9th Workshop on Social Media Mining for Health Research and Applications Workshop & Shared Tasks (SMM4H 2024)

Bangkok, Thailand
15 August 2024

ISBN: 979-8-3313-0177-4

Printed from e-media with permission by:

Curran Associates, Inc.
57 Morehouse Lane
Red Hook, NY 12571

Some format issues inherent in the e-media version may also appear in this print version.

Printed with permission by Curran Associates, Inc. (2025)

For permission requests, please contact the Association for Computational Linguistics at the address below.

Association for Computational Linguistics
209 N. Eighth Street
Stroudsburg, Pennsylvania 18360

Phone: 1-570-476-8006
Fax: 1-570-476-0860

acl@aclweb.org

Additional copies of this publication are available from:

Curran Associates, Inc.
57 Morehouse Lane
Red Hook, NY 12571 USA
Phone: 845-758-0400
Fax: 845-758-2633
Email: curran@proceedings.com
Web: www.proceedings.com

SMM4H 2024

The 9th Social Media Mining for Health Research and Applications (SMM4H 2024) Workshop and Shared Tasks

Proceedings of the Workshop

August 15, 2024

Association for Computational Linguistics (ACL)
317 Sidney Baker St. S
Suite 400 - 134
Kerrville, TX 78028
USA
Tel: +1-855-225-1962
acl@aclweb.org

Preface

Welcome to the 9th Social Media Mining for Health (#SMM4H) Research and Applications Workshop and Shared Tasks, co-located with the 62nd Annual Meeting of the Association for Computational Linguistics (ACL 2024). This year, #SMM4H will be a hybrid event, continuing our tradition of connecting data mining researchers who focus on leveraging social media data for health informatics. For #SMM4H 2024, we received 9 regular workshop paper submissions and 40 shared task paper submissions. Each submission underwent a rigorous review process: regular workshop papers were reviewed by at least two program committee members, while shared task papers were reviewed by at least one shared task committee member chosen for their expertise. Based on the reviewers' feedback, we accepted one long paper. Ultimately, we accepted 1 workshop paper and 38 shared task system description papers. The event will take place on August 15, 2024, in Thailand, offering both virtual and in-person attendance options.

The #SMM4H 2024 shared tasks aimed to advance the use of user-generated social media data for pharmacovigilance, epidemiology, patient-centered outcomes, and tracking the impacts of nonmedical substance use. This iteration of shared tasks included one re-run task about extraction and normalization of adverse drug events in English tweets (Task 1), and six new tasks: cross-lingual few-shot relation extraction for pharmacovigilance in French, German, and Japanese (Task 2), multi-class classification of effects of outdoor spaces on social anxiety symptoms in Reddit (Task 3), extraction of the clinical and social impacts of non-medical substance use from Reddit (Task 4), binary classification of English tweets reporting children's medical disorders (Task 5), self-reported exact age classification with cross-platform evaluation in English (Task 6), and identification of whether an LLM or a human domain expert annotated data in the context of health-related applications (Task 7). These tasks required methods for multi-class classification, named entity recognition, and normalization. The #SMM4H shared tasks attracted significant interest, with 84 teams from 22 countries registering and 45 teams submitting at least one set of predictions. Among the 38 accepted system description papers, 7 teams were invited for oral presentations.

We hope you find the workshop papers insightful and inspiring. We extend our gratitude to the shared task committee, program committee, additional reviewers of the system description papers, ACL 2024 organizers (especially the workshop chairs), annotators of the shared task data, and everyone who submitted a paper or participated in the shared tasks. #SMM4H 2024 would not have been possible without their contributions.

#SMM4H-2024 Workshop Chairs and Shared Task Committee

Organizing Committee

General Chairs

Graciela Gonzalez-Hernandez, Cedars-Sinai Medical Center, USA
Dongfang Xu, Cedars-Sinai Medical Center, USA
Ari Klein, University of Pennsylvania, USA

Shared Task Committee

Graciela Gonzalez-Hernandez, Cedars-Sinai Medical Center, USA
Dongfang Xu, Cedars-Sinai Medical Center, USA
Ari Klein, University of Pennsylvania, USA
Lisa Raithel, BIFOLD – Berlin Institute for the Foundations of Learning and Data, Germany
Roland Roller, German Research Center for Artificial Intelligence (DFKI), Germany
Philippe Thomas, German Research Center for Artificial Intelligence (DFKI), Germany
Eiji Aramaki, Nara Institute of Science and Technology, Japan
Shoko Wakamiya, Nara Institute of Science and Technology, Japan
Shuntaro Yada, Nara Institute of Science and Technology, Japan
Pierre Zweigenbaum, Université Paris-Saclay, France
Karen O'Connor, University of Pennsylvania, USA
Sai Tharuni Samineni, Cedars-Sinai Medical Center, USA
Yao Ge, Emory University, USA
Swati Rajwal, Emory University, USA
Sudeshna Das, Emory University, USA
Abeed Sarker, Emory University, USA
Ana Lucia Schmidt, Roche Innovation Center, Switzerland
Vishakha Sharma, Roche Diagnostics, USA
Raul Rodriguez-Esteban, Roche Innovation Center, Switzerland
Juan M. Banda, Stanford Health Care, USA
Ivan Flores Amaro, Cedars-Sinai Medical Center, USA

Workshop Committee

Graciela Gonzalez-Hernandez, Cedars-Sinai Medical Center, USA
Dongfang Xu, Cedars-Sinai Medical Center, USA
Davy Weissenbacher, Cedars-Sinai Medical Center, USA
Ari Klein, University of Pennsylvania, USA
Karen O'Connor, University of Pennsylvania, USA

Publication Chair

Dongfang Xu, Cedars-Sinai Medical Center, USA

Program Committee

Program Chairs

Juan M Banda

Yao Ge

Ari Klein

Karen O'Connor

Lisa Raithel, Roland Roller

Ana Lucia Schmidt

Philippe Thomas

Dongfang Xu

Program Committee

Natalia Grabar

Thierry Hamon

Antonio Jimeno Yepes

Robert Leaman

Rajesh Piryani, Thierry Poibeau

Nicolas Turenne

Karin Verspoor

Pierre Zweigenbaum

Reviewers

Harika Abburi, Falwah Alhamed, Bizhan Alipourpijani, João Rafael Almeida, Thushari Atapattu

Juan M Banda, Guntis Barzdins, Jacob S. Berkowitz

Steffen Castle, Harshita Chandwani, Manav Chaudhary

Liza Dahiya, Sudeshna Das, Andrew S. Davis, Sobha Lalitha Devi, Bin Dong

Ahmed El-Sayed

Yuming Fan, Sumam Francis

Helena Gomez Adorno, Anubhav Gupta

Thierry Hamon, Leon Hecht, Miguel E. Hernandez

Antonio Jimeno Yepes

Ram Mohan Rao Kadiyala, Ram Mohan Rao Kadiyala, Ramakanth Kavuluru, Yuanzhi Ke, Ari Klein

Paloma Martínez, Dana Moukheiber, Eduards Mukans, Hasan Murad

Nona Naderi, B Rahul Naik, Neha Nair

Karen O'Connor, Jesús Vázquez Osorio

Rajesh Piryani, Beatrice Portelli

Lisa Raithel, Swati Rajwal, Raul Rodriguez-Esteban, Roland Roller

Rana Salama, Ana Lucia Schmidt, Hsuan-Lei Shao, Vishakha Sharma, Kriti Singhal, Zafi Sherhan Syed

Noha Tawfik, Philippe Thomas, Giuliano Tortoreto, Nicolas Turenne

Karin Verspoor, V.G.Vinod Vydiswaran

Azmine Toushik Wasi, Angela M Wiley

Dongfang Xu

Yosuke Yamagishi, Zhai Yu

Yifan Zheng, Pierre Zweigenbaum

Keynote Talk
Social Media Mining for Substance Use Research

Abeed Sarker
Emory University

Abstract: The epidemic of substance use (SU) and substance use disorder (SUD) in the United States has been evolving for decades. Both prescription and illicit drugs have been involved in overdose deaths over the years, with notable increases in synthetic opioids (eg., fentanyl & analogs) and psychostimulants (eg., methamphetamine) in recent years. The emergence of high-potency novel psychoactive substances (NPSs), such as fentanyl analogs, have drastically contributed to rising deaths, and adversely impacted treatment engagement and response. A key element to tackling the crisis is improved surveillance. Specifically, there is a need for establishing novel approaches to provide timely insights about the trends, distributions, and trajectories of the SUD epidemic, as traditional surveillance approaches involve considerable lags. Many recent studies have identified social media (SM) as useful resources for conducting SU/SUD surveillance. Many people use SM to discuss personal experiences, provide advice, or seek answers to questions regarding SU/SUD, resulting in the generation of an abundance of information. Such information can be characterized, aggregated and analyzed to obtain population- or subpopulation-level insights, at low cost and in near real time. However, converting SM data into timely, actionable knowledge is non-trivial since the data is big, complex, and noisy, requiring the development of advanced, automated artificial intelligence methods. In this talk, I will highlight our ongoing and past work on developing NLP and machine learning methods for effectively leveraging social media data for substance use research.

Bio: Dr. Sarker (he/him) is an Associate Professor and the Vice Chair for Research at the Department of Biomedical Informatics, School of Medicine, Emory University. He leads several large-scale projects focusing on the application of NLP for health-related tasks, particularly those involving vulnerable populations such as people with substance use disorders, victims of intimate partner violence, and people at risk of self-harm and suicide. His research is primarily funded by the National Institutes of Health (NIH) and Centers for Disease Control and Prevention (CDC). Dr. Sarker's research has been covered by various national and international media outlets such as the Wall Street Journal, Forbes, and Scripps National News.

Table of Contents

Program

Thursday, August 15, 2024

08:55 - 09:00 *Welcome and Opening Remarks – Dongfang Xu*

09:00 - 09:15 *Workshop Introduction – Graciela Gonzalez-Hernandez*

09:15 - 09:40 *Overview of #SMM4H 2024 -– Task 3 Multi-class classification of effects of outdoor spaces on social anxiety symptoms in Reddit.*

IMS_medicALY at #SMM4H 2024: Detecting Impacts of Outdoor Spaces on Social Anxiety with Data Augmented Ensembling
Amelie Wuehrl, Lynn Greschner, Yarik Menchaca Resendiz and Roman Klinger

09:40 - 10:10 *Overview of #SMM4H 2024 — Task 5 Binary classification of English tweets reporting childrens medical disorders.*

CTYUN-AI@SMM4H-2024: Knowledge Extension Makes Expert Models
Yuming Fan, Dongming Yang and Lina Cao

10:10 - 10:35 *Overview of #SMM4H 2024 — Task 7 Identification of LLM or human domain-expert data annotations in the context of health-related applications.*

712forTask7 at #SMM4H 2024 Task 7: Classifying Spanish Tweets Annotated by Humans versus Machines with BETO Models
Hafizh Rahmatdianto Yusuf, David Belmonte, Dalton Simancek and V.G.Vinod Vydiswaran

10:35 - 11:00 *Coffee Break*

11:00 - 11:45 *Keynote – Social Media Mining for Substance Use Research*

11:45 - 12:15 *Overview of #SMM4H 2024 — Task 4 Extraction of the clinical and social impacts of nonmedical substance use from Reddit.*

UKYNLP@SMM4H2024: Language Model Methods for Health Entity Tagging and Classification on Social Media (Tasks 4 & 5)
Motasem S. Obeidat, Vinu H Ekanayake, Md Sultan Al Nahian and Ramakanth Kavuluru

12:15 - 12:45 *Lunch Break*

12:45 - 14:00 *Poster Presentation Session*

Thursday, August 15, 2024 (continued)

14:00 - 14:30 *Overview of #SMM4H 2024 -- Task 2 Cross-Lingual Few-Shot Relation Extraction for Pharmacovigilance in French, German, and Japanese*

Team Yseop at #SMM4H 2024: Multilingual Pharmacovigilance Named Entity Recognition and Relation Extraction
Anubhav Gupta

14:30 - 15:00 *Overview of #SMM4H 2024 -- Task 1 Extraction and normalization of adverse drug events (ADEs) in English tweets.*

SRCB at #SMM4H 2024: Making Full Use of LLM-based Data Augmentation in Adverse Drug Event Extraction and Normalization
Hongyu Li, Yuming Zhang, Yongwei Zhang, Shanshan Jiang and Bin Dong

15:00 - 15:30 *Overview of #SMM4H 2024 -- Task 6 Self-reported exact age classification with cross-platform evaluation in English.*

UTRad-NLP at #SMM4H 2024: Why LLM-Generated Texts Fail to Improve Text Classification Models
Yosuke Yamagishi and Yuta Nakamura

15:30 - 15:40 *Conclusion and Closing Remarks – Dongfang Xu*

ThangDLU at #SMM4H 2024: Encoder-decoder models for classifying text data on social disorders in children and adolescents

Hoang-Thang Ta
Dalat University
thangth@dlu.edu.vn

Abu Bakar Siddiqur Rahman
University of Nebraska at Omaha
abubakarsiddiqurra@unomaha.edu

Lotfollah Najjar
University of Nebraska at Omaha
lnajjar@unomaha.edu

Alexander Gelbukh
Instituto Politécnico Nacional (IPN), Mexico
gelbukh@cic.ipn.mx

Abstract

This paper describes our participation in Task 3 and Task 5 of the #SMM4H (Social Media Mining for Health) 2024 Workshop, explicitly targeting the classification challenges within tweet data. Task 3 is a multi-class classification task centered on tweets discussing the impact of outdoor environments on symptoms of social anxiety. Task 5 involves a binary classification task focusing on tweets reporting medical disorders in children. We applied transfer learning from pre-trained encoder-decoder models such as BART-base and T5-small to identify the labels of a set of given tweets. We also presented some data augmentation methods to see their impact on the model performance. Finally, the systems obtained the best F1 score of 0.627 in Task 3 and the best F1 score of 0.87 in Task 5.

1 Introduction

Social disorders are significantly influencing a large proportion of young people globally. Social anxiety disorder (SAD) typically emerges during early adolescence and is characterized by excessive anxiety in social situations (Rao et al., 2007). Although spending time outdoors in green or blue environments has been shown to alleviate symptoms of various anxiety disorders, limited research has explored its impact specifically on SAD. Meanwhile, numerous children receive diagnoses of conditions that can significantly affect their daily functioning and persist into adulthood. The commonly diagnosed childhood disorders are attention-deficit/hyperactivity disorder (ADHD) (Kidd, 2000), autism spectrum disorders (ASD) (Matson et al., 2009), speech delay, and asthma.

Datasets related to these disorders are usually extracted from tweets or user posts on social platforms such as Twitter and Reddit. Users describe their disorders daily and receive feedback from the community or comment on what they have passed. As a text classification problem, the most advanced and popular methods of social disorder identification use deep learning networks such as BERT (Devlin et al., 2018), RoBERTa (Liu et al., 2019), Transformer models (Vaswani et al., 2017), and their variants.

In participating in the #SMM4H 2024 workshop (Xu et al., 2024), we apply transfer learning from two pre-trained models (BART-base (Lewis et al., 2019) and T5-small (Raffel et al., 2020)), which follow the architecture of Transformer and sequence-to-sequence. Additionally, we exploited two data augmentation methods — (1) false inferred data and (2) paraphrased data extracted from ChatGPT — to supplement the training set and see their impact on model performance. Because the organizers hide the final ranking table, we can only present our results compared to the mean and median values by metrics (F1, precision, recall, and accuracy) they provided.

2 Methodology

After conducting several initial experiments on training the dataset with and without preprocessing steps, we observed that training without any preprocessing yielded better performance. It is assumed that every token within the data can positively impact the model's performance. Therefore, we fed raw data directly into the model in the training process.

We applied transfer learning from two pre-trained models based on the architecture of the Transformer and sequence-to-sequence: BART-base (Vaswani et al., 2017) and T5-small (Raffel et al., 2020) for Task 3 and Task 5, respectively. They are encoder-decoder language models, producing outputs that can be unknown labels. From a given text, the models must detect its label, and any out-of-scope label will be set automatically to a default label, which we choose as "0" for both tasks. The best models were saved based on their

Proceedings of The 9th Social Media Mining for Health Applications (#SMM4H) Workshop and Shared Tasks, pages 1–4
August 15, 2024 ©2024 Association for Computational Linguistics

performance on the validation set by F1-macro for Task 3 and the F1 score on the positive class for Task 5.

Two data augmentation methods complement new data to improve mode performance. First, the false inferred data in the validation set were utilized when inferring them in a trained model with the default training data. Second, paraphrased data by ChatGPT is taken from the training and validation sets at an insignificant cost.

3 Tasks & Datasets

3.1 Task 3

This task involves classifying Reddit posts mentioning predetermined keywords related to outdoor spaces into one of four categories: ("1") positive effect, ("2") neutral/no effect, ("3") negative effect, and ("4") unrelated [1]. The dataset comprises 3,000 annotated posts from the `r/socialanxiety` subreddit, filtered for users aged 12-25, and keywords related to green or blue spaces. 80% of the data will be used for training/validation, and 20% for evaluation. Evaluation will be based on the macro-averaged F1-score across all categories. Data will be provided in CSV format with fields: `post_id`, `keyword`, `text`, and `label`. The distribution of subsets follows a ratio 6:2:2, in which training, validation, and testing sets take 1800, 600, and 600 posts correspondingly. However, the organizer provided a test set with 1200 posts to hide the real ones.

3.2 Task 5

This task involves automatically classifying tweets from users who reported pregnancy on Twitter. It distinguishes tweets reporting children with ADHD, autism, delayed speech, or asthma ("1") from those merely mentioning a disorder ("0")[2]. The goal is to enable large-scale epidemiologic studies and explore parents' experiences for targeted support interventions. The dataset includes 7398 training tweets, 389 validation tweets, and 1947 test tweets. Like Task 3, the organizer gave a new test with 10000 tweets to hide the actual data.

4 Experiments

4.1 Task 3

The organizer limited each team to 3 submissions. Therefore, we used BART-base to train 3 models over 3 different training data sets.

- `Training`: The model was trained over the original training set offered by the organizer.
- `Training + Paraphrased`: We extracted the paraphrased data based on the validation set by ChatGPT. Then, we added this new data to the training set.
- `Training + Validation`: We added a validation set to the training set and then used this new set for training the model.

The training has the same parameters for all models, including `epochs = 10`, `batch_size = 4`, and `max_source_length = 768`. For any input with a length over 768 tokens, we process it to take its first 256 tokens and its last 512 tokens.

Table 1 shows the results of our team and the mean and median performance of all teams by metrics: F1, precision, recall, and accuracy. It is clear our models outperformed the mean and median overall metrics. Our best model was trained on the `Training + Validation` data and obtained an F1 value of 0.627, while the model with `Training + Paraphrased` data takes slightly lower performance. While paraphrased data helps improve the model, it is better to collect actual data to obtain the best performance. The low F1 value indicates the task's difficulty and the need for adding more training data to improve the model performance.

4.2 Task 5

The task limits each team to only 2 submissions. Therefore, we pick 2 trained T5-small models on two training sets for participation. In the post-eval phase, we also trained the other 2 models with different training sets. Finally, we have 4 models with 4 training sets, which are:

- `Training`: The model was trained over the original training set offered by the organizer.
- `Training + Validation`: We added a validation set to the training set and then used this new set for training the model.

[1]https://codalab.lisn.upsaclay.fr/competitions/18305
[2]https://codalab.lisn.upsaclay.fr/competitions/17310

#Submission	Data	F1	Precision	Recall	Accuracy
1	Training	0.595	0.589	0.615	0.631
2	Training + Paraphrased	0.601	0.592	0.622	0.640
3	**Training + Validation**	**0.627**	**0.620**	**0.644**	**0.670**
Compared to other teams					
-	Mean	0.518	0.564	0.537	0.574
-	Median	0.579	0.630	0.588	0.627

Table 1: F1-macro, Precison-macro, and Recall-macro values of BART-base models on Task 3, which were trained over different data combinations.

#Submission	Data	F1	Precision	Recall
1	Training + False inferred + Paraphrased	0.841	0.844	0.839
2	Training + False inferred	0.829	0.803	0.856
3*	Training + Validation	0.870	0.869	0.867
4*	Training	0.820	0.809	0.831
Compared to other teams				
-	Mean	0.822	0.818	0.838
-	Median	0.901	0.885	0.917
Other works				
-	**RoBERTa-Large** (Klein et al., 2024)	**0.930**	-	-
*Our extra participation in the post-eval phase.				

Table 2: The metrics of T5-small models on Task 5, which were trained over different data combinations.

- `Training + False inferred`: First, we used the model trained on the original training set to infer the labels of inputs in the validation set. Then, we collect false inferred texts with their labels (41 examples) and add them to the original training set to form a new one.
- `Training + False inferred + Paraphrased`: Similar to `Training + False inferred`, we add more paraphrased data to the training set. First, we used BM25 (Robertson et al., 1995) to take similar texts in the training set based on the validation set. Not that the new set's size equals the validation set's size (389 examples). Then, we used ChatGPT APIs[3] to extract paragraphed texts and add them to the training set.

The training has the same parameters for all models, including `epochs = 20`, `batch_size = 4`, and `max_source_length = 128`. Table 2 shows the results of our team and the mean and median performance of all teams by metrics: F1, precision, and recall. All our models have metric values that are better than the mean but lower than the median. Especially, our metric values are significantly lower than the benchmark F1 (Klein et al., 2024) when using RoBERTa-Large. It can be explained that we only use a small-scale pre-trained model like T5-small for the classification task.

Due to the small size of false inferred data, the model performance is not much better. However, we realize that the paraphrased data contributes positively to the model performance even though with a subset. Unfortunately, we can not experiment with training on more paraphrased data based on the full validation and the training sets, but we expect the model will be much better. Our best model was trained on the `Training + Validation` with an F1 value of 0.87, indicating that the more data, the better model performance.

5 Conclusion

This paper introduced our approach, utilizing pre-trained encoder-decoder models with two data augmentation methods to address Task 3 and Task 5 of the #SMM4H 2024 workshop. Our findings underscore the advantages of encoder-decoder models in text classification problems when they offer a strong baseline performance. Furthermore, it is beneficial to exploit data augmentation methods to enhance the model's performance by comple-

[3]https://platform.openai.com/docs/overview

menting paraphrased texts from ChatGPT. In the experiments, we achieved the highest F1 score of 0.627 for Task 3 and 0.87 for Task 5. In the future, we will investigate further how large language models' outputs like ChatGPT can positively impact downstream classification tasks' performance.

References

Jacob Devlin, Ming-Wei Chang, Kenton Lee, and Kristina Toutanova. 2018. Bert: Pre-training of deep bidirectional transformers for language understanding. *arXiv preprint arXiv:1810.04805*.

Parris M Kidd. 2000. Attention deficit/hyperactivity disorder (adhd) in children: rationale for its integrative management. *Alternative medicine review*, 5(5):402–428.

Ari Z Klein, José Agustín Gutiérrez Gómez, Lisa D Levine, and Graciela Gonzalez-Hernandez. 2024. Using longitudinal twitter data for digital epidemiology of childhood health outcomes: an annotated data set and deep neural network classifiers. *Journal of Medical Internet Research*, 26:e50652.

Mike Lewis, Yinhan Liu, Naman Goyal, Marjan Ghazvininejad, Abdelrahman Mohamed, Omer Levy, Ves Stoyanov, and Luke Zettlemoyer. 2019. Bart: Denoising sequence-to-sequence pre-training for natural language generation, translation, and comprehension. *arXiv preprint arXiv:1910.13461*.

Yinhan Liu, Myle Ott, Naman Goyal, Jingfei Du, Mandar Joshi, Danqi Chen, Omer Levy, Mike Lewis, Luke Zettlemoyer, and Veselin Stoyanov. 2019. Roberta: A robustly optimized bert pretraining approach. *arXiv preprint arXiv:1907.11692*.

Michael L Matson, Sara Mahan, and Johnny L Matson. 2009. Parent training: A review of methods for children with autism spectrum disorders. *Research in Autism Spectrum Disorders*, 3(4):868–875.

Colin Raffel, Noam Shazeer, Adam Roberts, Katherine Lee, Sharan Narang, Michael Matena, Yanqi Zhou, Wei Li, and Peter J Liu. 2020. Exploring the limits of transfer learning with a unified text-to-text transformer. *Journal of machine learning research*, 21(140):1–67.

Patricia A Rao, Deborah C Beidel, Samuel M Turner, Robert T Ammerman, Lori E Crosby, and Floyd R Sallee. 2007. Social anxiety disorder in childhood and adolescence: Descriptive psychopathology. *Behaviour Research and Therapy*, 45(6):1181–1191.

Stephen E Robertson, Steve Walker, Susan Jones, Micheline M Hancock-Beaulieu, Mike Gatford, et al. 1995. Okapi at trec-3. *Nist Special Publication Sp*, 109:109.

Ashish Vaswani, Noam Shazeer, Niki Parmar, Jakob Uszkoreit, Llion Jones, Aidan N Gomez, Łukasz Kaiser, and Illia Polosukhin. 2017. Attention is all you need. *Advances in neural information processing systems*, 30.

Dongfang Xu, Guillermo Lopez Garcia, Lisa Raithel, Rolland Roller, Philippe Thomas, Eiji Aramaki, Shuntaro Yada, Pierre Zweigenbaum, Karen O'Connor, Yao Ge, Sudeshna Das, Abeed Sarker, Ari Klein, Lucia Schmidt, Vishakha Sharma, Raul Rodriguez-Esteban, Juan Banda, Ivan Flores Amaro, Sai Tharuni Samineni, and Graciela Gonzalez-Hernandez. 2024. Overview of the 9th social media mining for health applications (#SMM4H) shared tasks at ACL 2024. In *Proceedings of the 9th Social Media Mining for Health Applications Workshop and Shared Task*, Bangkok, Thailand. Association for Computational Linguistics.

CTYUN-AI@SMM4H-2024: Knowledge Extension Makes Expert Models

Yuming Fan* and **Dongming Yang** *† and **Lina Cao**
{fanym, yangdm1, caoln}@chinatelecom.cn
China Telecom Cloud Technology Co., Ltd

Abstract

This paper explores the potential of social media as a rich source of data for understanding public health trends and behaviors, particularly focusing on emotional well-being and the impact of environmental factors. We employed large language models (LLMs) and developed a suite of knowledge extension techniques to analyze social media content related to mental health issues, specifically examining 1) effects of outdoor spaces on social anxiety symptoms in Reddit, 2) tweets reporting children's medical disorders, and 3) self-reported ages in posts of Twitter and Reddit. Our knowledge extension approach encompasses both supervised data (i.e., sample augmentation and cross-task fine-tuning) and unsupervised data (i.e., knowledge distillation and cross-task pretraining), tackling the inherent challenges of sample imbalance and informality of social media language. The effectiveness of our approach is demonstrated by the superior performance across multiple tasks (i.e., Task 3, 5 and 6) at the SMM4H-2024. Notably, we achieved the best performance in all three tasks, underscoring the utility of our models in real-world applications.

1 Introduction

In recent years, the surge in social media usage has transformed these platforms into valuable repositories of public health attitudes and behaviors. Users not only share snippets of their daily lives but also discuss a variety of health issues, drug reactions, and treatment outcomes, providing a wealth of real-time data for medical and health research. Social media plays an especially crucial role in monitoring adverse drug reactions, tracking diseases, and facilitating public discussions on health conditions. This data aids healthcare organizations and researchers in understanding disease trends and patient needs, enhancing drug safety monitoring and optimizing treatment plans.

Against this backdrop, the significance of the 9th Social Media Mining for Health Research and Applications Workshop (SMM4H-2024)(Xu et al., 2024) is particularly pronounced. This workshop brings together researchers, developers, and medical professionals from around the world to address the challenges of automating the extraction and analysis of health information from social media. The conference not only serves as a platform for sharing the latest research findings and cutting-edge technologies but also organizes multiple shared tasks targeting specific practical application problems, attracting numerous teams. These tasks are designed to enhance data processing capabilities across languages and cultural contexts, aiming to more accurately parse and utilize health-related information from social media, thereby supporting global health research and public health surveillance.

In this workshop, our goal is to construct and enhance the given limited data, including both unsupervised and supervised data, to achieve maximum knowledge extension for training large language models. This will enable us to turn the models into specialized experts for each individual task. The core points are summarized as follows:

- Sample Augmentation: by performing self-augmentation on long-tail samples, we aim to address the issue of sample imbalance in the tasks.
- Knowledge Distillation: we utilized ensemble learning with multiple models to process unsupervised data, thereby generating supervised samples that are beneficial for model training.
- Cross-Task Training: we applied both unsupervised and supervised data from one task to train the model for another task, in order to

*Equal Contribution.
†Corresponding Author.

Proceedings of The 9th Social Media Mining for Health Applications (#SMM4H) Workshop and Shared Tasks, pages 5–9
August 15, 2024 ©2024 Association for Computational Linguistics

expand the model's background knowledge in that task.

These strategies not only resolved issues related to uneven class distribution and data scarcity but also refined the sentiment analysis process. Finally, experimental results confirm the effectiveness of our strategies, as our team (i.e., CTYUN-AI) achieved **best performance** in tasks 3, 5, and 6 among all participating teams.

2 Related Work

Recent advances in natural language processing (NLP) have significantly enhanced the ability to analyze health-related discussions on social media. Zanwar et al. utilized advanced NLP techniques alongside psycholinguistic features to effectively detect chronic stress expressions on social media, addressing data imbalance issues in the process (Zanwar et al., 2022). Liu et al. demonstrated the use of multiple pre-trained models to detect adverse drug reactions on Twitter, showcasing methods that tackle the complexities of social media data (Liu et al., 2022). Additionally, Tamayo et al. developed a transfer learning approach with post-processing enhancements to accurately extract disease mentions from Spanish tweets, improving the robustness of disease monitoring across different languages (Tamayo et al., 2022). These contributions highlight the evolving capabilities of NLP to provide valuable insights into public health from social media content.

Concurrently, generative models in the realm of NLP, exemplified by the GPT (Brown et al., 2020) series, have exhibited remarkable abilities in comprehending and producing natural language. Bai et al. (Bai et al., 2023) developed the Qwen models, which excel at multiple tasks. Consequently, we employ the Qwen models as our base model to cultivate further specialized experts.

3 System Overview

In this section, we systematically explicate the knowledge extension strategies employed by our team for the respective sub-tasks. We commence with a descriptive analysis of the datasets pertinent to each sub-task, followed by an exposition of our methodologies, specifically devised and optimized in accordance with the task-specific data characteristics.

3.1 Task 3: Classification of reported effects of outdoor spaces on social anxiety symptoms

Task 3 is centered on categorizing Reddit posts by individuals aged 12 to 25 discussing the effects of green or blue spaces on symptoms of Social Anxiety Disorder (SAD). The dataset comprises 3,000 annotated posts, divided into 1,800 for training, 600 for validation, and 600 for testing.

Considering the long-tail distribution of the task data, which poses a challenge in achieving satisfactory performance for certain classes on the test set and detrimentally impacts the overall F1 score, we created a knowledge extension approach relying on random shuffling to alleviate this concern, as illustrated in Figure 1.

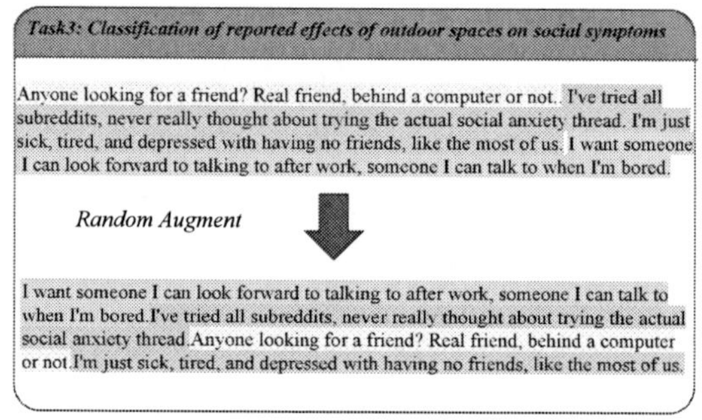

Figure 1: Example of random shuffling.

Unlike formal articles, Reddit posts are less structured, and some level of sequence disorder can still reflect the sentiment analysis inherent in social media text. Therefore, we utilized commas and periods as delimiters to randomly shuffle and augment the split data. We then balanced the dataset by augmenting the less frequent classes to match the quantity of the most populous 'positive effect' category.

3.2 Task 6: Self-reported exact age classification with cross-platform evaluation in English

Task 6 focuses on identifying precise self-reported ages from social media posts on Twitter and Reddit. This enables the analysis of health-related observational studies by determining the age of users directly from their posts. The dataset comprises 8,800 labeled tweets and 100,000 unlabeled Reddit posts, with the F1-score for the positive class serving as the evaluation metric.

To effectively extends professional knowledge

of the LLM, we employed the unlabeled data for pre-training, thereby strengthening the model's capacity to learn domain-specific features pertinent to social media text. Using the unlabeled data for pre-training involves treating each post sample directly as a training example.

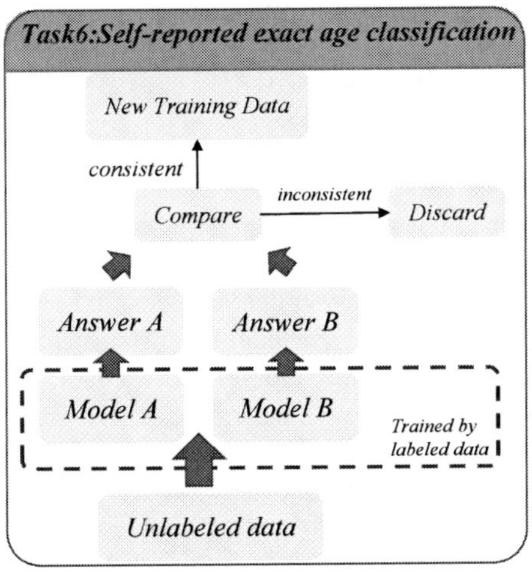

Figure 2: Example of random shuffling.

Furthermore, we introduced an ensemble model voting strategy to further mine professional knowledge from unsupervised data. Specifically, we first trained two additional models, qwen72b and qwen1.5 72b(Bai et al., 2023), using train set of the task and inferred the unlabeled data employing the trained LLMs. Then, the inference results from the two models are compared to see if they are consistent. Next, the samples with consistent inference results are retained as extended fine-tuning data, as shown in Figure 2. Finally, the participating model was fine-tuned using a combination of the original training data and augmented training data. In summary, by integrating responses from multiple models, we significantly expanded the training dataset and enhanced the robustness and generalization capability of our model across different textual contexts in social media.

3.3 Task 5: Binary classification of tweets reporting children's medical disorders

Task 5 focuses on binary classification of tweets to determine whether they report a child's medical condition, such as Attention Deficit/Hyperactivity Disorder (ADHD). The dataset consists of 7,398 training tweets, 389 validation tweets, and 1,947 test tweets, differentiating between tweets that directly report children's disorders from those that merely mention such conditions. The performance is assessed using the F1-score for tweets that substantively report on a child's medical disorder.

As task 6 is also a binary classification task focused on social media content analysis, task 5 and 6 demonstrated promising potential for transferability due to the similarities in task nature and data characteristics. Thus, our cross-task training strategy utilized the unlabeled and labeled data from task 6, which involved analyzing Twitter and other social media content, to improve the model at task 5. Building on this, we adopted the supervised fine-tuned model from Task 6 as the base model and further fine-tune the model using the train set of task 5. The experiments have proven that cross-task training not only deepens the model's comprehension of the domain-specific data but also improves its generalization capabilities in practical applications.

It is important to note that while cross-task training strategy have achieved notable success, there are still some limitations. The effectiveness of this method largely depends on the correlation between the source task and the target task. If there is a significant difference in data characteristics or objectives between the two tasks, the performance of the pre-trained model may be substantially compromised. This means that selecting tasks with high relevance for pre-training is a critical issue in practical applications. Additionally, the decline in model performance may be more pronounced when there are significant differences in the nature of the tasks, data distribution, or language style.

4 Experiment

4.1 Implement Detail

We employed the Qwen-72B-Chat(Bai et al., 2023) as our base model, upon which post pre-traing and fine-tuning of all parameters was carried out. The computational experiments were executed on an Nvidia A800 GPU, equipped with 80GB of VRAM. In the training phase, we configured the model to handle sequences with a maximum of 2048 tokens, a batch size of 8, and accumulated the gradient after every training step. The training initiated with a learning rate of 5e-6, adopting a cosine decay schedule, and spanned across three complete epochs. For the inference process, the model's built-in default parameters were utilized.

4.2 Result

In this section, we present the evaluation results of our participation in Tasks 3, 5, and 6 at the SMM4H-2024, comparing our system's performance against the mean and median scores of all participating teams. According to the organizers' assessment, our CTYUN-AI team achieved the best performance across these tasks.

Table 1: Evaluation Result on SMM4H Task 3.

Task 3 Result	F1-score	P	R	Acc
CTYUN-AI	0.692	0.704	0.686	0.726
Mean	0.5186	0.5649	0.5379	0.5746
Median	0.5795	0.63	0.5885	0.627

For Task 3, which involved classifying social media posts about the impact of social anxiety disorder, we achieved an F1 score of 0.692, as shown in Table 1. This performance significantly surpassed the mean F1 score of 0.5186 and the median of 0.5795, demonstrating our model's robust capability in accurately classifying relevant posts. The precision and recall were 0.704 and 0.686 respectively, with an accuracy of 0.726.

Table 2: Evaluation Result on SMM4H Task 5.

Task 5 Result	F1-score	P	R
CTYUN-AI	0.956	0.954	0.959
Mean	0.822	0.818	0.838
Median	0.901	0.885	0.917

In Task 5, aimed at identifying tweets reporting on children's medical conditions, our model demonstrated exceptional effectiveness with an F1 score of 0.956, considerably outperforming the mean score of 0.822 and the median score of 0.901, as shown in Table 2. This indicates that our model was highly precise (P=0.954) and sensitive (R=0.959) in identifying relevant tweets.

Table 3: Evaluation Result on SMM4H Task 6.

Task 6 Result	F1-score	P	R
CTYUN-AI	0.970	0.976	0.963
Mean	0.924	0.924	0.926
Median	0.936	0.934	0.949

Finally, in Task 6, our approach achieved an F1 score of 0.970, which is notably higher than the average F1 score of 0.924 and the median of 0.936 reported by other teams, as shown in Table 3. Our model exhibited a precision of 0.976 and a recall of 0.963, indicating superior performance in accurately identifying and classifying age-related information from the posts.

These results across different tasks highlight the efficacy of our approaches and underline the potential of our model configurations in effectively handling diverse and complex social media datasets.

5 Conclusion

In this work, we employed large language models and crafted a comprehensive set of knowledge extension techniques for the purpose of analyzing social media content pertaining to mental health concerns. We have provided a detailed account of how to leverage knowledge extension techniques to maximize the utilization of limited data, enabling us to train a general large language model into a domain-specific expert model. Specifically, our approach encompasses both supervised data and unsupervised data, including sample augmentation, knowledge distillation and cross-task training. We achieved the best performance at multiple SMM4H-2024 tasks (i.e., Task 3, 5 and 6), validating the effectiveness of our approach.

References

Jinze Bai, Shuai Bai, Yunfei Chu, Zeyu Cui, and et al. 2023. Qwen technical report. *arXiv:2309.16609*.

Tom B. Brown, Benjamin Mann, Nick Ryder, Melanie Subbiah, and et al. 2020. Language models are few-shot learners. *Preprint*, arXiv:2005.14165.

Xi Liu, Han Zhou, and Chang Su. 2022. Pingantech at smm4h task1: Multiple pre-trained model approaches for adverse drug reactions. In *Proceedings of The Seventh Workshop on Social Media Mining for Health Applications, Workshop & Shared Task*, pages 4–6.

Antonio Tamayo, Alexander Gelbukh, and Diego A Burgos. 2022. Nlp-cic-wfu at socialdisner: Disease mention extraction in spanish tweets using transfer learning and search by propagation. In *Proceedings of The Seventh Workshop on Social Media Mining for Health Applications, Workshop & Shared Task*, pages 19–22.

Dongfang Xu, Guillermo Lopez Garcia, Lisa Raithel, Rolland Roller, Philippe Thomas, Eiji Aramaki, Shuntaro Yada, Pierre Zweigenbaum, Karen O'Connor, Yao Ge, Sudeshna Das, Abeed Sarker, Ari Klein, Lucia Schmidt, Vishakha Sharma, Raul Rodriguez-Esteban, Juan Banda, Ivan Flores Amaro, Sai Tharuni Samineni, and Graciela Gonzalez-Hernandez. 2024. Overview of the 9th social media mining for health applications (smm4h) shared tasks at acl 2024. *arXiv preprint arXiv:2405.02994*.

Sourabh Zanwar, Daniel Wiechmann, Yu Qiao, and Elma Kerz. 2022. Mantis at smm4h'2022: pre-trained language models meet a suite of psycholinguistic features for the detection of self-reported chronic stress. In *Proceedings of the Seventh Workshop on Social Media Mining for Health Applications, Workshop & Shared Task*, pages 16–18.

A Appendix

Additionally, we will report the improvements in model performance on the validation set. It is important to note that the results shown in the previous tables are from the final test sets, which differ from the results presented here. Specifically, we initially defined the baseline method by directly feeding the labeled training data into the qwen-72b-chat model without applying any augmentation strategies.

In Task-6, starting from the initial baseline model, we achieved the highest score of 94.78 by incorporating unsupervised data and semantic alignment processing. In contrast, the baseline model using only labeled training data scored 92.29. This indicates that the strategies of using unsupervised data and semantic processing significantly improved the model's performance, especially when the accuracy was already at a high level.

In Task-5, we experimented with different learning rates and model bases. The initial baseline model scored 93.92. By adjusting the learning rate and employing a pre-trained model with unsupervised data, we improved the score to 95.88. This result was achieved using the model from Task-6 as the base, demonstrating that selecting the appropriate pre-trained model and fine-tuning parameters can significantly enhance classification performance.

In Task-3, we improved the model's classification performance through a mix of data augmentation strategies. The initial baseline model scored 57. After applying class-wise random exchange augmentation, the score increased to 61. Further, by enhancing the smaller classes, the model score rose to 64. This shows that appropriate training strategies and data augmentation techniques play a crucial role in improving multi-class classification task performance.

In summary, by integrating unsupervised data, using pretrained model from other tasks, and employing data augmentation strategies, we achieved significant performance improvements across these three tasks, validating the effectiveness of these methods in enhancing model performance.

DILAB at #SMM4H 2024: RoBERTa Ensemble for Identifying Children's Medical Disorders in English Tweets

Azmine Toushik Wasi
Shahjalal University of Science and Technology
Sylhet, Bangladesh
azmine32@student.sust.edu

Sheikh Ayatur Rahman
BRAC University
Dhaka, Bangladesh
sheikh.ayatur.rahman@g.bracu.ac.bd

Abstract

This paper details our system developed for the 9th Social Media Mining for Health Research and Applications Workshop (SMM4H 2024), addressing Task 5 focused on binary classification of English tweets reporting children's medical disorders. Our objective was to enhance the detection of tweets related to children's medical issues. To do this, we use various pre-trained language models, like RoBERTa and BERT. We fine-tuned these models on the task-specific dataset, adjusting model layers and hyperparameters in an attempt to optimize performance. As we observe unstable fluctuations in performance metrics during training, we implement an ensemble approach that combines predictions from different learning epochs. Our model achieves promising results, with the best-performing configuration achieving F1 score of 93.8% on the validation set and 89.8% on the test set.

1 Introduction

Health informatics research often involves analyzing social media data from platforms like Twitter, Facebook, and Reddit to understand public sentiment on health-related topics. The 9th edition of the Social Media Mining for Health Research and Applications Workshop (SMM4H 2024) (Xu et al., 2024) is dedicated to advancing this area. Researchers at SMM4H 2024 aim to contribute by exploring topics like deriving health trends from social media, classifying health-related messages, identifying health-related or medical terms and monitoring diseases using social media content.

We decided to engage in Task 5 of this workshop, focusing on the binary classification of English tweets reporting children's medical disorders. This task aims to refine methods for accurately identifying and categorizing tweets related to pediatric health conditions. It aligns with broader efforts in health informatics research to understand public sentiment on pediatric health topics through social media analysis, inspired by Klein et al.'s work on using longitudinal Twitter data for digital epidemiology of childhood health outcomes.

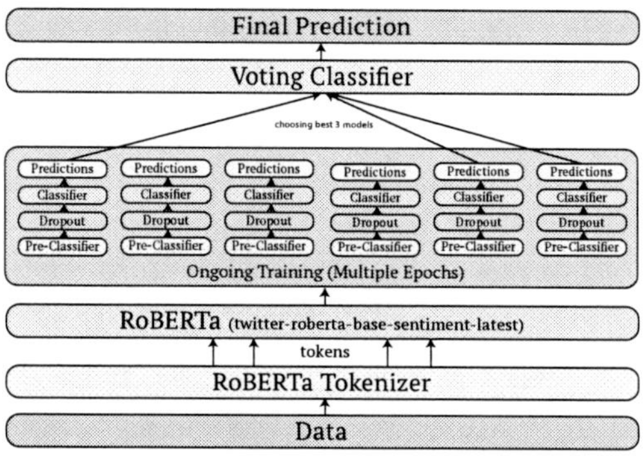

Figure 1: Model architecture, containing tokenizer, pre-trained RoBERTa, classifier and other components

2 System Description

2.1 Problem and Dataset

This binary classification task involves automatically distinguishing tweets indicating a user reported pregnancy on Twitter and subsequently mentioned having a child with specific disorders (ADHD, ASD, delayed speech, or asthma) labeled as "1", from tweets that simply mention a disorder (labeled as "0"). This task enables large-scale use of Twitter for epidemiologic studies and to understand parents' experiences with targeted support interventions (Klein et al., 2024). The dataset includes 7398 training tweets, 389 validation tweets, and 1947 test tweets.

2.2 RoBERTa Model

Our work uses a pre-trained RoBERTa model, namely "*cardiffnlp/twitter-roberta-base-sentiment-latest*" (R-TRBSL, in short) from HuggingFace[1], originally developed by Loureiro et al.

[1] https://huggingface.co/cardiffnlp/twitter-roberta-base-sentiment-latest

Proceedings of The 9th Social Media Mining for Health Applications (#SMM4H) Workshop and Shared Tasks, pages 10–12
August 15, 2024

Model	MLP	Dropout	T F1	T Precision	T Recall	V F1	V Precision	V Recall
BERT-base	786, 786	0.3	88.92	87.32	89.12	86.67	82.55	89.31
BERT-base	786, 786	0.5	89.42	87.84	89.58	86.97	83.12	89.64
RoBERTa-base	786, 786	0.3	92.42	92.72	93.32	90.34	86.53	92.42
RoBERTa-base	786, 786, 512	0.3 3	92.22	92.27	92.78	90.02	87.46	92.24
RoBERTa-base	786, 786	0.5	92.12	92.22	**93.46**	90.13	87.55	92.23
RoBERTa-base	786, 786, 512	0.5	92.36	92.24	93.02	91.24	88.32	91.72
RoBERTa-large	786, 786	0.3	91.23	91.23	91.36	91.44	91.43	91.34
RoBERTa-large	786, 786	0.5	91.34	91.55	92.33	91.34	88.34	92.26
R-TRBSL	786, 786	0.3	**92.81**	**92.89**	92.69	93.42	93.43	93.27
R-TRBSL	786, 786	0.5	92.23	92.34	92.33	92.82	92.76	92.81
R-TRBSL Top3 En.[t]	786, 786	0.3	92.75	93.06	92.71 6	93.21	93.26	93.16
R-TRBSL Best S.[t]	786, 786	0.3	92.93	93.03	92.83	**93.52**	**93.41**	**93.31**

Table 1: Evaluation results by our models on the training and validation set on different setups.
[t]Test set submissions, T=Train, V=Validation, E=Epoch, En.=Ensemble, S.=Single

Models Submitted	F1	Precision	Recall
1. R-TRBSL Top3 En.	0.898	0.883	0.914
2. R-TRBSL Best S.	0.892	0.866	**0.921**
Task Mean	0.822	0.818	0.838
Task Median	**0.901**	**0.885**	0.917

Table 2: Evaluation results by our models on the test set, together with the mean and median results of the task.

(TimeLMs) and fine-tuned by Camacho-collados et al. (TweetNLP) on TweetEval dataset, developed by Barbieri et al. (2020). We also use the original tokenizer (*RobertaTokenizer* class) used to get embeddings for our model, from HuggingFace transformers library (Wolf et al., 2020).

2.3 Implementation Details

The *RobertaTokenizer* uses a maximum length of 256 tokens with lowercase text processing. Both the pre-classifier and classifier are MLPs with 768 nodes, employing a Sigmoid activation function and ReLU between layers, with 0.3 dropout. Training batch size is 8, validation, and test batch sizes are 4. Learning rate is $1e-05$, using Binary Cross Entropy loss and Adam optimizer (Kingma and Ba, 2017). All the random seeds used are 101. In R-TRBSL, we train the model for 10 epochs and combines predictions from the best 3 epochs using a voting classifier as shown in Figure 1. Other models are also trained for 10 epochs with early stopping if over-fitted. No additional data is used.

3 Evaluation

In Table 1, we can see both train and test results of different model setups. Among BERT (Devlin et al., 2018), RoBERTa-base (Liu et al., 2019), RoBERTa-large (Conneau et al., 2019) and R-TRBSL (Camacho-collados et al., 2022), R-TRBSL works the best. MLP setup with (786,786) consistently performs the best, and a dropout probability of 0.3 generally outperforms 0.5 in most cases. Overall, the best single model score is 93.52 on F1-Macro score, with 92.82% accuracy in validation set. While Accuracy and F1 scores remain consistent, precision and recall exhibit variability. Certain models excel in precision but show lower recall, while others demonstrate strong recall but lower precision. Despite these variations, the combined Macro-F1 score remains stable across different configurations. Also, R-TRBSL performs better than other models, probably because it is previously trained on a twitter sentiment dataset.

The results on the test set are outlined in Table 2. The two models we sent for evaluation, the single best model (R-TRBSL Best S.+ MLP (786, 786) + Dropout p=0.3, 2 Epochs) and the ensemble of top 3 epochs obtained almost similar results, with a 89.8% F1 for the former and 89.2% for the later. This puts our solution 6.4% F1 above the mean task score. Though the ensemble model performs weakly in validation set than single model (see Table 1, 93.82 vs 93.16 on F1 score), it works better in the test data (89.8 vs 89.2); showing our approach to keep the model stable worked.

4 Conclusion

Studying social media data continues to be vital in health informatics research, providing valuable insights into public sentiment on health-related topics. Using pre-trained language models like RoBERTa and BERT, we obtain text embeddings and fine-tuned them using MLP classifiers on the task dataset with strategic adjustments to model layers and hyperparameters, enhancing performance. Additionally, we applied an ensemble approach on epochs to ensure performance stability. Our top models achieved 0.9316 and 0.9382 F1-macro on validation and 0.898 and 0.892 on the test set, demonstrating its effectiveness.

References

Francesco Barbieri, Jose Camacho-Collados, Luis Espinosa-Anke, and Leonardo Neves. 2020. TweetEval:Unified Benchmark and Comparative Evaluation for Tweet Classification. In *Proceedings of Findings of EMNLP*.

Jose Camacho-collados, Kiamehr Rezaee, Talayeh Riahi, Asahi Ushio, Daniel Loureiro, Dimosthenis Antypas, Joanne Boisson, Luis Espinosa Anke, Fangyu Liu, and Eugenio Martínez Cámara. 2022. TweetNLP: Cutting-edge natural language processing for social media. In *Proceedings of the 2022 Conference on Empirical Methods in Natural Language Processing: System Demonstrations*, pages 38–49, Abu Dhabi, UAE. Association for Computational Linguistics.

Alexis Conneau, Kartikay Khandelwal, Naman Goyal, Vishrav Chaudhary, Guillaume Wenzek, Francisco Guzmán, Edouard Grave, Myle Ott, Luke Zettlemoyer, and Veselin Stoyanov. 2019. Unsupervised cross-lingual representation learning at scale. *CoRR*, abs/1911.02116.

Jacob Devlin, Ming-Wei Chang, Kenton Lee, and Kristina Toutanova. 2018. BERT: pre-training of deep bidirectional transformers for language understanding. *CoRR*, abs/1810.04805.

Diederik P. Kingma and Jimmy Ba. 2017. Adam: A method for stochastic optimization. *Preprint*, arXiv:1412.6980.

Ari Z Klein, José Agustín Gutiérrez Gómez, Lisa D Levine, and Graciela Gonzalez-Hernandez. 2024. Using longitudinal twitter data for digital epidemiology of childhood health outcomes: An annotated data set and deep neural network classifiers. *Journal of Medical Internet Research*, 26:e50652.

Yinhan Liu, Myle Ott, Naman Goyal, Jingfei Du, Mandar Joshi, Danqi Chen, Omer Levy, Mike Lewis, Luke Zettlemoyer, and Veselin Stoyanov. 2019. Roberta: A robustly optimized BERT pretraining approach. *CoRR*, abs/1907.11692.

Daniel Loureiro, Francesco Barbieri, Leonardo Neves, Luis Espinosa Anke, and Jose Camacho-collados. 2022. TimeLMs: Diachronic language models from Twitter. In *Proceedings of the 60th Annual Meeting of the Association for Computational Linguistics: System Demonstrations*, pages 251–260, Dublin, Ireland. Association for Computational Linguistics.

Thomas Wolf, Lysandre Debut, Victor Sanh, Julien Chaumond, Clement Delangue, Anthony Moi, Pierric Cistac, Tim Rault, Rémi Louf, Morgan Funtowicz, Joe Davison, Sam Shleifer, Patrick von Platen, Clara Ma, Yacine Jernite, Julien Plu, Canwen Xu, Teven Le Scao, Sylvain Gugger, Mariama Drame, Quentin Lhoest, and Alexander M. Rush. 2020. Huggingface's transformers: State-of-the-art natural language processing. *Preprint*, arXiv:1910.03771.

Dongfang Xu, Guillermo Lopez Garcia, Lisa Raithel, Rolland Roller, Philippe Thomas, Eiji Aramaki, Shuntaro Yada, Pierre Zweigenbaum, Karen O'Connor, Yao Ge, Sudeshna Das, Abeed Sarker, Ari Klein, Lucia Schmidt, Vishakha Sharma, Raul Rodriguez-Esteban, Juan Banda, Ivan Flores Amaro, Sai Tharuni Samineni, and Graciela Gonzalez-Hernandez. 2024. Overview of the 9th social media mining for health applications (#SMM4H) shared tasks at ACL 2024. In *Proceedings of the 9th Social Media Mining for Health Applications Workshop and Shared Task*, Bangkok, Thailand. Association for Computational Linguistics.

DILAB at #SMM4H 2024: Analyzing Social Anxiety Effects through Context-Aware Transfer Learning on Reddit Data

Sheikh Ayatur Rahman
BRAC University
Dhaka, Bangladesh
sheikh.ayatur.rahman@g.bracu.ac.bd

Azmine Toushik Wasi
Shahjalal University of Science and Technology
Sylhet, Bangladesh
azmine32@student.sust.edu

Abstract

This paper illustrates the system we design for Task 3 of the 9th Social Media Mining for Health (SMM4H 2024) shared tasks. The task presents posts made on the Reddit social media platform, specifically the *r/SocialAnxiety* subreddit, along with one or more outdoor activities as pre-determined keywords for each post. The task then requires each post to be categorized as either one of *positive*, *negative*, *no effect*, or *not outdoor activity* based on what effect the keyword(s) have on social anxiety. Our approach focuses on fine-tuning pre-trained language models to classify the posts. Additionally, we use fuzzy string matching to select only the text around the given keywords so that the model only has to focus on the contextual sentiment associated with the keywords. Using this system, our peak score is 0.65 macro-F1 on the validation set and 0.654 on test set.

1 Introduction

Analyzing health-related topics from social media data, such as Twitter, Facebook, and Reddit, to gauge public sentiment is an area of significant research interest. The SMM4H 2024 (Xu et al., 2024) shared tasks encourage researchers to address some of these research problems.

We decided to take part in Task 3. The focus of Task 3 is on **S**ocial **A**nxiety **D**isorder (SAD) (Leigh and Clark, 2018), and the motivation behind it is that while a significant number of people may experience SAD in their lives, they may experience its symptoms for much longer before actually seeking professional help. However, people often turn to social media to discuss symptoms of SAD such as the *r/socialanxiety* subreddit. In particular, this task aims to understand the effects of outdoor activities on the symptoms of SAD.

Our approach to this involved stripping text from each post so that only the context surrounding the keywords was fed into our model. Our model consisted of a RoBERTa backbone, followed a dense layer, a batch normalization layer, a dropout layer, and finally a classification layer, which will classify the post into one of four categories depending on the effect of the outdoor activity on symptoms of SAD : *positive*, *negative*, *no effect*, or *keyword is not an outdoor activity*. We then proceeded to fine-tune this model on that data provided.

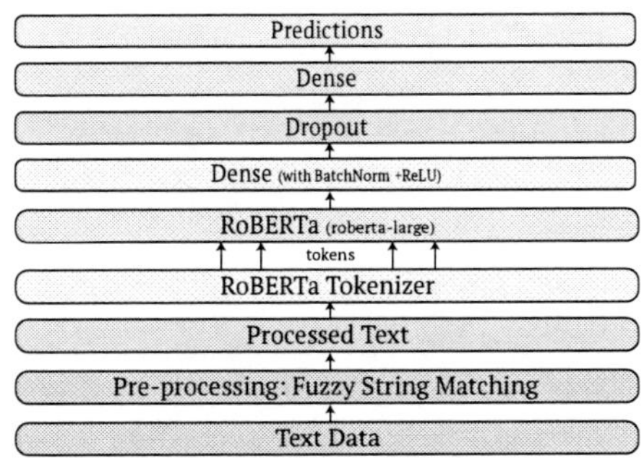

Figure 1: Model architecture

2 System Description

Dataset. The data provided consisted of posts made on the *r/socialanxiety* subreddit, along with one or more outdoor activity keywords, and a class label for each post. In total, there were 1,800 posts in the training set, 600 posts in the validation set, and 600 posts in the test set.

Keyword	Text	Class
run	21/m. I want to experience young love, but I've never had a relationship before... (continued)	0

Table 1: Example of a data point

Pre-processing with Fuzzy String Matching. Using the *fuzzywuzzy* library, we select all instances of each keyword that appeared in the posts using Fuzzy String Matching (FSM) using Levenshtein Distance (Miller et al., 2009). To limit false positives, we only select matched keywords whose length was at least 3, and whose *similarity_score* returned from the *fuzzywuzzy.process.extract* func-

Proceedings of The 9th Social Media Mining for Health Applications (#SMM4H) Workshop and Shared Tasks, pages 13–16
August 15, 2024 ©2024 Association for Computational Linguistics

Model	FSM	Dropout	Tr Ac.	Tr F1	Tr Pr.	Tr Rc.	Val Ac.	Val F1	Val Pr.	Val Rc.
RoBERTa-base	N	0.3	**0.993**	**0.992**	**0.992**	**0.991**	0.728	0.569	0.585	0.556
RoBERTa-base	N	0.4	0.986	0.980	0.979	0.980	0.728	0.577	0.610	0.563
RoBERTa-base	Y	0.3	0.981	0.969	0.974	0.965	0.773	**0.653**	0.644	0.665
RoBERTa-base^t	Y	0.4	0.977	0.966	0.965	0.967	0.762	0.637	0.620	0.662
RoBERTa-Large^t	N	0.3	0.966	0.940	0.948	0.934	0.750	0.624	0.6200	0.632
RoBERTa-Large	N	0.4	0.977	0.967	0.971	0.977	0.747	0.598	0.625	0.581
RoBERTa-Large	Y	0.3	0.961	0.963	0.961	0.969	0.750	0.641	0.619	**0.687**
RoBERTa-Large^t	Y	0.4	0.983	0.968	0.965	0.972	**0.777**	0.651	**0.656**	0.656

Table 2: Evaluation results by our models on the training and validation set on different setups. ^{t}Test submissions, Tr=Train, Val=Validation,Ac.=Accuracy Pr=Precision Rc =Recall Y=Yes N=No

Submission	F1	Prec.	Rec.	Acc.
RoBERTa-base	0.590	0.587	0.620	0.633
RoBERTa-L	0.631	0.617	0.657	0.670
RoBERTa-L-FSM	**0.654**	**0.654**	**0.661**	**0.693**
Task Mean	0.5186	0.5649	0.5379	0.5746
Task Median	0.5795	0.6300	0.5885	0.6270

Table 3: Test set performance

tion is at least 90. For each instance of a matched keyword, we only select the sentence containing the keyword, the sentence preceding it, and the sentence following it. This is used as a means to ensure the model focused only on the outdoor activity and the contextual sentiment associated with it in order to perform a classification. We also keywords at the beginning of each post to give them more impact on the model's final classification.

Model. Our model relies on a RoBERTa[1] backbone (Liu et al., 2019), from HuggingFace transformers library (Wolf et al., 2020). Each sentence is first passed into the RoBERTa backbone. Since, we treat the task as a simple sequence classification task, we perform mean-pooling on the 768-dimensional embeddings (1024 if RoBERTa-large) generated by the RoBERTa model. Then, we pass the pooled embedding into 768-dimensional (1024 if RoBERTa-large) dense layer followed by a batch normalization and ReLU layer. We then pass the output through a dropout layer before passing it through a dense layer, which classifies it as one of the four categories.

Implementation Details. Our tokenizer employs a maximum token length of 256 tokens with lowercase text processing. We fine-tune models for 20 epochs using a batch size of 8, a learning rate of $1e^{-5}$ with the Adam optimizer (Kingma and Ba, 2017) and dropout p=0.4, and the cross entropy loss function. All trials use a fixed random seed of 42, and no additional data is used.

[1]https://huggingface.co/FacebookAI/roberta-base

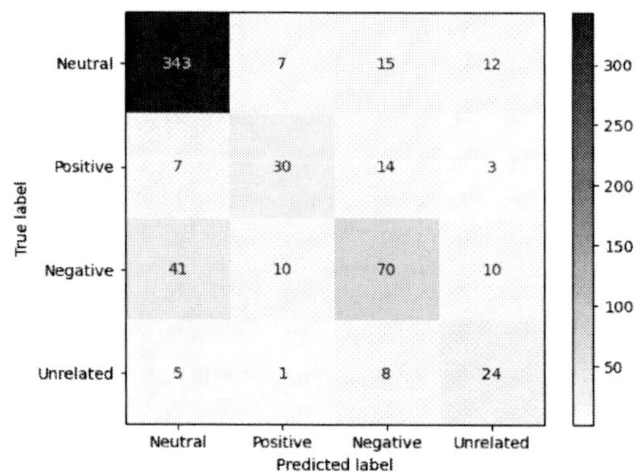

Figure 2: Raw confusion matrix of the output of RoBERTa-large with FSM on validation data

3 Results and Discussion

3.1 Training and Validation Results.

Table 2 presents the performance of different models on both training and validation data. We trained two models, RoBERTa-base and RoBERTa-large. Among these, RoBERTa-base without FSM but with dropout probability $p = 0.3$ performed best during training, achieving higher scores across all metrics. However, it showed signs of overfitting as it did not generalize well to the validation set. For RoBERTa-base, dropout probability $p = 0.3$ yielded better average scores. Conversely, for RoBERTa-large, $p = 0.4$ led to better average scores. In terms of validation data, the top-performing model was RoBERTa-large with FSM and dropout probability $p = 0.4$, achieving 77.7% accuracy and 65.6% precision. Meanwhile, the model with the highest F1 score was RoBERTa-base with FSM and dropout probability $p = 0.3$, achieving a macro F1 score of 0.653.

3.2 Impact of Text Pre-processing with Fuzzy String Matching (FSM).

In Table 2, we observe a consistent trend: models without text pre-processing using Fuzzy String

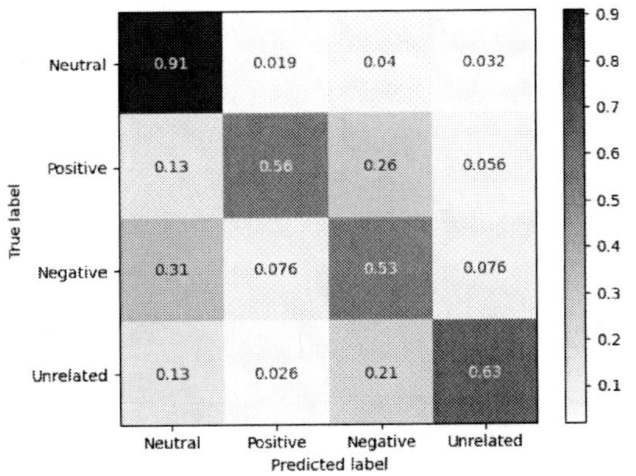

Figure 3: Normalized confusion matrix of the output of RoBERTa-large with FSM on validation data

Matching (FSM) perform better on the training data across all metrics but exhibit lower scores on the validation data. This suggests that without FSM, the models tend to overfit on noisy, unrelated data. However, with FSM, the models can focus on essential context to improve prediction accuracy.

3.3 Submissions and Test Results.

In Table 3, the test results are summarized. Among our three submissions RoBERTa-base with FSM, RoBERTa-Large with and without FSM—RoBERTa-Large with Fuzzy String Matching (FSM) performed the best, achieving an F1 score of 0.654 and 69.3% accuracy. All our submissions surpassed the mean and median task scores across all metrics.

3.4 Error analysis of our best performing model

We performed error analysis on the output of our best performing model - RoBERTa large with FSM trained using a dropout of 0.3 - on the validation set. The raw and normalized confusion matrices of the model's outputs are given in 2 and 3 respectively. We can observe that the *neutral* class (the outdoor activity has no effect), is the class on which the model performs the best. This can be explained by the imbalanced nature of the dataset. In the training set, of the 1800 training examples, 1131 belong to the *neutral* class, 160 to the *positive* class, 395 to the *negative* class, and 114 to the *unrelated* class (the keyword is not intended as an outdoor activity in the text). Due to the low numbers in the other classes, the scope for more granular analysis is small. However, a possible source inconsistency in the dataset may arise due to the fact that for examples, the correct label may be subjective. For instance consider the text - *Be a night shift stocker at Walmart. You don't really have to run the register and interact* - for which the keyword mentioned was *run*. The original label for this example was *neutral*. However, it can argued that since "run" in this scenario is not referring to the outdoor activity of running as in one would in a field or park, it should be labelled as *unrelated*. However, in this instance, "running a register" was likely considered an outdoor social activity. Subjectivity in assigning a label can lead to inconsistencies even if one person is labelling, as being consistent across multiple examples can be challenging. As such, this may lead to noisy labels and affect the model's ability to recognize patterns.

4 Conclusion

In this work, we study Social Anxiety Disorder using Reddit data to identify individuals who may experience its symptoms for a prolonged period before seeking professional help. By applying fuzzy string matching to find contexts and transfer learning with RoBERTa, we achieve a validation set macro-F1 score of 0.65 and a test set score of 0.654, outperforming the task mean and median scores. We also have analyzed errors in our best performing model with possible explanations and reasoning.

5 Limitations

While our method outperforms the mean macro-F1 score for the competition, there are some limitations to our approaches which can be addressed in future works. Some of them are listed below:

- Since we are only selecting sentences containing the keyword along with the ones preceding and succeeding them, context relevant to the keyword that are far away from the keyword is being lost. This may result in information relevant to the keyword being lost. It may be possible to design more intelligent ways of restricting context.
- We employed a very simple method to make the model focus more on the keywords - by prepending the text a few times with the keyword. It may be worthwhile to explore other ways of integrating keyword information.

References

Diederik P. Kingma and Jimmy Ba. 2017. Adam: A method for stochastic optimization. *Preprint*, arXiv:1412.6980.

Eleanor Leigh and David M Clark. 2018. Understanding social anxiety disorder in adolescents and improving treatment outcomes: Applying the cognitive model of clark and wells (1995). *Clin. Child Fam. Psychol. Rev.*, 21(3):388–414.

Yinhan Liu, Myle Ott, Naman Goyal, Jingfei Du, Mandar Joshi, Danqi Chen, Omer Levy, Mike Lewis, Luke Zettlemoyer, and Veselin Stoyanov. 2019. Roberta: A robustly optimized BERT pretraining approach. *CoRR*, abs/1907.11692.

Frederic P. Miller, Agnes F. Vandome, and John McBrewster. 2009. *Levenshtein Distance: Information theory, Computer science, String (computer science), String metric, Damerau?Levenshtein distance, Spell checker, Hamming distance*. Alpha Press.

Thomas Wolf, Lysandre Debut, Victor Sanh, Julien Chaumond, Clement Delangue, Anthony Moi, Pierric Cistac, Tim Rault, Rémi Louf, Morgan Funtowicz, Joe Davison, Sam Shleifer, Patrick von Platen, Clara Ma, Yacine Jernite, Julien Plu, Canwen Xu, Teven Le Scao, Sylvain Gugger, Mariama Drame, Quentin Lhoest, and Alexander M. Rush. 2020. Huggingface's transformers: State-of-the-art natural language processing. *Preprint*, arXiv:1910.03771.

Dongfang Xu, Guillermo Lopez Garcia, Lisa Raithel, Rolland Roller, Philippe Thomas, Eiji Aramaki, Shuntaro Yada, Pierre Zweigenbaum, Karen O'Connor, Yao Ge, Sudeshna Das, Abeed Sarker, Ari Klein, Lucia Schmidt, Vishakha Sharma, Raul Rodriguez-Esteban, Juan Banda, Ivan Flores Amaro, Sai Tharuni Samineni, and Graciela Gonzalez-Hernandez. 2024. Overview of the 9th social media mining for health applications (#SMM4H) shared tasks at ACL 2024. In *Proceedings of the 9th Social Media Mining for Health Applications Workshop and Shared Task*, Bangkok, Thailand. Association for Computational Linguistics.

Dolomites@#SMM4H 2024
Helping LLMs "*Know The Drill*" in Low-Resource Settings A Study on Social Media Posts

Giuliano Tortoreto †, **Seyed Mahed Mousavi** ‡
† WISECode LLC
‡ Signals and Interactive Systems Lab, University of Trento, Italy
gtortoreto@wisecode.ai, mahed.mousavi@unitn.it

Abstract

The amount of data to fine-tune LLMs plays a crucial role in the performance of these models in downstream tasks. Consequently, it is not straightforward to deploy these models in low-resource settings. In this work, we investigate two new multi-task learning data augmentation approaches for fine-tuning LLMs when little data is available: "In-domain Augmentation" of the training data and extracting "*Drills*" as smaller tasks from the target dataset. We evaluate the proposed approaches in three natural language processing settings in the context of SMM4H 2024 competition tasks: multi-class classification, entity recognition, and information extraction. The results show that both techniques improve the performance of the models in all three settings, suggesting a positive impact from the knowledge learned in multi-task training to perform the target task.

1 Introduction

Collecting an adequate amount of data to fine-tune Large Language Models (LLM) in low-resource settings is an expensive practice. To address the lack of enough data in such settings, data augmentation techniques have been commonly used in different natural language processing tasks (Wei and Zou, 2019; Feng et al., 2021) as a low-cost solution. Such techniques focus on generating new examples close to the original samples and data distribution, thus increasing the number of training samples.

More recently, data augmentation has been studied from the perspective of multi-task learning. In Multi-Task Learning Data Augmentation (MTL-DA), the dataset of the target task is augmented with the data of other auxiliary tasks that can improve the model performance, despite having different characteristics (Sánchez-Cartagena et al., 2021). The model is then jointly optimized in a multi-task manner on the augmented data, resulting in more robust performance compared to traditional augmentation techniques (Wei et al., 2021).

In this work, we investigate two new MTL-DA techniques to fine-tune LLMs for low-resource tasks: **a) In-domain Augmentation** where we augment the original target dataset with the data (and tasks) collected from the same source (e.g. Reddit Social Forum); and **b) Drills** where we extract from the target task a set of smaller tasks and replicate the dataset to cover the corresponding samples for the extracted drills. We study the proposed techniques on three language processing tasks in the context of SMM4H 2024 challenge[1]: *I) Task 3*: four-class classification of Reddit social forum posts; *II) Task 4*: an entity recognition task on Reddit posts to identify two entity types (Ge et al., 2024); *III) Task 6*: information extraction from tweets and Reddit posts. A complete description of the tasks is presented in §A.

The results indicate the effectiveness of both augmentation techniques by achieving higher F_1 scores on the validation sets, compared to target task fine-tuning (note that the competition test sets are not publicly available), as well as scoring two to five points higher than overall scores in SMM4H competition.

2 Approach

2.1 Proposed Techniques

We experiment with two new MTL-DA techniques to fine-tune LLMs for low-resource settings:

a) In-domain Augmentation We jointly fine-tune the model on an augmented dataset, consisting of the data of the target task and other tasks collected from the same source. Here, we leverage the fact that Reddit is the common data source across all tasks. We then optimize/evaluate the performance of the model on each target task.

b) Drills We extract smaller tasks for each tar-

[1]SMM4H 2024: "The 9th Social Media Mining for Health Research and Applications Workshop and Shared Tasks" (Xu et al.)

Proceedings of The 9th Social Media Mining for Health Applications (#SMM4H) Workshop and Shared Tasks, pages 17–22
August 15, 2024

Approach	Task 3			Task 4			Task 6		
	F_1	P	R	F_1	P	R	F_1	P	R
GPT-4 *(In-Context Learning)*									
0-shot	46.4	47.4	55.5	57.2	63.3	57.6	88.6	86.2	91.2
5-shot	50.1	51.1	62.4	59.4	56.5	67.6	85.6	80.7	91.2
Mistral *(fine-tuning)*									
Target Task	52.7	60.3	50.3	50.5	56.3	52.6	**89.5**	95.0	84.6
Target Task + Drills	46.6	55.5	44.6	55.4	65.9	51.2	86.2	92.0	81.1
In-domain	52.9	61.7	50.2	**58.5**	62.9	55.7	87.5	94.2	81.8
In-domain + Drills	**53.8**	57.1	52.5	56.0	60.1	53.5	85.8	94.1	78.8

Table 1: Evaluation of the proposed MTL-DA techniques (In-domain Augmentation and Drills) on the validation set of each task, compared to In-Context Learning (ICL) in {*zero,few*}-shot settings (the test set is not publicly available). While ICL with GPT-4 achieves the highest ***R***ecall in all tasks, MTL-DA approaches generally achieve higher ***P***recision and thus competitive $\boldsymbol{F_1}$ scores. In-domain augmentation of the training set achieves competitive results in all tasks and manages to obtain the highest $\boldsymbol{F_1}$ score in Task 4, the highest score in Task 3 by additional Drills, and the second best $\boldsymbol{F_1}$ score in Task 6.

get task to augment the dataset. Regarding Task 3, we extracted two drills of "*sentiment classification: Pos|Neg|Neut.*" and "*post relevance classification: Y|N*". As a result, the training set was replicated twice to include the additional training examples. Regarding Task 4, we extracted two binary classification drills as "*includes social impact: Y|N*" and "*includes clinical: Y|N*", as well as two sequence extraction drills as "*social impact extraction*" and "*clinical impact extraction*". As a result, the training set was replicated four times to include the additional training examples. Concerning Task 6, we extracted one binary classification drill, in addition to the main task, as "*Source Platform Classification:Reddit|Twitter*", and duplicated the training set.

2.2 Model

We experimented with Mistral (7B) (Jiang et al., 2023), a decoder-only LLM. We applied QLoRA (Dettmers et al., 2024) to fine-tune the model, due to the limited GPU memory. Details of our implementation are provided in §B.

3 Evaluation

Since the test sets of the competition tasks are not publicly available, we analyze the performance of the model and the impact of the MTL-DA techniques on the validation set. We consider the *vanilla fine-tuning* of the model for the target downstream task as the baseline in our analysis. Furthermore, we compare the performance of the model with GPT-4[2] via In-Context Learning (ICL) in {0,5}-shot settings in each task as a low-cost alternative. We then present the scores obtained by the best-performing models and MTL-DA techniques in SMM4H competition, compared to the overall mean and median scores obtained by all participants.

3.1 Validation Set

The performance of Mistral with different combinations of the proposed augmentation techniques on the validation set is presented in Table 1. MTL-DA techniques result in considerable improvements in the model performance for Tasks 3 and 4. However, vanilla fine-tuning of the model achieves the highest score in Task 6, suggesting that the relatively bigger training data for Task 6 is adequate to train the model without additional augmentation techniques. In-domain Augmentation shows promising results in all tasks, suggesting the positive impact of joint multi-task learning. Introducing the Drills improved the model in Tasks 3 and 4 when combined with In-domain Augmentation. Regarding ICL, 5-shot learning achieves the highest recall score in all tasks. However, fine-tuned models manage to obtain higher precision and, accordingly, outperforming F_1 scores.

3.2 Error Analysis

To gain better insight into the impact of MTL-DA techniques, we manually controlled the fraction of the validation set in which models with different

[2]GPT-4-Turbo (gpt-4-0125-preview)

Approach	Task 3			Task 4			Task 6		
	F_1	P	R	Rel F_1	Str F_1	Tok F_1	F_1	P	R
SMM4H Competition									
Overall Mean	52.7	57.2	54.4	40.7	14.6	41.7	92.4	92.4	92.6
Overall Median	59.6	63.1	60.1	40.4	16.4	48.9	93.6	93.4	94.9
Our Scores									
Target Task	/	/	/	/	/	/	91.4	97.7	86.0
In-domain	55.8	66.3	53.0	**44.9**	**20.1**	49.6	/	/	/
In-domain + Drills	**64.2**	67.0	62.3	36.8	16.4	37.5	**95.7**	96.5	94.9

Table 2: The performance of the proposed approaches on the test sets of each task, in addition to the overall mean and median of all participants in the SMM4H 2024 competition. The results indicate the positive impact of MTL-DA techniques by achieving two to five points higher than overall scores. Note that the metrics used for Task 4 are Relaxed F_1 (For Ranking), Strict F_1, Token-level F_1.

MTL-DA techniques provided different predictions (disagreements). Regarding **Task 3** we observed that the model with vanilla fine-tuning on the target task excels at correctly predicting the label for emotionally nuanced samples. Meanwhile, the augmentation techniques further improve the model's ability in correctly predicting the label for neutral/unrelated samples. In **Task 4**, we noticed that the model with MTL-DA techniques performs better on samples that include keywords for medical side-effects/symptoms and their associated impacts such as "*brain zaps*", "*difficult urinating*" and "*hair loss*". Regarding **Task 6**, we observed that the model with augmentation techniques is more effective in identifying explicit age mentions. However, the model with vanilla fine-tuning on the target task shows more robustness in predicting cases lacking exact age mentions, which is frequent in discussions focused on age-related health concerns. More details are presented in §D.

3.3 SMM4H Test Sets

Given that the submissions for the competition are limited to 3, we applied Borda Count to determine the three top models in the validation set. The results of the best-performing models on each task, presented in Table 2, indicate the positive impact of augmentation techniques. Similar to the performance observed on the validation set, the model with *In-domain augmentation* and with *In-domain augmentation + Drills* achieved the highest performance in Tasks 3 and 4, respectively, each scoring approximately four points higher than the overall median of the competition. Interestingly, while vanilla fine-tuning achieved the highest performance on the validation set in Task 6, it is outperformed on the test set by the model with *In-domain augmentation + Drills* by approximately four points. This result suggests that augmentation techniques can increase the model's robustness to handle unseen samples.

4 Conclusion

We studied two new Multi-Task Learning Data Augmentation techniques to address the lack of adequate data to fine-tune LLMs in low-resource settings. We evaluated the proposed techniques in three tasks in the context of the SMM4H 2024 competition, and we observed a) the positive impact of augmentation techniques in improving model performance; and b) the potential of in-context learning as a low-cost surrogate to fine-tuning. Nevertheless, besides further ablation studies, there are a few unanswered questions: a) "*What is the best set of drills given a task?*"; b) "*When should a dataset be considered in-domain?*"; c) "*Would the same task on different domains also contribute positively?*"; d) "*How does the introduced approach compare to existing data augmentation techniques?*"; and e) "*Does the replication of the dataset potentially lead to overfitting?*". We aim to explore these questions in future work.

Acknowledgement

We acknowledge the support of the MUR PNRR project FAIR - Future AI Research (PE00000013) funded by the NextGenerationEU.

References

Tim Dettmers, Artidoro Pagnoni, Ari Holtzman, and Luke Zettlemoyer. 2024. Qlora: Efficient finetuning of quantized llms. *Advances in Neural Information Processing Systems*, 36.

Steven Y. Feng, Varun Gangal, Jason Wei, Sarath Chandar, Soroush Vosoughi, Teruko Mitamura, and Eduard Hovy. 2021. A survey of data augmentation approaches for NLP. In *Findings of the Association for Computational Linguistics: ACL-IJCNLP 2021*, pages 968–988, Online. Association for Computational Linguistics.

Yao Ge, Sudeshna Das, Karen O'Connor, Mohammed Ali Al-Garadi, Graciela Gonzalez-Hernandez, and Abeed Sarker. 2024. Reddit-impacts: A named entity recognition dataset for analyzing clinical and social effects of substance use derived from social media. *Preprint*, arXiv:2405.06145.

Albert Q Jiang, Alexandre Sablayrolles, Arthur Mensch, Chris Bamford, Devendra Singh Chaplot, Diego de las Casas, Florian Bressand, Gianna Lengyel, Guillaume Lample, Lucile Saulnier, et al. 2023. Mistral 7b. *arXiv preprint arXiv:2310.06825*.

Colin Raffel, Noam Shazeer, Adam Roberts, Katherine Lee, Sharan Narang, Michael Matena, Yanqi Zhou, Wei Li, and Peter J. Liu. 2020. Exploring the limits of transfer learning with a unified text-to-text transformer. *Journal of Machine Learning Research*, 21(140):1–67.

Víctor M. Sánchez-Cartagena, Miquel Esplà-Gomis, Juan Antonio Pérez-Ortiz, and Felipe Sánchez-Martínez. 2021. Rethinking data augmentation for low-resource neural machine translation: A multi-task learning approach. In *Proceedings of the 2021 Conference on Empirical Methods in Natural Language Processing*, pages 8502–8516, Online and Punta Cana, Dominican Republic. Association for Computational Linguistics.

Jason Wei, Chengyu Huang, Shiqi Xu, and Soroush Vosoughi. 2021. Text augmentation in a multi-task view. In *Proceedings of the 16th Conference of the European Chapter of the Association for Computational Linguistics: Main Volume*, pages 2888–2894, Online. Association for Computational Linguistics.

Jason Wei and Kai Zou. 2019. EDA: Easy data augmentation techniques for boosting performance on text classification tasks. In *Proceedings of the 2019 Conference on Empirical Methods in Natural Language Processing and the 9th International Joint Conference on Natural Language Processing (EMNLP-IJCNLP)*, pages 6382–6388, Hong Kong, China. Association for Computational Linguistics.

Dongfang Xu, Guillermo Lopez Garcia, Lisa Raithel, Rolland Roller, Philippe Thomas, Eiji Aramaki, Shuntaro Yada, Pierre Zweigenbaum, Sai Tharuni Samineni, Karen O'Connor, Yao Ge, Sudeshna Das, Abeed Sarker, Ari Klein, Lucia Schmidt, Vishakha Sharma, Raul Rodriguez-Esteban, Juan Banda, Ivan Flores Amaro, Davy Weissenbacher, and Graciela Gonzalez-Hernandez. Overview of the 9th social media mining for health applications (#SMM4H) shared tasks at ACL 2024. In *Proceedings of The 9th Social Media Mining for Health Research and Applications Workshop and Shared Tasks", month = Aug, year =*.

Appendix

A SMM4H 2024 Tasks

We evaluate the proposed techniques on three tasks: **Task 3: Multi-class classification of effects of outdoor spaces on social anxiety symptoms** consisting of categorizing posts on the Reddit platform into four classes `positive effect`, `negative effect`, `neutral`, and `unrelated`. A total of 3000 annotated posts are provided into sets of 1800, 600, and 600 posts as training, validation, and test sets, respectively. **Evaluation:** To evaluate the model, we compute the macro-average F1-score over all 4 classes as done in the competition.

Task 4: Extraction of the clinical and social impacts of nonmedical substance use from Reddit consisting of an entity recognition tasks on Reddit posts for two entity types `clinical impact` and `social impact`. A total of 1380 posts are provided into sets of 843, 259, and 278 posts as training, validation, and test sets respectively. **Evaluations:** The model is tasked to predict the relevant spans. We automatically post-process the predicted tag for each token in the predicted span for both social and clinical impact. The prediction is then evaluated as a traditional entity tagging task (Note that it is different than the evaluation set-up deployed by SMM4H competition organizers).

Task 6: Self-reported exact age classification with cross-platform evaluation in English consists of extracting self-reported exact age from posts on X (Twitter) and Reddit. The training data consists of 8.8k tweets as well as 100k unannotated Reddit posts, while the validation set consists of 2.2k tweets and 1000 Reddit posts. The test data for this task includes 2.2k tweets and 14.4k Reddit posts. **Evaluations:** To evaluate the model, we compute the macro-average F1-score for the self-report class (post contains exact age) as done in the Task 6 competition.

B Implementation Details

QLoRA (Dettmers et al., 2024) introduces additional adapter layers that are fine-tuned for the downstream task. We trained the model with two different GPUs on a cloud hosting platform: an A100 (80GB) and an A6000 (48GB). Considering our limitation of GPU power, we applied quantization to the LORA layers. We used 100 warm-up steps and, as suggested in QLora (Dettmers et al., 2024), we kept a ratio of $\alpha = 2 \times rank$ with rank=4,8,32 for 30 epochs. We applied LoRA adapters to every linear layer *"q_proj", "k_proj", "v_proj", "o_proj","gate_proj", "up_proj", "down_proj"*, and *"lm_head"*. We used a decaying learning rate that starts from 2e-4. We experimented with 2 parameters for the max sequence length, 796 and 1496. The best-performing max sequence length was 1496. This is probably due to the fact that Reddit posts' length required a longer number of tokens. To reduce GPU memory occupation, we halved batch size from two to one and doubled gradient accumulation steps from 16 to 32. To keep the memory footprint small in QLoRA, we enabled model loading in 4 bits and the computation type in `bfloat16`. To keep the memory footprint of the optimizer small, we used Paged AdamW 8bit (a version of Adam Optimizer that leverages CUDA paging). Each entry is packed with multiple sequences, as in T5 (Raffel et al., 2020). The hyperparameter configuration selected is rank=32, alpha=64 LoRA dropout = 0.1.

C Fine-Tuning Costs

Using the A100 GPU, the longest training run took 24 hours on the model that included task drills and tasks for all three SMM4H tasks, summing up to a cost of $76. At the end of the study, the total cost, including all experiments, was $1600. The cost per hour of A100 GPU (80GB) and the A6000 (48GB) was respectively $3.18 and $1.89.

D Error analysis

D.1 Task 6

In Task 6, the Target model proved to be more effective in predicting the absence of the exact age in posts. We can observe this especially in discussions focused on health conditions broadly associated with age ranges ("cataracts with age...", "diagnosed in my late 20s...") rather than specific numerical ages. Conversely, the In-domain + drills model better recognized posts where age is explicitly stated ("I'm 28 going on 29...", "I'm 19..."). This indicates that the Target model tends to predict more effectively the lack of exact self-reported age, while the In-domain + drills model is more reliable when exact ages are directly provided within the text.

D.2 Task 3

The target task model performs better in recognizing the emotional nuances of outdoor experiences.

It accurately predicts posts expressing anxiety, reporting social difficulties ("Crushed by SA...") or highlighting positive achievements ("I GOT TO RIDE IN A TESLA...").

In contrast, the In-domain + Drills model demonstrates strength in identifying neutral activities involving outdoor spaces ("Went outside for the first time in awhile...") and posts where the outdoor setting is peripheral to the core discussion ("Social anxiety has ruined my diet..."). This suggests that the In-domain + Drills model is better at recognizing when the impact of outdoor spaces is primarily neutral or indirect.

D.3 Task 4

In Task 4, the in-domain model outperforms the target model in recognizing specialized medical terminology and its associated impacts. It accurately identifies clinically significant terms such as "brain zaps" and "hair loss" (potential side effects), as well as the severity of "cravings for heroin" (addiction). The in-domain model also demonstrates a better grasp of the nuanced social impacts related to health. Although the in-domain model's higher recall leads to some incorrectly predicted social impacts, in general it achieves higher performance in the identification and classification of clinically relevant keywords.

RIGA at SMM4H-2024 Task 1: Enhancing ADE discovery with GPT-4

Eduards Mukans
eduards.mukans@lu.lv
University of Latvia, Faculty of Computing

Guntis Barzdins
guntis.barzdins@lumii.lv
University of Latvia, IMCS

Abstract

The following is a description of the RIGA team's submissions for the SMM4H-2024 Task 1: Extraction and normalization of adverse drug events (ADEs) in English tweets. Our approach focuses on utilizing Large Language Models (LLMs) to generate data that enhances the fine-tuning of classification and Named Entity Recognition (NER) models. Our solution significantly outperforms mean and median submissions of other teams. The efficacy of our ADE extraction from tweets is comparable to the current state-of-the-art solution, established as the task baseline. The code for our method is available on GitHub[1].

1 Introduction

The SMM4H-2024 Task 1, as outlined in the overview (Xu et al., 2024), challenged participants to extract and normalize ADEs to MedDRA high-level term identifiers (HLTIs).

Our submission aims to harness the capabilities of large language models (LLMs) to enhance performance. Additionally, we compare the performance of the off-the-shelf submission, which did not involve model training, with a fine-tuned model that combines the original input with the output generated by GPT.

2 Related work

The baseline system (Magge et al., 2021) utilizes a pipeline method for solving the task. The pipeline involves 3 components and are executed sequentially: (1) the ADE classifier for identifying tweets containing ADE mentions, (2) the ADE span extractor or named entity recognition (NER) for extracting ADE mentions, and (3) the ADE normalizer, which maps the extracted ADE mention to MedDRA HLT identifiers. In our submission we utilize the same pipeline components.

[1]https://github.com/emukans/smm4h2024-riga

Dataset	Train	Dev	Test
Full	18185	965	11799
Contain ADEs	1239	65	N/A

Table 1: Dataset size distribution

The paper concentrates on integrating the GPT model generation with the original text. A comparable methodology was employed in SemEval-2023 (Mukans and Barzdins, 2023), where the task involved token classification with highly specific tags. To streamline the process, the RIGA team utilized GPT as a knowledge database for individuals, entities, food items, and other relevant entities mentioned in the text.

According to the LLM for Generative Information Retrieval Survey (Xu et al., 2023), our method can be classified as a form of data augmentation. Similar approaches have been independently employed in several studies (Amalvy et al., 2023; Chen and Feng, 2023; Li et al., 2023)

3 Data

In contrast to the previous version of the task, the new challenge in the most recent dataset lies in the inclusion of negative samples (falling outside MedDRA categories) in each data split.

As presented in Table 1, the data is highly imbalanced. The amount of tweets containing any ADE is 6.8% for train data split and 6.7% for dev data split.

4 Methodology

In order to address the issue of high data imbalance, our pipeline includes tweet classification as the initial step to filter out the majority of the tweets. Subsequently, for the filtered tweets, we conduct NER to extract the precise spans that contain an ADEs. In the final step, we generate a sentence embedding for the span and identify the nearest

Proceedings of The 9th Social Media Mining for Health Applications (#SMM4H) Workshop and Shared Tasks, pages 23–27
August 15, 2024 ©2024 Association for Computational Linguistics

Submission	F1-Norm	P-Norm	R-Norm	F1-NER	P-NER	R-NER	F1-Norm-Unseen
GPT few-shot	**31.8**	**29.5**	**34.6**	40.3	37.7	43.4	21.2
Custom + GPT	10.3	12.1	9	**47.9**	**52.5**	**44.1**	6.5
Baseline	43.9	39.3	49.8	48.1	43.1	54.3	32.3
Mean	28.264	29.244	33.388	32.672	35.625	34.032	20.936
Median	29.3	33.9	32.6	37.6	43.7	37.4	14.1

Table 2: The performance of our submissions

HLTIs using cosine similarity.

We used four Tesla v100 16GB GPUs, provided by our institution, for conducting these experiments.

4.1 Tweet classification

According to Table 1, more than 93% of the tweets do not contain any ADEs. To filter out these tweets, we developed a binary classification model to identify the presence of ADEs in the input tweets. This model is based on a language model fine-tuned from RoBERTa-large (Antypas et al., 2023; Liu et al., 2019).

Before the classification model fine-tuning, all tweets are preprocessed with GPT-4 Turbo (OpenAI et al., 2024) prompt engineering (Brown et al., 2020) to extract mentioned ADEs in the text. The generative model simply needs to mention all ADEs from the provided text in a free-form manner. The prompt used in our submission is detailed in Appendix A.

The GPT output is then concatenated with the original tweet in the following format and used as input to a binary classification model:

`{tweet} <sep> {ADE extracted with GPT}`.

4.2 ADE span extraction

All categorized tweets with ADEs are forwarded to the span extraction stage. We employ a BIO-tagging schema with only three tags: B-ADE, I-ADE, and O.

In this stage, we also incorporate GPT output as additional context for downstream fine-tuning. The prompt utilized in our submission is detailed in Appendix B.

As shown in Table 2, the ADEs generated by `GPT few-shot` demonstrate strong performance in comparison to the mean and median scores. However, a notable limitation of GPT is its verbosity and propensity for hallucinations. Often, the generated spans contain verbs that contribute to coherent sentence structures but are not directly pertinent to ADEs.

Furthermore, the model may generate text that deviates from the original text. For instance, it might produce ADE expressions that do not exactly match the words in the given tweet. This issue goes beyond minor discrepancies, such as differences in American and British spelling, and highlights a broader challenge in utilizing generative models for extracting ADEs from tweets. The foundational model's training datasets, like C4, which predominantly feature texts with American dialects, contribute to this bias (Dodge et al., 2021).

To fine-tune DeBERTaV3 for span extraction (He et al., 2021), we adopt a similar input structure as in the classification step. However, since tweets may contain multiple ADEs, we separate each ADE in the input using the `sep` token.

The output generated by the fine-tuned model `Custom + GPT`, using the following input format is less noisy compared to the original GPT results.

4.3 Span mapping to MedDRA HLTIs

In total, MedDRA contains 23,389 HLTIs, but the training and development data only contain 319 unique identifiers. This indicates that the majority of HLTIs are not present in our dataset.

Training a classifier to map the spans to the HLTIs using the provided data would be futile due to the high variety of HLTIs. Additionally, the test data includes unseen categories that the trained classifier would not be able to identify.

In our submission, we utilized an off-the-shelf solution by leveraging OpenAI's Embedding API. Initially, we computed an embedding representation for all MedDRA HLTIs, followed by doing the same for each ADE span. Subsequently, we identified the closest HLTIs by calculating the cosine similarity between the embeddings.

Unfortunately, we ran out of resources and time to achieve a higher F1-Norm score for the `Custom + GPT` model. Despite using the same approach as the `GPT few-shot` model, the `Custom + GPT` model's performance on F1-Norm suffered.

5 Results

In Table 2, we compare our solutions with the current state-of-the-art solution, which serves as a baseline for the task. The competition evaluates performance using two metrics: F1-Norm and F1-NER. Our primary focus was on the F1-NER metric, where the `Custom + GPT` model demonstrates performance comparable to the baseline and significantly higher than both the mean and median. The `GPT few-shot` submission also achieved results above both the mean and median for both metrics.

Acknowledgments

This work has been supported by the EU Recovery and Resilience Facility projects Language Technology Initiative (No 2.3.1.1.i.0/1/22/I/CFLA/002) and Latvian Quantum Initiative (No. 2.3.1.1.i.0/1/22/I/CFLA/001).

References

Arthur Amalvy, Vincent Labatut, and Richard Dufour. 2023. Learning to rank context for named entity recognition using a synthetic dataset. In *Proceedings of the 2023 Conference on Empirical Methods in Natural Language Processing*, pages 10372–10382, Singapore. Association for Computational Linguistics.

Dimosthenis Antypas, Asahi Ushio, Francesco Barbieri, Leonardo Neves, Kiamehr Rezaee, Luis Espinosa-Anke, Jiaxin Pei, and Jose Camacho-Collados. 2023. Supertweeteval: A challenging, unified and heterogeneous benchmark for social media nlp research. In *Findings of the Association for Computational Linguistics: EMNLP 2023*.

Tom B. Brown, Benjamin Mann, Nick Ryder, Melanie Subbiah, Jared Kaplan, Prafulla Dhariwal, Arvind Neelakantan, Pranav Shyam, Girish Sastry, Amanda Askell, Sandhini Agarwal, Ariel Herbert-Voss, Gretchen Krueger, Tom Henighan, Rewon Child, Aditya Ramesh, Daniel M. Ziegler, Jeffrey Wu, Clemens Winter, Christopher Hesse, Mark Chen, Eric Sigler, Mateusz Litwin, Scott Gray, Benjamin Chess, Jack Clark, Christopher Berner, Sam McCandlish, Alec Radford, Ilya Sutskever, and Dario Amodei. 2020. Language models are few-shot learners. In *Proceedings of the 34th International Conference on Neural Information Processing Systems*, NIPS '20, Red Hook, NY, USA. Curran Associates Inc.

Feng Chen and Yujian Feng. 2023. Chain-of-thought prompt distillation for multimodal named entity recognition and multimodal relation extraction. *Preprint*, arXiv:2306.14122.

Jesse Dodge, Ana Marasovic, Gabriel Ilharco, Dirk Groeneveld, Margaret Mitchell, and Matt Gardner. 2021. Documenting large webtext corpora: A case study on the colossal clean crawled corpus. In *Conference on Empirical Methods in Natural Language Processing*.

Pengcheng He, Jianfeng Gao, and Weizhu Chen. 2021. Debertav3: Improving deberta using electra-style pretraining with gradient-disentangled embedding sharing. *Preprint*, arXiv:2111.09543.

Jinyuan Li, Han Li, Zhuo Pan, Di Sun, Jiahao Wang, Wenkun Zhang, and Gang Pan. 2023. Prompting ChatGPT in MNER: Enhanced multimodal named entity recognition with auxiliary refined knowledge. In *Findings of the Association for Computational Linguistics: EMNLP 2023*, pages 2787–2802, Singapore. Association for Computational Linguistics.

Yinhan Liu, Myle Ott, Naman Goyal, Jingfei Du, Mandar Joshi, Danqi Chen, Omer Levy, Mike Lewis, Luke Zettlemoyer, and Veselin Stoyanov. 2019. Roberta: A robustly optimized BERT pretraining approach. *CoRR*, abs/1907.11692.

Arjun Magge, Elena Tutubalina, Zulfat Miftahutdinov, Ilseyar Alimova, Anne Dirkson, Suzan Verberne, Davy Weissenbacher, and Graciela Gonzalez-Hernandez. 2021. DeepADEMiner: a deep learning pharmacovigilance pipeline for extraction and normalization of adverse drug event mentions on Twitter. *Journal of the American Medical Informatics Association*, 28(10):2184–2192.

Eduards Mukans and Guntis Barzdins. 2023. RIGA at SemEval-2023 task 2: NER enhanced with GPT-3. In *Proceedings of the 17th International Workshop on Semantic Evaluation (SemEval-2023)*, pages 331–339, Toronto, Canada. Association for Computational Linguistics.

OpenAI, Josh Achiam, Steven Adler, Sandhini Agarwal, Lama Ahmad, Ilge Akkaya, Florencia Leoni Aleman, Diogo Almeida, Janko Altenschmidt, Sam Altman, Shyamal Anadkat, Red Avila, Igor Babuschkin, Suchir Balaji, Valerie Balcom, Paul Baltescu, Haiming Bao, Mohammad Bavarian, Jeff Belgum, Irwan Bello, Jake Berdine, Gabriel Bernadett-Shapiro, Christopher Berner, Lenny Bogdonoff, Oleg Boiko, Madelaine Boyd, Anna-Luisa Brakman, Greg Brockman, Tim Brooks, Miles Brundage, Kevin Button, Trevor Cai, Rosie Campbell, Andrew Cann, Brittany Carey, Chelsea Carlson, Rory Carmichael, Brooke Chan, Che Chang, Fotis Chantzis, Derek Chen, Sully Chen, Ruby Chen, Jason Chen, Mark Chen, Ben Chess, Chester Cho, Casey Chu, Hyung Won Chung, Dave Cummings, Jeremiah Currier, Yunxing Dai, Cory Decareaux, Thomas Degry, Noah Deutsch, Damien Deville, Arka Dhar, David Dohan, Steve Dowling, Sheila Dunning, Adrien Ecoffet, Atty Eleti, Tyna Eloundou, David Farhi, Liam Fedus, Niko Felix, Simón Posada Fishman, Juston Forte, Isabella Fulford, Leo Gao, Elie Georges, Christian Gibson, Vik Goel, Tarun Gogineni, Gabriel Goh, Rapha Gontijo-Lopes, Jonathan Gordon, Morgan Grafstein, Scott Gray, Ryan Greene, Joshua Gross, Shixiang Shane

Gu, Yufei Guo, Chris Hallacy, Jesse Han, Jeff Harris, Yuchen He, Mike Heaton, Johannes Heidecke, Chris Hesse, Alan Hickey, Wade Hickey, Peter Hoeschele, Brandon Houghton, Kenny Hsu, Shengli Hu, Xin Hu, Joost Huizinga, Shantanu Jain, Shawn Jain, Joanne Jang, Angela Jiang, Roger Jiang, Haozhun Jin, Denny Jin, Shino Jomoto, Billie Jonn, Heewoo Jun, Tomer Kaftan, Łukasz Kaiser, Ali Kamali, Ingmar Kanitscheider, Nitish Shirish Keskar, Tabarak Khan, Logan Kilpatrick, Jong Wook Kim, Christina Kim, Yongjik Kim, Jan Hendrik Kirchner, Jamie Kiros, Matt Knight, Daniel Kokotajlo, Łukasz Kondraciuk, Andrew Kondrich, Aris Konstantinidis, Kyle Kosic, Gretchen Krueger, Vishal Kuo, Michael Lampe, Ikai Lan, Teddy Lee, Jan Leike, Jade Leung, Daniel Levy, Chak Ming Li, Rachel Lim, Molly Lin, Stephanie Lin, Mateusz Litwin, Theresa Lopez, Ryan Lowe, Patricia Lue, Anna Makanju, Kim Malfacini, Sam Manning, Todor Markov, Yaniv Markovski, Bianca Martin, Katie Mayer, Andrew Mayne, Bob McGrew, Scott Mayer McKinney, Christine McLeavey, Paul McMillan, Jake McNeil, David Medina, Aalok Mehta, Jacob Menick, Luke Metz, Andrey Mishchenko, Pamela Mishkin, Vinnie Monaco, Evan Morikawa, Daniel Mossing, Tong Mu, Mira Murati, Oleg Murk, David Mély, Ashvin Nair, Reiichiro Nakano, Rajeev Nayak, Arvind Neelakantan, Richard Ngo, Hyeonwoo Noh, Long Ouyang, Cullen O'Keefe, Jakub Pachocki, Alex Paino, Joe Palermo, Ashley Pantuliano, Giambattista Parascandolo, Joel Parish, Emy Parparita, Alex Passos, Mikhail Pavlov, Andrew Peng, Adam Perelman, Filipe de Avila Belbute Peres, Michael Petrov, Henrique Ponde de Oliveira Pinto, Michael, Pokorny, Michelle Pokrass, Vitchyr H. Pong, Tolly Powell, Alethea Power, Boris Power, Elizabeth Proehl, Raul Puri, Alec Radford, Jack Rae, Aditya Ramesh, Cameron Raymond, Francis Real, Kendra Rimbach, Carl Ross, Bob Rotsted, Henri Roussez, Nick Ryder, Mario Saltarelli, Ted Sanders, Shibani Santurkar, Girish Sastry, Heather Schmidt, David Schnurr, John Schulman, Daniel Selsam, Kyla Sheppard, Toki Sherbakov, Jessica Shieh, Sarah Shoker, Pranav Shyam, Szymon Sidor, Eric Sigler, Maddie Simens, Jordan Sitkin, Katarina Slama, Ian Sohl, Benjamin Sokolowsky, Yang Song, Natalie Staudacher, Felipe Petroski Such, Natalie Summers, Ilya Sutskever, Jie Tang, Nikolas Tezak, Madeleine B. Thompson, Phil Tillet, Amin Tootoonchian, Elizabeth Tseng, Preston Tuggle, Nick Turley, Jerry Tworek, Juan Felipe Cerón Uribe, Andrea Vallone, Arun Vijayvergiya, Chelsea Voss, Carroll Wainwright, Justin Jay Wang, Alvin Wang, Ben Wang, Jonathan Ward, Jason Wei, CJ Weinmann, Akila Welihinda, Peter Welinder, Jiayi Weng, Lilian Weng, Matt Wiethoff, Dave Willner, Clemens Winter, Samuel Wolrich, Hannah Wong, Lauren Workman, Sherwin Wu, Jeff Wu, Michael Wu, Kai Xiao, Tao Xu, Sarah Yoo, Kevin Yu, Qiming Yuan, Wojciech Zaremba, Rowan Zellers, Chong Zhang, Marvin Zhang, Shengjia Zhao, Tianhao Zheng, Juntang Zhuang, William Zhuk, and Barret Zoph. 2024. Gpt-4 technical report. *Preprint*, arXiv:2303.08774.

Derong Xu, Wei Chen, Wenjun Peng, Chao Zhang, Tong Xu, Xiangyu Zhao, Xian Wu, Yefeng Zheng, and Enhong Chen. 2023. Large language models for generative information extraction: A survey. *Preprint*, arXiv:2312.17617.

Dongfang Xu, Guillermo Lopez Garcia, Lisa Raithel, Rolland Roller, Philippe Thomas, Eiji Aramaki, Shuntaro Yada, Pierre Zweigenbaum, Sai Tharuni Samineni, Karen O'Connor, Yao Ge, Sudeshna Das, Abeed Sarker, Ari Klein, Lucia Schmidt, Vishakha Sharma, Raul Rodriguez-Esteban, Juan Banda, Ivan Flores Amaro, Davy Weissenbacher, and Graciela Gonzalez-Hernandez. 2024. Overview of the 9th social media mining for health applications (#SMM4H) shared tasks at ACL 2024. In *Proceedings of The 9th Social Media Mining for Health Research and Applications Workshop and Shared Tasks*, Bangkok, Thailand. Association for Computational Linguistics.

A Classification prompt

For tweet classification we used the following prompt:

```
   You will be provided with a tweet.
Summarise it into a brief sentence and
highlight already happened adverse drug
events (ADE) if there are any related to
drugs.
Format:
Summary: {text}
ADE: {text or null}
_
Tweet:
"""
{tweet}
"""
```

The model generates two lines in the output: "Summary" and "ADE." In our submission, we utilize only the "ADE" field. The intention behind the "Summary" field was to classify summarized tweets instead of the original text, potentially simplifying the task by producing summaries in a unified language and style. Unfortunately, this hypothesis did not hold. GPT likely omits important keywords common to many ADE-containing tweets, or the semantics of the generated text do not match the original tweet. It is probable that using a "rewrite" instruction instead of "summarize" would have been more effective.

B ADE extraction prompt

For mining text spans containing ADEs we used the following prompt:

```
You will be provided with a tweet.
Your task is to identify and highlight
any adverse drug events (ADEs) mentioned
in relation to drug use. Only the
exact phrases describing the ADEs should
be outputted, without including any
additional context. Each ADE should
be listed on a new line. If the same
ADE is mentioned multiple times, each
occurrence should be listed separately.
If multiple different ADEs are identified
within the same tweet, they should be
listed on separate lines. If no ADEs are
found, output "null".
-
Format:
SPAN: {text or null}
-
Samples:
Tweet:
"""
 user
if avelox has hurt your liver, avoid
tylenol always, as it further damages
liver, eat grapefruit unless taking
cardiac drugs
"""
SPAN: hurt your liver
-
Tweet:
"""
losing it. could not remember the word
power strip. wonder which drug is doing
this memory lapse thing. my guess the
cymbalta. helps
"""
SPAN: not remember
SPAN: memory lapse
Tweet:
"""
is adderall a performance enhancing drug
for mathletes?
"""
SPAN: null
-
Tweet:
"""
{tweet}
"""
```

Since the most of tweets will be filtered out during the classification step, and token classification is more complex task, than sequence classification, the prompt contains more instructions and output samples.

Golden_Duck at #SMM4H 2024: A Transformer-based Approach to Social Media Text Classification

Md. Ayon Mia, Mahshar Yahan, Hasan Murad, Muhammad Ibrahim Khan
Department of Computer Science and Engineering
Chittagong University of Engineering and Technology, Bangladesh
u1804{128, 007}@student.cuet.ac.bd, hasanmurad@cuet.ac.bd, muhammad_ikhan@cuet.ac.bd

Abstract

In this paper, we have addressed Task 3 on social anxiety disorder identification and Task 5 on mental illness recognition organized by the SMM4H 2024 workshop. In Task 3, a multi-classification problem has been presented to classify Reddit posts about outdoor spaces into four categories: Positive, Neutral, Negative, or Unrelated. Using the pre-trained RoBERTa-base model along with techniques like Mean pooling, CLS, and Attention Head, we have scored an F1-Score of 0.596 on the test dataset for Task 3. Task 5 aims to classify tweets into two categories: those describing a child with conditions like ADHD, ASD, delayed speech, or asthma (class 1), and those merely mentioning a disorder (class 0). Using the pre-trained RoBERTa-large model, incorporating a weighted ensemble of the last 4 hidden layers through concatenation and mean pooling, we achieved an F1 Score of 0.928 on the test data for Task 5.

1 Introduction

In the past few years, social media has become the primary source of communication. Unfortunately, millions of social media users suffer from mental illnesses such as social anxiety, attention-deficit or hyperactivity disorder (ADHD), autism spectrum disorders (ASD), delayed speech or asthma. The World Health Organization (WHO) estimates that around 450 million individuals are impacted by these conditions. According to WHO, 1 in 7 adolescents from age 10 to 19 experience mental health issues, many of which go unnoticed and untreated.[1].

In this research work, we have presented our findings for the two shared tasks on social anxiety disorder identification (Task 3) and mental illness recognition (Task 5) organized under SMM4H 2024 workshop (Xu et al., 2024). By employing transformer-based pre-trained models alongside different techniques, we obtained F1 scores of 0.596 for Task 3 and 0.928 for Task 4.

By examining user-generated content from social media platforms like Reddit and Twitter, this research aims to shed light on the potential therapeutic benefits of outdoor environments and enhance our understanding of childhood developmental disorders.

2 Related Work

Automatic mental illness detection using social media data has been a key research area in literature. The key focus of this study (Klein et al., 2022) is to detect and identify conditions such as ADHD, ASD, delayed speech, and asthma in children using data from Twitter. It suggests Twitter data could offer a high-performance method for this epidemiological research. They have achieved notable performance using several models: SVM with an F1 score of 0.74, BERT-Base-Uncased achieving an F1 score of 0.85, and BERTweet-Large scoring an F1 score of 0.92. Mukherjee et al. (2023) discuss a system using traditional machine learning models for the early detection of Autism Spectrum Disorder (ASD). They have utilized various machine learning techniques, including SVM (accuracy of 0.71), Logistic Regression (accuracy of 0.71), K-nearest neighbor (accuracy: 0.62), and Random Forest (accuracy: 0.69). Additionally, they have explored the stacked deep learning models like CNN+SVM (accuracy of 0.916) and RNN+SVM (accuracy of 0.866). Ta et al. (2024) focus on classifying text data related to social disorders in children and adolescents using encoder-decoder models and data augmentation methods, achieving the best F1 scores of 0.627 in Task 3 and 0.841 in Task 5. The research highlights the importance of employing sophisticated deep learning networks to identify social disorders from social media data. According

[1]https://www.who.int/news-room/fact-sheets/detail/adolescent-mental-health

Proceedings of The 9th Social Media Mining for Health Applications (#SMM4H) Workshop and Shared Tasks, pages 28–31
August 15, 2024 ©2024 Association for Computational Linguistics

to Ameer et al. (2022), the paper explores the application of social media for mental health communication. It emphasizes the significance of automating the detection of mental illness using RoBERTa (achieving an accuracy and F1 score of 0.83) and BERT (with accuracy and F1 score of 0.78 and 0.80 respectively) models for classifying mental health disorders based on Reddit posts.

3 Task and Dataset Description

These shared tasks have been organized by the SMM4H 2024 workshop (Xu et al., 2024), including two tasks related to identifying and classifying medical information from social media data.

Task 3, entitled "Classification of reported effects of outdoor spaces on social anxiety symptoms", entails a multi-class classification challenge. Its objective is to sort Reddit posts containing specific keywords related to outdoor environments into four distinct categories: (i) Positive effect (0), (ii) Neutral or no effect (1), (iii) Negative effect (2), and (iv) Unrelated (keywords not pertaining to actual outdoor spaces or activities) (3).

Task 5, named "Binary classification of tweets reporting children's medical disorders", introduces a binary classification challenge. The goal of this task is to automatically identify tweets from users who disclosed their pregnancy on Twitter and report a child with ADHD, ASD, delayed speech, or asthma (labeled as "1"), versus tweets that just mention these disorders (labeled as "0"). In Task 3, the data is divided into 1800 training samples, 600 validation samples, and 1200 test samples, with the test data not publicly provided. Similarly, in Task 5, there are 7398 training samples, 389 validation samples, and 10,000 test samples, with the test data also not publicly provided.

Sets	**Classes**				**Total**
	0	1	2	3	
Train	1131	160	395	114	1800
Valid	377	54	131	38	600

Table 1: Dataset statistics of Task 3.

Sets	**Classes**		**Total**
	0	1	
Train	5118	2280	7398
Valid	254	135	389

Table 2: Dataset statistics of Task 5.

4 Methodology

4.1 Preprocessing

We have applied similar preprocessing steps for both tasks: removed URLs, hashtags, and user mentions from text, removed emojis, lowered all English letters, and removed duplicate punctuations. For Task 5, we have conducted additional preprocessing by expanding contractions like "wouldn't" to "would not".

4.2 Augmentation

In Task 3, as the classes are imbalanced, we have employed random oversampling using the scikit-learn library. For minority classes 1, 2, and 3, instances have been resampled with replacements to a target count of 350 each.

4.3 Fine-tuning Process

We have explored different transformer models for both shared tasks. For Task 3, we have fine-tuned the following transformer models:

- XLM-RoBERTa-base, followed by Mean pooling of the output
- RoBERTa-base with the concatenation of Mean pooling, CLS, and Attention Head
- RoBERTa-base with additional Attention Head on top of the model
- BERT-base-uncased, with weighted layer pooling from the last 4 layers

For Task 5, we have employed ensemble methods combining multiple output representations from large pre-trained language models. Specifically, we have fine-tuned following models:

- RoBERTa-large (Liu et al., 2019) utilizes a weighted ensemble approach that concatenates the last 4 hidden layers and employs mean pooling.
- RoBERTa-large uses a combination of weighted ensemble by concatenating the last 4 hidden layers, weighted pooling of these layers, and mean pooling.
- BERT-large uses a weighted ensemble by concatenating the last 4 hidden layers and utilizes mean pooling.

Model	F1 Score	Precision	Recall	Accuracy
XLM-RoBERTa-base (Mean pooling)	0.540	0.543	0.560	0.697
RoBERTa-base: Concat of Mean pooling, CLS, Attention Head	**0.562**	0.563	0.573	0.692
RoBERTa-base + Attention Head	0.333	0.533	0.332	0.67
BERT-base-uncased: WeightedLayer-Pooling (last 4 layers)	0.226	0.253	0.266	0.64

Table 3: Validation set outcomes for Task 3 for different models

Model	F1 Score	Precision	Recall	Accuracy
XLM-RoBERTa-base (Mean pooling)	0.59	0.586	0.612	0.623
RoBERTa-base: Concat of Mean pooling, CLS, Attention Head	**0.596**	0.603	0.601	0.618
RoBERTa-base + Attention Head	0.361	0.577	0.376	0.505
Mean	0.5186	0.5649	0.5379	0.5746
Median	0.5795	0.63	0.5885	0.627

Table 4: Results of various models in Task 3 on the official test set

Model	F1 Score	Precision	Recall
RoBERTa-large: Weighted Ensemble of LastConcat, LastPooled, Mean Pooling	0.912	0.899	0.926
RoBERTa-large: Weighted Ensemble of LastConcat and Mean Pooling	**0.923**	0.913	0.933
BERT-large: Weighted Ensemble of LastConcat, Mean Pooling	0.864	0.834	0.896

Table 5: Performance of different models in task 5 on the validation set where LastConcat, and LastPooled denote "Concatenation of last 4 hidden layers" and "Weighted pooling of last 4 layers" respectively

Model	F1 Score	Precision	Recall
RoBERTa-large: Weighted Ensemble of LastConcat, LastPooled, Mean Pooling	0.908	0.902	0.914
RoBERTa-large: Weighted Ensemble of LastConcat and Mean Pooling	**0.928**	0.919	0.937
Mean	0.822	0.818	0.838
Median	0.901	0.885	0.917

Table 6: Official results in Task 5 on the test set using different models

For the ensemble models, the weights have been treated as hyperparameters and tuned on the validation set to maximize the F1 score. First, we have evaluated each output representation independently on the validation data. Next, the top performing representations have been linearly combined using tuned weights, effectively creating an ensemble of multiple complementary representations.

The textual data has been preprocessed as per subsection 4.1 and tokenized with maximum lengths of 512 and 100 for Tasks 3 and 5, respectively, using a subword tokenizer.The hyperparameters used for both tasks include the AdamW optimizer, with learning rates set to 2e-5 for Task 3 and 1e-5 for Task 5. Batch sizes were 16 for Task 3 and 32 for Task 5, with training conducted over 10 epochs and early stopping determined by validation performance. A dropout rate of 0.3 was applied for regularization purposes. The whole experiment has been run on the Kaggle environment with the infrastructure including the NVIDIA Tesla P100 GPU with 16GB VRAM, 29GB RAM, and 4 CPU cores.

5 Results Analysis

We have evaluated the model's performance on the validation and test sets for both Task 3 and Task 5. Specifically, Table 3 and Table 4 present the results for Task 3 on the validation and test sets, respectively. On the validation set, the RoBERTa-base model with the concatenation of Mean pooling, CLS, and Attention Head has achieved the best F1 score of 0.562. This model with the same configuration has obtained the highest F1 score of 0.596, precision of 0.603, recall of 0.601, and accuracy of 0.618 on the test case. For Task 5, Table 5 and Table 6 present the validation and test set performance respectively. On the validation set, the RoBERTa-large model with a weighted ensemble of the last 4 hidden layers concatenation and Mean pooling has attained the best F1-score of 0.923. This model configuration has shown the best performance on the test dataset, with an F1-score of 0.928, precision of 0.919, and recall of 0.937.

6 Conclusion

In this paper, we outline our contributions to Task 3 and Task 5 for the 2024 SMM4H workshop on Social Media Mining for Health Applications. We utilized several transformer models to perform multi-class classification on Reddit posts and binary classification on tweets. Utilizing RoBERTa-base with Mean Pooling, CLS, and Attention Head concatenation, we have reached an F1 score of 0.596 for Task 3. For Task 5, we have employed RoBERTa-large with a weighted ensemble of the last 4 hidden layers concatenation and Mean Pooling, obtaining an F1-score of 0.928. Our experiments demonstrated our system's reliability and adaptability, and we are working on further improvements.

References

Iqra Ameer, Muhammad Arif, Grigori Sidorov, Helena Gòmez-Adorno, and Alexander Gelbukh. 2022. Mental illness classification on social media texts using deep learning and transfer learning. *arXiv preprint arXiv:2207.01012*.

Ari Z Klein, José Agustín Gutiérrez Gómez, Lisa D Levine, and Graciela Gonzalez-Hernandez. 2022. Automatically identifying childhood health outcomes on twitter for digital epidemiology in pregnancy. *medRxiv*, pages 2022–11.

Yinhan Liu, Myle Ott, Naman Goyal, Jingfei Du, Mandar Joshi, Danqi Chen, Omer Levy, Mike Lewis, Luke Zettlemoyer, and Veselin Stoyanov. 2019. Roberta: A robustly optimized bert pretraining approach. *arXiv preprint arXiv:1907.11692*.

Prasenjit Mukherjee, Sourav Sadhukhan, Manish Godse, and Baisakhi Chakraborty. 2023. Early detection of autism spectrum disorder (asd) using traditional machine learning models. *International Journal of Advanced Computer Science and Applications*, 14(6).

Hoang-Thang Ta, Abu Bakar Siddiqur Rahman, Lotfollah Najjar, and Alexander Gelbukh. 2024. Thangdlu at# smm4h 2024: Encoder-decoder models for classifying text data on social disorders in children and adolescents. *arXiv preprint arXiv:2404.19714*.

Dongfang Xu, Guillermo Lopez Garcia, Lisa Raithel, Rolland Roller, Philippe Thomas, Eiji Aramaki, Shuntaro Yada, Pierre Zweigenbaum, Karen O'Connor, Yao Ge, Sudeshna Das, Abeed Sarker, Ari Klein, Lucia Schmidt, Vishakha Sharma, Raul Rodriguez-Esteban, Juan Banda, Ivan Flores Amaro, Sai Tharuni Samineni, and Graciela Gonzalez-Hernandez. 2024. Overview of the 9th social media mining for health applications (#SMM4H) shared tasks at ACL 2024. In *Proceedings of the 9th Social Media Mining for Health Applications Workshop and Shared Task*, Bangkok, Thailand. Association for Computational Linguistics.

SRCB at #SMM4H 2024: Making Full Use of LLM-based Data Augmentation in Adverse Drug Event Extraction and Normalization

Hongyu Li[1], Yuming Zhang[1], Yongwei Zhang[1], Shanshan Jiang[1], and Bin Dong[1]

[1]Ricoh Software Research Center (Beijing) Co., Ltd
{Hongyu.Li, Yuming.Zhang1, Yongwei.Zhang, Shanshan Jiang, Bin Dong}@cn.ricoh.com

Abstract

This paper reports on the performance of SRCB's system in the Social Media Mining for Health (#SMM4H) 2024 Shared Task 1: extraction and normalization of adverse drug events (ADEs) in English tweets. We develop a system composed of an ADE extraction module and an ADE normalization module which further includes a retrieval module and a filtering module. To alleviate the data imbalance and other issues introduced by the dataset, we employ 4 data augmentation techniques based on Large Language Models (LLMs) across both modules. Our best submission achieves an F1 score of 53.6 (49.4 on the unseen subset) on the ADE normalization task and an F1 score of 52.1 on ADE extraction task.

1 Introduction

The Social Media Mining for Health (#SMM4H) Workshop has served as a competitive platform aimed at promoting the development and evaluation of advanced natural language processing (NLP) systems for the detection, extraction and normalization of health related information in social media texts (tweets, reviews and Reddit posts). Among the shared tasks in #SMM4H 2024 (Xu et al., 2024), ADE extraction and normalization has been the longest running task, which requires participants first extract spans of adverse drug events (ADEs) expressions from tweets and then normalize the spans to MedDRA[1] ontology's preferred terms (PTs). This task is evaluated in the following order of priority: the F1 score of ADE normalization, the F1 score of ADE normalization on the unseen subset[2] and F1 score of ADE extraction.

The challenges of this task lie in: (1) The dataset exhibits extreme imbalance between samples containing ADEs (positive) and those not containing ADEs (negative). (2) The validation and test sets contain ADEs to which the corresponding PTs are not seen during training. (3) The given texts are tweets with frequent use of informal grammar as well as irregular vocabulary.

In recent years, LLMs have been widely used for data augmentation (Cai et al., 2023; Whitehouse et al., 2023; Zhang et al., 2024) to improve data quality, especially for the unbalanced and noisy data. To this end, we propose to make full use of LLM-based data augmentation using GLM-4[3] and GPT-3.5 across both ADE extraction and ADE normalization modules. For ADE extraction, we first enrich the training set from both mention-level and context-level by prompting the LLMs to rewrite ADE spans and the other parts of the positive samples. To cover more ADEs with unseen preferred terms, we prompt LLMs to generate synthetic tweets and obtain pseudo ADE annotations by our previously trained ADE extraction models. For ADE normalization, we ask the LLMs to rewrite the given tweets into formal written texts. In addition, we also ask the LLMs to give an explanation for each lowest level term (LLT) and preferred term (PT) in MedDRA dictionary. These LLM-based data augmentation techniques show varying degrees of performance improvement on our system. Finally, our system obtained the highest ADE normalization F1 score in task 1.

2 System Description

we employ a relatively comprehensive pipeline for data pre-processing detailed in Appendix A. Our system includes an ADE extraction module which extracts spans of ADEs from the given tweets and an ADE normalization module which normalizes the extracted spans to PTs. We further decompose ADE normalization into two steps, namely MedDRA term (LLTs, PTs) retrieval and MedDRA

[1]https://www.meddra.org/

[2]The ADEs to which the corresponding preferred terms are not seen during training.

[3]https://open.bigmodel.cn/

Proceedings of The 9th Social Media Mining for Health Applications (#SMM4H) Workshop and Shared Tasks, pages 32–37
August 15, 2024

term filtering. MedDRA term retrieval involves retrieving up to 20 MedDRA terms by conducting similarity search bewteen the embeddings of each extracted span and the LLTs and PTs. Given the input tweet, an extracted span and each of the retrieved LLTs or PTs, a MedDRA term filtering model is asked to classify whether the retrieved LLT or PT happens as an ADE in the tweet or not. The base models we use are detailed in Appendix B.

3 Data Augmentation Using LLMs

Since the dataset poses challenges such as data unbalance, unseen PTs and domain mismatch as illustrated before, we utilize 4 different LLM-based (GLM-4, GPT-3.5) data augmentation techniques across ADE extraction and ADE normalization. More details are provided in D

3.1 ADE Mention Rewriting and Context Rewriting

To address the issue of unbalanced data, we increase the number of positive samples by leveraging LLMs to rewrite ADE mentions and their contexts separately, and then combining them. The process includes: (1) ADE Mention Rewriting: We prompt GLM-4 to rewrite each ADE mention while keeping the remaining parts unchanged three times. This gives us four versions of each ADE mention (the original plus three new ones). (2) Context Rewriting: We mask the ADE mentions in the tweets and prompt GLM-4 to rewrite the remaining parts, creating three new versions of each tweet. This results in four different tweet templates for each original tweet (the original plus three new ones). (3) Combining Rewritten Mentions and Contexts: We sample up to three combinations[4] of ADE mentions and insert each combination into a randomly chosen rewritten tweet template, generating new positive samples. By using 3 different random seeds during the sampling process, we create three distinct augmented training sets, leading to more model candidates trained with different datasets during ensemble.

3.2 LLM Synthetic Tweets with Pseudo ADE Annotations

We note that there are ADEs of unseen PTs in the validation and test sets. Therefore we prompt the LLMs to generate synthetic tweets with more ADE expressions which may be corresponding to unseen PTs. We first extract all potential drugs in the train and validation sets using GLM-4 and then validate whether the extracted drugs are genuine drugs with GPT-3.5 to get rid of the noise. Meanwhile, we request GPT-3.5 to list the common side effects of each validated drug. We then prompt GPT-3.5 to generate 3 tweets based on a given drug and one or two sampled side effects of it. We sample the side effects for 3 times, thus obtain 9 tweets for each drug ideally. Finally, we get pseudo ADE annotations by conducting majority voting over the predictions of 27 ADE extraction models, which are trained with data augmented by ADE mention rewriting and context rewriting[5].

3.3 Tweet rewriting

In the provided tweets, informal grammar and irregular vocabulary are frequently used. This typically results in performance degradation in language models pre-trained with a general-domain corpus such as BERT (Devlin et al., 2019). Therefore, we prompt GLM-4 to rewrite the tweets into formal written texts easier to understand. The rewritten tweets are used during MedDRA filtering.

3.4 MedDRA Term Explanation

Considering explanation of the MedDRA terms (LLTs, PTs) may benefit context understanding in MedDRA filtering, we prompt GPT-3.5 to give an explanation for each MedDRA term. During MedDRA term filtering, we append the description of the given MedDRA term to the end of the input sequence.

4 Results

4.1 Experiment Results

ADE extraction The evaluation results on the validation set of our models are illustrated in Table 1. The ADE extraction F1 scores shown in the table are averaged over models trained with random seeds of 42, 21 and 1, and also averaged over all models trained with 3 different MR-CR augmented training sets with different sampling seeds. A significant performance improvement is observed for both techniques of LLM-based data augmentation. And the scores are further improved

[4] We did not utilize all combinations, as doing so would bias the data distribution towards samples with a higher number of entities.

[5] 27 models: using 3 different augmented training sets, fine-tuned based on RoBERTa-large, BERTweet-large and DeBERTa-v3-large with random seeds of 42, 21 and 1

Traing data	PLM	ADE Extraction F1 (Dev)
ORIG†	RoBERTa-large	67.16
	BERTweet-large	67.51
	DeBERTa-v3-large	71.97
ORIG+MR-CR	RoBERTa-large	70.82
	Bertweet-large	71.16
	DeBERTa-v3-large	74.43
ORIG+SynTweet	RoBERTa-large	71.93
	BERTweet-large	72.88
	DeBERTa-v3-large	73.07
ORIG+MR-CR +SynTweet	Roberta-large	72.10
	BERTweet-large	72.01
	DeBERTa-v3-large	75.39

Table 1: The ADE extraction F1 scores on validation set for our ADE extraction models. ORIG represents the original training set. † represents the baseline models. MR-CR denotes the augmented data created by ADE Mention Rewriting and Context Rewriting. SynTweet denotes the LLM-generated tweets.

through combining both techniques on RoBERTa-large and DeBERTa-v3-large. Among the 3 pre-trained language models used, DeBERTa-v3-large outperforms the other two by a considerable margin.

ADE normalization We select the ensemble ADE extraction results with highest F1 score as the input for ADE normalization to retrieve MedDRA terms. The MedDRA term retrieval model using fine-tuned text embedding model achieves a recall of 85.88, higher than 78.82 without fine-tuning. The evaluation results of our MedDRA term filtering models are illustrated in Table 2. By introducing Tweet Rewriting and MedDRA Term Explanation, the performance is improved in most of our experiments. Furthermore, the models using only rewritten tweets achieved a more impressive improvement than the ones that include original tweets in their inputs. This indicates that the informal grammar as well as irregular vocabulary in the original tweets hinder model learning, and converting the original tweets into formal written texts can effectively address this issue.

4.2 Test Results

Table 3 shows the submission results of our system. Among the submission files, **submission-2** uses the ensemble ADE extraction result over all models trained from DeBERTa-v3-large except for the baseline models. We observe a smaller proportion of positive samples in the ensemble ADE extraction result on the test set, compared to the training and validation sets. Therefore, we add more ADE predictions for **submission-1** and **submission-3** by including the ensemble result of a smaller number of model candidates (9 models based on DeBERTa-v3-large w/ ORIG+MR-CR+SynTweet). In addition, **submission-1** and **submission-2** are ensembled with MedDRA filtering models achieving an F1 score over 77. For **submission-3**, we use the model combination that obtains the highest F1 score on validation set out of all possible combinations.

Input sequence	PLM	ADE Normalization F1 (Dev)
ORIG†	RoBERTa-large	74.73
	BERTweet-large	73.63
	DeBERTa-v3-large	75.82
ORIG&TE	RoBERTa-large	75.14
	Bertweet-large	74.03
	DeBERTa-v3-large	76.24
ORIG&TR	RoBERTa-large	74.87
	BERTweet-large	73.91
	DeBERTa-v3-large	76.50
ORIG&TR&TE	Roberta-large	77.53
	BERTweet-large	76.24
	DeBERTa-v3-large	77.35
TR&TE	Roberta-large	78.89
	BERTweet-large	77.53
	DeBERTa-v3-large	79.33

Table 2: The ADE normalization F1 scores on validation set for our ADE normalization models. The first column compares the differences in input sequences other than the extracted spans and retrieved MedDRA terms. ORIG represents using the original training tweets. † represents the baseline models. & represents the concatenation operation of sequences. TR denotes the Tweet Rewriting sequence. TE denotes the Term Explanation sequence.

Submission	ADE Normalization F1	ADE Normalization F1 (Unseen)	ADE Extraction F1
1	53.6	49.4	52.1
2	52.7	48.9	51.8
3	53.3	49.1	52.1
Official Baseline	43.9	32.3	48.1
Mean	28.3	20.9	32.7
Median	29.3	14.1	37.6

Table 3: Submission results on test set

5 Conclusion

In this work, we propose to use 4 different LLM-based data augmentation techniques for Task 1, including ADE mention rewriting and context rewriting, synthetic tweets with psuedo annotations, tweet rewriting and MedDRA term explanation. As a result, our system, including an ADE extraction module and an ADE normalization module, achieves the highest F1 score (53.6) in Task 1 among all the teams. For future work, modification on the models will be studied to achieve more model-level improvements. Additionally, pipelines based solely on LLMs will be further explored.

References

Alham Fikri Aji, Made Nindyatama Nityasya, Haryo Akbarianto Wibowo, Radityo Eko Prasojo, and Tirana Fatyanosa. 2021. Bert goes brrr: a venture towards the lesser error in classifying medical self-reporters on twitter. In *Proceedings of the Sixth Social Media Mining for Health (# SMM4H) Workshop and Shared Task*, pages 58–64.

Christos Baziotis, Nikos Pelekis, and Christos Doulkeridis. 2017. Datastories at semeval-2017 task 4: Deep lstm with attention for message-level and topic-based sentiment analysis. In *Proceedings of the 11th International Workshop on Semantic Evaluation (SemEval-2017)*, pages 747–754, Vancouver, Canada. Association for Computational Linguistics.

Xunxin Cai, Meng Xiao, Zhiyuan Ning, and Yuanchun Zhou. 2023. Resolving the imbalance issue in hierarchical disciplinary topic inference via llm-based data augmentation. In *2023 IEEE International Conference on Data Mining Workshops (ICDMW)*, pages 1424–1429. IEEE.

Jacob Devlin, Ming-Wei Chang, Kenton Lee, and Kristina Toutanova. 2019. BERT: Pre-training of deep bidirectional transformers for language understanding. In *Proceedings of the 2019 Conference of the North American Chapter of the Association for Computational Linguistics: Human Language Technologies, Volume 1 (Long and Short Papers)*, pages 4171–4186, Minneapolis, Minnesota. Association for Computational Linguistics.

Matthijs Douze, Alexandr Guzhva, Chengqi Deng, Jeff Johnson, Gergely Szilvasy, Pierre-Emmanuel Mazaré, Maria Lomeli, Lucas Hosseini, and Hervé Jégou. 2024. The faiss library.

Jeff Johnson, Matthijs Douze, and Hervé Jégou. 2019. Billion-scale similarity search with GPUs. *IEEE Transactions on Big Data*, 7(3):535–547.

Kenton Lee, Luheng He, Mike Lewis, and Luke Zettlemoyer. 2017. End-to-end neural coreference resolution. In *Proceedings of the 2017 Conference on Empirical Methods in Natural Language Processing*. Association for Computational Linguistics.

Xianming Li and Jing Li. 2023. Angle-optimized text embeddings. *arXiv preprint arXiv:2309.12871*.

Xi Liu, Han Zhou, and Chang Su. 2022. Pingantech at smm4h task1: Multiple pre-trained model approaches for adverse drug reactions. In *Proceedings of The Seventh Workshop on Social Media Mining for Health Applications, Workshop & Shared Task*, pages 4–6.

Yi Luan, Luheng He, Mari Ostendorf, and Hannaneh Hajishirzi. 2018. Multi-task identification of entities, relations, and coreference for scientific knowledge graph construction. In *Proceedings of the 2018 Conference on Empirical Methods in Natural Language Processing*. Association for Computational Linguistics.

Yi Luan, Dave Wadden, Luheng He, Amy Shah, Mari Ostendorf, and Hannaneh Hajishirzi. 2019. A general framework for information extraction using dynamic span graphs. In *Proceedings of the 2019 Conference of the North*. Association for Computational Linguistics.

Nils Reimers and Iryna Gurevych. 2019. Sentence-bert: Sentence embeddings using siamese bert-networks. In *Proceedings of the 2019 Conference on Empirical Methods in Natural Language Processing*. Association for Computational Linguistics.

Andrey Sakhovskiy, Zulfat Miftahutdinov, and Elena Tutubalina. 2021. Kfu nlp team at smm4h 2021 tasks: Cross-lingual and cross-modal bert-based models for adverse drug effects. In *Proceedings of the Sixth Social Media Mining for Health (# SMM4H) Workshop and Shared Task*, pages 39–43.

Darius Koenig Julius Lipp Sean Lee, Aamir Shakir. 2024. Open source strikes bread - new fluffy embeddings model.

Chenxi Whitehouse, Monojit Choudhury, and Alham Fikri Aji. 2023. Llm-powered data augmentation for enhanced cross-lingual performance. In *The 2023 Conference on Empirical Methods in Natural Language Processing*.

Dongfang Xu, Guillermo Lopez Garcia, Lisa Raithel, Rolland Roller, Philippe Thomas, Eiji Aramaki, Shuntaro Yada, Pierre Zweigenbaum, Sai Tharuni Samineni, Karen O'Connor, Yao Ge, Sudeshna Das, Abeed Sarker, Ari Klein, Lucia Schmidt, Vishakha Sharma, Raul Rodriguez-Esteban, Juan Banda, Ivan Flores Amaro, Davy Weissenbacher, and Graciela Gonzalez-Hernandez. 2024. Overview of the 9th social media mining for health applications (#SMM4H) shared tasks at ACL 2024. In *Proceedings of The 9th Social Media Mining for Health Research and Applications Workshop and Shared Tasks*, Bangkok, Thailand. Association for Computational Linguistics.

Meishan Zhang, Gongyao Jiang, Shuang Liu, Jing Chen, and Min Zhang. 2024. Llm–assisted data augmentation for chinese dialogue–level dependency parsing. *Computational Linguistics*, pages 1–24.

Zexuan Zhong and Danqi Chen. 2021. A frustratingly easy approach for entity and relation extraction. In *Proceedings of the 2021 Conference of the North American Chapter of the Association for Computational Linguistics: Human Language Technologies*, pages 50–61.

A Data Pre-Processing

We follow similar data pre-processing steps with (Aji et al., 2021; Sakhovskiy et al., 2021; Liu et al., 2022). We utilize Ekphrasis[6] (Baziotis et al., 2017) package as well as some customized processing steps. In addition to lowercase, (1) we remove the "@USER", "HTTPURL" and the following placeholders. (2) We remove emojis, but convert the emoticons into natural language (e.g. ":)" to "<smile>"). (3) We convert the slangs into natural language (e.g. "lol" to "laugh out loud"). (4) We replace the HTML entities into corresponding symbols (e.g. "&" to "&" and "<" to "<"). We then convert all ampersand symbols "&" into "and". (5) We conduct hashtag unpacking, which performs word segmentation on the content following a "#" symbol. (6) We employ spell correction.

B Base Models

B.1 ADE Extraction Models

The ADE extraction task can be treated as a Named Entity Recognition (NER) task with only one entity label, namely "ADE". Therefore, we implemented a span-based NER model following prior works (Lee et al., 2017; Luan et al., 2018, 2019; Zhong and Chen, 2021). We use a pre-trained language model (PLM: RoBERTa, BERTweet or DeBERTa) as encoder to obtain contextualized representation $\mathbf{X}_t$ for each input token $x_t \in X$. Consequently, the span representation $\mathbf{h}_e(s_i)$ for each potential span $s_i \in S$ is denoted as:

$$\mathbf{h}_e(s_i) = \left[\mathbf{X}_{START(i)}; \mathbf{X}_{END(i)}; \phi(s_i)\right],$$

where $\phi(s_i) \in \mathbb{R}^{d_F}$ denotes the length embeddings of s_i. The span representation $\mathbf{h}_e(s_i)$ is then fed into a softmax layer to predict the probability distribution of "the span is an ADE" and "the span is not an ADE".

B.2 ADE Normalization Models

MedDRA term retrieval We implement a basic dense retrieval model with LangChain[7] and Faiss[8] (Douze et al., 2024; Johnson et al., 2019) libraries for efficient similarity search of dense vectors. We use a fine-tuned mxbai-embed-large-v1[9] (Sean Lee, 2024; Li and Li, 2023) to obtain text embeddings for both queries (extracted spans) and MedDRA terms. Details of text embedding model fine-tuning could be found at Appendix C. We decide to retrieve up to 20 LLTs and PTs, which strike a balance between the performance of retrieval model and the subsequent filtering model.

MedDRA term filtering We simply utilize a PLM-based (RoBERTa, BERTweet or DeBERTa) binary classification model to classify whether each retrieved MedDRA term happens as an ADE in the given tweet. The basic inputs of our models is the concatenation of the given tweet, a extracted span as an potential ADE and one of the retrieved MedDRA terms. We collect the training samples for MedDRA term filtering models by using the MedDRA term retrieval model to retrieve 20 MedDRA terms for each gold-annotated ADE mention. We add the gold MedDRA term to the retrieved ones if it is not within them. Besides, we collect more negative samples by adding the ADE predictions from our earlier 5-fold cross-validation experiments.

C Text Embedding Model Fine-tuning

We use the example training script for NLI tasks provided by Sentence-Transformers (Reimers and Gurevych, 2019) to fine-tune mxbai-embed-large-v1. To construct the fine-tuning data, we first group the LLTs and PTs with the same PTID (ID of a preferred term). We obtain the "entailment" samples by pairing each gold ADE span in the training set with each of the LLTs and PTs belonging to the same group with its gold LLT or PT. Then for each "entailment" sample, we construct a "contradiction" sample by pairing the gold ADE span with a PT or LLT belonging to another randomly sampled group.

D Details of LLM-based Data Augmentation

Figure 4 shows an example of LLM-based data augmentation using mention rewriting and context rewriting. For ADE mention rewriting, we first convert the ADE mentions in the given tweets into JSON-formatted data indicating the gold LLTs or PTs (e.g. "withdrawal" to {"withdrawal syndrome": "withdrawal"}). We then prompt GLM-4 to output the entire input tweet while only replacing the ADE mentions in the JSON-formatted data for 3 times. These measures ensure that GLM-4 does not change the PT corresponding to each ADE

[6] https://github.com/cbaziotis/ekphrasis

[7] https://github.com/langchain-ai/langchain

[8] https://github.com/facebookresearch/faiss

[9] https://huggingface.co/mixedbread-ai/mxbai-embed-large-v1

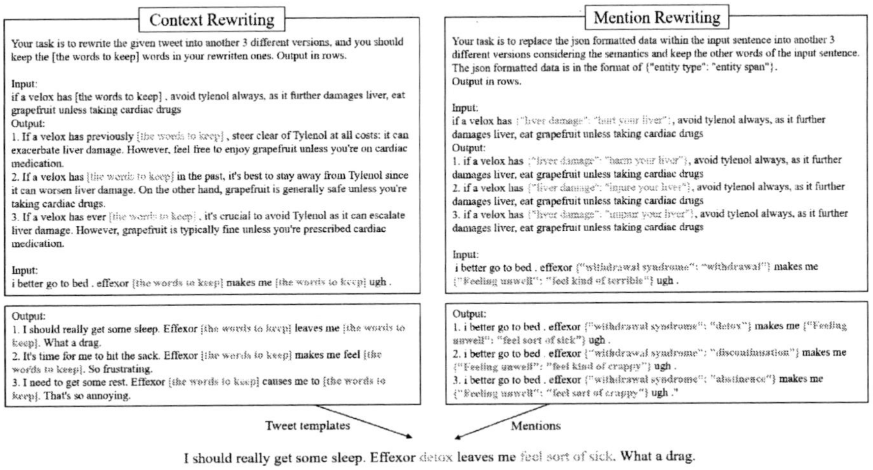

Figure 1: An example of LLM-based data augmentation using mention rewriting and context rewriting

Assuming you are a Twitter user, your task is to write at least 3 tweets with less than 30 words complaining the side effects of "[side_effects]" of "[drug_name]" that happens to you.
The expressions for the side effects should be colloquial and varied, and you are also recommended to use words other than the original terms.

Examples:
1. I found the humira to fix all my crohn's issues, but cause other issues. i went off it due to issues w nerves/muscle spasms
2. ciprofloxacin: how do you expect to sleep when your stomach is a cement mixer?
3. debating on taking a trazodone and literally passing out for the day.
4. well played tysabri...kicking butt #nosleep.

Output:

Figure 2: Prompt template for LLM-based data augmentation using LLM synthetic tweets

Your task is to rewrite the input tweet for better flow within 30 words. You should make it easier to understand and you should not use any abbreviations.

Input: if # avelox has hurt your liver , avoid tylenol always , as it further damages liver . eat grapefruit unless taking cardiac drugs.
Output: If Avelox has caused liver damage, always avoid Tylenol as it can further harm the liver. Also, avoid grapefruit if taking cardiac drugs.

Input: apparently , baclofen greatly exacerbates the " " ad " " part of my attention deficit hyperactivity disorder . average length of focus today : about 30 seconds .
Output: Baclofen makes my attention deficit hyperactivity disorder worse. I can only focus for around 30 seconds.

Input: [tweet]
Output:

Figure 3: Prompt template for LLM-based data augmentation using tweet rewriting

Your task is to explain the given medical term in one sentence using simple and understandable vocabulary.

Input: Abnormal weight gain
Output: Abnormal weight gain refers to gaining an unusually large amount of weight, which may be a sign of health issues or imbalances in the body.

Input: [LLT or PT]
Output:

Figure 4: Prompt template for LLM-based data augmentation using MedDRA term explanation

mention and maintains the coherence of texts after ADE mention rewriting. For context rewriting, we use "[the words to keep]" to mask the ADE mentions and ask GLM-4 to rewrite the other parts of the given tweets into another 3 versions. In this way, GLM-4 will pay more attention to the part-of-speech of the masked words while rewriting the contexts, thus ensuring the smoothness of the rewritten contexts when inserting the rewritten ADE mentions. Prompt templates used in synthetic tweets, tweet rewriting and MedDRA term explanation are illustrated in Figure 2,3 and 4.

E LLM-based ADE Extraction

We abandoned LLM-based ADE Extraction for the following reasons: (1) The extracted ADEs are often absent from the original texts and include a large number of false positives, making it impossible to ensemble with other models. (2) Despite filtering the LLM extraction results with other models', the retrieval module achieves a lower recall of 82.35 compared to 85.88 achieved on the ensemble results of PLM-based ADE extraction models.

F Model Ensemble

We employ majority voting for both ADE extraction and ADE normalization. For ADE extraction, we set the threshold at $\frac{1}{3}$ of the number of model candidates. For ADE normalization, we collect all LLTs or PTs classified into "happens as an ADE" by at least one model candidate for each extracted span. We then map the LLTs to PTs and the span is labeled with the specific PT that has the highest counts.

LT4SG@SMM4H'24: Tweets Classification for Digital Epidemiology of Childhood Health Outcomes Using Pre-Trained Language Models

Dasun Athukoralage
NirvanaClouds
dasun@nirvanaclouds.com

Thushari Atapattu
University of Adelaide
thushari.atapattu@adelaide.edu.au

Menasha Thilakaratne
University of Adelaide
menasha.thilakaratne@adelaide.edu.au

Katrina Falkner
University of Adelaide
katrina.falkner@adelaide.edu.au

Abstract

This paper presents our approaches for the SMM4H'24 Shared Task 5 on the binary classification of English tweets reporting children's medical disorders. Our first approach involves fine-tuning a single RoBERTa-large model, while the second approach entails ensembling the results of three fine-tuned BERTweet-large models. We demonstrate that although both approaches exhibit identical performance on validation data, the BERTweet-large ensemble excels on test data. Our best-performing system achieves an F1-score of 0.938 on test data, outperforming the benchmark classifier by 1.18%. Our code is available on Github[1].

1 Introduction & Motivation

Chronic childhood disorders like attention-deficit/hyperactivity disorder (ADHD), autism spectrum disorders (ASD), delayed speech, and asthma significantly impact a child's development and well-being, often extending into adulthood. Approximately 1 in 6 (17%) children aged 3-17 years in the United States experience a developmental disability, with ADHD, ASD, and others contributing to this statistic (Zablotsky et al., 2019). In previous studies (Guntuku et al., 2019; Hswen et al., 2019; Edo-Osagie et al., 2019), Twitter data have been utilized to identify self-reports of the aforementioned disorders; however, the identification of reports concerning these disorders in users' children has not been explored. It may be of interest to explore Twitter's potential in continuing to collect users' tweets postpartum, enabling the detection of outcomes in childhood.

2 Task and Data Description

2.1 Task

The SMM4H-2024 workshop and shared tasks have a special focus on Large Language Models (LLMs) and generalizability for natural language processing (NLP) in social media. We participated in Task 5, which is 'Binary classification of English tweets reporting children's medical disorders'. The objective is to automatically differentiate tweets from users who have disclosed their pregnancy on Twitter and mention having a child with ADHD, ASD, delayed speech, or asthma (annotated as "1"), from tweets that merely refer to a disorder (annotated as "0").

2.2 Data

There were three different datasets provided: training, validation, and test datasets. The training and validation datasets were labeled while the test dataset was not. All datasets are composed entirely of tweets posted by users who had reported their pregnancy on Twitter, that report having a child with a disorder and tweets that merely mention a disorder. The training, validation, and test sets contain 7398 tweets, 389 tweets, and 1947 tweets, respectively.

3 Methodology

3.1 Baseline

A benchmark classifier, based on a RoBERTa-large model (Liu et al., 2019), has achieved an F1-score of 0.927 for the 'positive' class (i.e., tweets that report having a child with a disorder) on the test data for Task 5 (Klein et al., 2024).

3.2 Models Used

We investigated three Transformer based models which are BioLinkBERT-large (Yasunaga et al., 2022), RoBERTa-large and BERTweet-large (Nguyen et al., 2020). BioLinkBERT was selected for its specialized understanding of biomedical NLP tasks, RoBERTa for its domain-independent NLP capabilities, and BERTweet for its superior performance in Tweet-specific NLP tasks. We fine-

[1]To access the code, please visit: [GitHub Source Code]

Proceedings of The 9th Social Media Mining for Health Applications (#SMM4H) Workshop and Shared Tasks, pages 38–41
August 15, 2024 ©2024 Association for Computational Linguistics

tuned each model with the training dataset and evaluated its performance using the validation dataset.

3.3 Training Regime

Experiments were conducted using Google Colab Pro+ equipped with an NVIDIA A100 Tensor Core GPU boasting 40 gigabytes of available GPU RAM. The Hugging Face Transformers Python library (Wolf et al., 2019) and its Trainer API facilitated training procedures. Each model was trained on the training datasets for 3 iterations and 10 epochs per iteration. We used HuggingFace's Trainer Class's default 'AdamW' and 'linear warmup with cosine decay' as the optimizer and scheduler respectively. The maximum sequence length for all models was set to 512. FP-16 mixed precision training was employed to enable larger batch sizes and expedited training. Primary hyperparameters including learning rate, weight decay, and batch size were determined as described in the subsequent section.

3.4 Hyperparameter Optimization

We conducted hyperparameter optimization that relied on HuggingFace's Trainer API with the Ray Tune backend (Liaw et al., 2018). We utilized Ray Tune's built-in "BasicVariantGenerator" algorithm[2] for hyperparameter search, paired with the First-In-First-Out (FIFO) scheduler. Since BasicVariantGenerator has the ability to dynamically generate hyperparameter configurations based on predefined search algorithms (e.g., random search, Bayesian optimization), it enables more efficient exploration of the search space. The hyperparameters optimized using BasicVariantGenerator are presented in Table 1.

Model	Learning Rate	Weight Decay	Batch Size
BioLinkBERT-large	6.10552e-06	0.00762736	16
RoBERTa-large	7.21422e-06	0.00694763	8
BERTweet-large	1.17754e-05	0.01976150	8

Table 1: Hyperparameters optimized via BasicVariantGenerator.

4 Preliminary Experiments

Each selected model was trained for 3 iterations, with 10 epochs per iteration. At the end of each epoch, its F1-score was recorded. The F1-score for each model was determined based on its performance with the validation dataset. We saved the best-performing epoch (i.e., the best F1-score for the positive class) for each model in each iteration. The results are shown in Table 2.

Model	1st run	2nd run	3rd run	Mean F1	SD
BioLinkBERT-large	0.855019	**0.875969**	0.863159	0.864716	0.010561
RoBERTa-large	0.931408	**0.945055**	0.931408	0.935957	0.007879
BERTweet-large	**0.940741**	0.934307	0.933824	**0.936291**	**0.003862**

Table 2: The F1 scores of the BioLinkBERT-large, RoBERTa-large, and BERTweet-large classifiers on the validation data. The mean F1 score and standard deviation are also provided.

As shown in Table 2, RoBERTa-large and BERTweet-large perform similarly on the validation dataset, and considerably better than BioLinkBERT-large, even though it has been pretrained on a large corpus of biomedical data. Therefore, we decided to remove BioLinkBERT-large to carry out further experiments for this task (Guo et al., 2020).

4.1 Ensembling Strategy

The issue arises when fine-tuning large-transformer models on small datasets: the classification performance varies significantly with slightly different training data and random seed values, even when using the same hyperparameter values (Dodge et al., 2020). To overcome this high variance and provide more robust predictions, we propose ensembles of multiple fine-tuned RoBERTa-large models and BERTweet-large models separately. We create two separate ensemble models using the best models corresponding to three iterations for each RoBERTa-large and BERTweet-large. All three iterations use the same hyperparameters, and only differ in the initial random seed. A hard majority voting mechanism combines the predictions of these models:

$$\hat{y} = \arg\max_{c} \sum_{i=1}^{n} \mathbf{1}(\hat{y}_i = c) \qquad (1)$$

where $\mathbf{1}(\cdot)$ represents the indicator function, which returns either '1' or '0' for the class label c predicted by the i-th classifier.

Classifier	F1-score	Precision	Recall
RoBERTa-large Ensemble	0.934783	0.914894	**0.955556**
BERTweet-large Ensemble	**0.945055**	**0.934783**	**0.955556**

Table 3: Performance results for ensemble classifiers on validation data.

As shown in Table 3, the BERTweet-large ensemble performs better than the RoBERTa-large

[2]https://docs.ray.io/en/latest/tune/api/doc/ray.tune.search.basic_variant.BasicVariantGenerator.html

ensemble. This is fundamentally because its performance variation is less for three iterations, as indicated in Table 2. Another noteworthy observation is that the performance of the BERTweet-large ensemble is identical to that of the best iteration (Table 2, 2nd run) of RoBERTa-large as shown in Table 4. Figure 1 shows the corresponding confusion matrices for both classifiers which are also identical.

Classifier	F1-score	Precision	Recall
RoBERTa-large best-run	**0.945055**	**0.934783**	**0.955556**
BERTweet-large Ensemble	**0.945055**	**0.934783**	**0.955556**

Table 4: Performance comparison of the RoBERTa-large best-run vs the BERTweet-large ensemble on validation data.

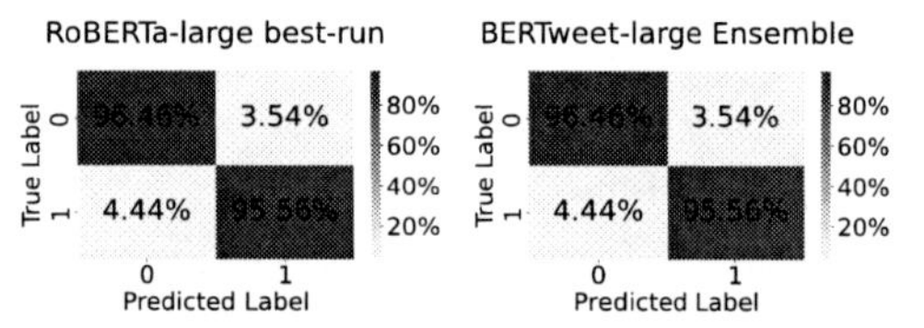

Figure 1: Confusion matrices of the RoBERTa-large best-run and BERTweet-large ensemble on the validation dataset.

5 Results and Conclusion

Since RoBERTa-large best-run and BERTweet-large ensemble are performing equally well on the validation data, we tested the performance of both classifiers on unseen, unlabeled test data. As shown in Table 5, the BERTweet-large ensemble classifier outperforms the mean and median performance on the test data among all teams' submissions by a considerable margin, as well as the benchmark classifier by 1.18%. Additionally, we can observe that even though both classifiers perform equally well on validation data, the BERTweet-large ensemble model performs significantly better on test data. One possible reason for this is that different runs of BERTweet-large might excel at capturing different aspects of the data or learning different patterns.

Previously, authors (Klein et al., 2024) have achieved an F1-score of 0.92 using a BERTweet-large classifier on the same dataset. We believe that our approach achieved better results for several fundamental reasons. First, we performed more thorough hyperparameter tuning compared to the previous authors. Better-optimized hyperparameters can significantly improve model performance. Second, by creating an ensemble of BERTweet-large models from the three best epochs of three runs, we captured more robust and generalized patterns in the data. Ensemble methods typically outperform individual models because they reduce variance and mitigate the risk of overfitting to specific subsets of data. Third, a likely reason is the diversity in model runs. Each run of our BERTweet-large model may have encountered slightly different initialization and training dynamics (e.g., random seed), resulting in diverse decision boundaries. Combining these diverse models helps in making more accurate predictions by leveraging the strengths of each individual model.

Model	F1-score	Precision	Recall
Baseline	0.927	0.923	0.940
Mean	0.822	0.818	0.838
Median	0.901	0.885	0.917
RoBERTa-large best-run	0.925	0.908	0.942
BERTweet-large Ensemble	**0.938**	**0.930**	**0.946**

Table 5: Results for our two proposed approaches on the test data, including the mean, median, and baseline scores.

When fine-tuning complex pre-trained language models, one issue on small datasets is the instability of the classification performance. To overcome this, we combined the predictions of multiple BERTweet-large models in an ensemble. By doing so, we achieved significantly better results in terms of F1-score for SMM4H'24 Task 5. For future work, it's interesting to investigate how the system performance varies when adding more BERTweet-large iterations (i.e., runs) to the ensemble.

References

Jesse Dodge, Gabriel Ilharco, Roy Schwartz, Ali Farhadi, Hannaneh Hajishirzi, and Noah Smith. 2020. Fine-tuning pretrained language models: Weight initializations, data orders, and early stopping. *arXiv preprint arXiv:2002.06305*.

Oduwa Edo-Osagie, Gillian Smith, Iain Lake, Obaghe Edeghere, and Beatriz De La Iglesia. 2019. Twitter mining using semi-supervised classification for relevance filtering in syndromic surveillance. *PLoS One*, 14(7):e0210689.

Sharath Chandra Guntuku, Jared R. Ramsay, Raina M. Merchant, and Lyle H. Ungar. 2019. Language of adhd in adults on social media. *Journal of Attention Disorders*, 23(12):1475–1485.

Yuting Guo, Xiangjue Dong, Mohammed Ali Al-Garadi, Abeed Sarker, Cécile Paris, and Diego Mollá-Aliod. 2020. Benchmarking of transformer-based pretrained models on social media text classification datasets. In *Workshop of the Australasian Language Technology Association*, pages 86–91.

Yulin Hswen, Ashwin Gopaluni, John S. Brownstein, and Jared B. Hawkins. 2019. Using twitter to detect psychological characteristics of self-identified persons with autism spectrum disorder: A feasibility study. *JMIR mHealth and uHealth*, 7(2):e12264.

Ari Klein, José Gutiérrez Gómez, Lisa Levine, and Graciela Gonzalez-Hernandez. 2024. Using longitudinal twitter data for digital epidemiology of childhood health outcomes: An annotated data set and deep neural network classifiers. *J Med Internet Res*, 26:e50652.

Richard Liaw, Eric Liang, Robert Nishihara, Philipp Moritz, Joseph E. Gonzalez, and Ion Stoica. 2018. Tune: A Research Platform for Distributed Model Selection and Training. *arXiv preprint arXiv:1807.05118*.

Yinhan Liu, Myle Ott, Naman Goyal, Jingfei Du, Mandar Joshi, Danqi Chen, Omer Levy, Mike Lewis, Luke Zettlemoyer, and Veselin Stoyanov. 2019. Roberta: A robustly optimized bert pretraining approach. *arXiv preprint arXiv:1907.11692*.

Dat Quoc Nguyen, Thanh Vu, and Anh Tuan Nguyen. 2020. Bertweet: A pre-trained language model for english tweets. In *Proceedings of the 2020 Conference on Empirical Methods in Natural Language Processing: System Demonstrations*, pages 9–14.

Thomas Wolf, Lysandre Debut, Victor Sanh, Julien Chaumond, Clement Delangue, Anthony Moi, Pierric Cistac, Tim Rault, Rémi Louf, Morgan Funtowicz, and et al. 2019. Huggingface's transformers: State-of-the-art natural language processing. *arXiv preprint arXiv:1910.03771*.

Makoto Yasunaga, Jure Leskovec, and Percy Liang. 2022. Linkbert: Pretraining language models with document links. In *Proceedings of the 60th Annual Meeting of the Association for Computational Linguistics (Volume 1: Long Papers)*, pages 8003–8016.

Benjamin Zablotsky, Lindsey I. Black, Matthew J. Maenner, Laura A. Schieve, Melanie L. Danielson, Rebecca H. Bitsko, et al. 2019. Prevalence and trends of developmental disabilities among children in the united states: 2009-2017. *Pediatrics*, 144(4):e20190811.

UTRad-NLP at #SMM4H 2024: Why LLM-Generated Texts Fail to Improve Text Classification Models

Yosuke Yamagishi
The University of Tokyo, Japan
yamagishi-yosuke0115@g.ecc.u-tokyo.ac.jp

Yuta Nakamura
The University of Tokyo, Japan
yutanakamura-tky@umin.ac.jp

Abstract

In this paper, we present our approach to addressing the binary classification tasks, Tasks 5 and 6, as part of the Social Media Mining for Health (SMM4H) text classification challenge. Both tasks involved working with imbalanced datasets that featured a scarcity of positive examples. To mitigate this imbalance, we employed a Large Language Model to generate synthetic texts with positive labels, aiming to augment the training data for our text classification models. Unfortunately, this method did not significantly improve model performance. Through clustering analysis using text embeddings, we discovered that the generated texts significantly lacked diversity compared to the raw data. This finding highlights the challenges of using synthetic text generation for enhancing model efficacy in real-world applications, specifically in the context of health-related social media data.

1 Introduction

In recent years, the burgeoning field of social media mining has opened new avenues for health-related research (Shakeri Hossein Abad et al., 2021), providing rich data sources for public health surveillance, including understanding public health trends and individual health behaviors. The Social Media Mining for Health (SMM4H) initiative, through its various tasks, aims to leverage these data sources to address pertinent health questions (Xu et al., 2024). This paper focuses on our approaches to Tasks 5 and 6, both of which present unique challenges and opportunities in the realm of text classification for health-related social media mining.

Task 5 targets the binary classification of tweets related to children's medical disorders, differentiating between tweets that report a genuine diagnosis and those that only mention these disorders without a diagnosis. Task 6 involves identifying the exact ages from social media posts, crucial for health research applications and enabling more accurate analysis of age-related health outcomes and behaviors in observational studies.

For both tasks, the challenge of imbalanced datasets is prominent. To address this, we employed the Large Language Model (LLM), GPT-4, aiming to augment our training data with synthetic positive examples to balance the dataset and enhance the performance of our binary classification models. Despite these efforts, our initial results were underwhelming, as the synthetic texts generated by GPT-4 lacked the diversity found in the raw data. This finding raises important questions about the practical challenges and limitations of using synthetic data augmentation in real-world applications, particularly in the nuanced field of health-related social media mining.

2 Dataset & Metrics

2.1 Task 5

The Task 5 dataset comprises tweets posted by users who reported their pregnancy on Twitter, used for binary classification. It includes 7,398 tweets for training, 389 tweets for validation, and 1,947 tweets for testing. The evaluation is based on the F1-score for tweets reporting a child with a disorder (annotated as '1').

2.2 Task 6

Task 6 focuses on extracting self-reported exact ages from posts on Twitter and Reddit. The dataset features 8,800 labeled tweets and 100,000 unlabeled Reddit posts from r/AskDocs containing 2-digit numbers for training; 2,200 tweets and 1,000 Reddit posts on dry eye disease for validation; and 2,200 tweets, 2,000 Reddit posts on dry eye disease, and 12,482 posts on social anxiety with ages 13 to 25 for testing. The evaluation uses the F1-score on the positive class ('1'), with micro-averaging.

Proceedings of The 9th Social Media Mining for Health Applications (#SMM4H) Workshop and Shared Tasks, pages 42–47
August 15, 2024 ©2024 Association for Computational Linguistics

2.3 Label Distribution

Both Tasks 5 and 6 have imbalanced datasets where Label 1 is less than half as numerous as Label 0. The distributions are documented in Table 1.

Table 1: Label Distribution in Tasks 5 and 6

Task	Label 0	Label 1
5	5,118 (69.1%)	2,280 (30.8%)
6	5,966 (67.8%)	2,834 (32.2%)

3 Methods

3.1 LLM Text Generation

We used the OpenAI API's GPT-4 ("gpt-4-0125-preview") with in-context learning techniques. We explained the label definitions and showed 5 randomly selected examples for each of the positive and negative labels. The model's temperature was initially set to 0, allowing for automatic adjustments to ensure a balance between determinism and diversity. We then generated 10 texts for the positive category, repeating this 100 times to produce 1,000 synthetic texts for both Tasks 5 and 6. The schematic diagram of the prompts is shown in the figure 1. Additionally, the examples of the full prompts and the examples of the generated texts are presented in the appendix A.1 and A.2.

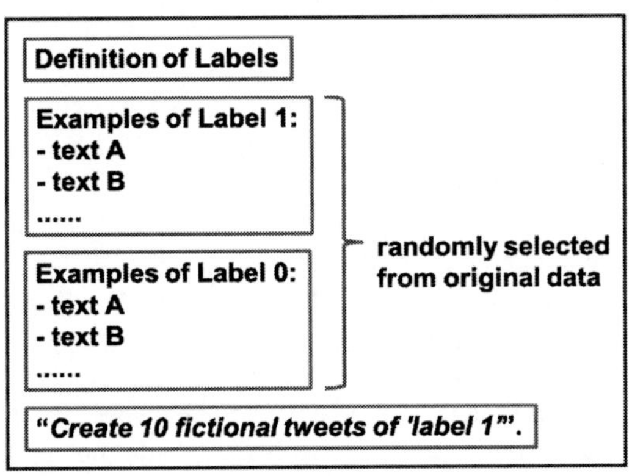

Figure 1: Schematic diagram of the prompt. Composed of the definition of labels, specific examples of texts, and instructions for generation.

3.2 Model Development

For the development of our text classification models, we utilized the DeBERTa v3 Large model (He et al., 2021b,a). For each task, we trained two versions of the model: one using synthetic data for training and the other without using synthetic data.

3.2.1 Training Procedure

Training was conducted over 10 epochs, starting with a learning rate of 5e-5. A scheduler was used to reduce the learning rate to zero towards the end of the training process.

3.2.2 Validation and Model Selection

The model's performance was evaluated on the validation dataset at the end of each epoch using the F1-score. The model that achieved the highest F1-score on the validation set was selected for inference on the test dataset.

3.3 LLM-Generated Text Assessment

3.3.1 Clustering of Texts Embeddings

In the text analysis using sentence transformers, we extracted embeddings using "all-MiniLM-L6-v2" available at https://github.com/UKPLab/sentence-transformers (Reimers and Gurevych, 2019). On top of this, we performed k-means clustering and t-distributed stochastic neighbor embedding (t-SNE) to visualize and evaluate the distribution of the synthetic data and the raw data.

3.3.2 *N*-gram Analysis

Additionally, we conducted the same n-gram analysis for ($n = 1, 2, 3$) on the raw data of Tasks 5 and 6 for comparison and discussion. To ensure a fair comparison with the GPT-4 generated text, which had a limited number of samples, we randomly selected 1,000 samples from each of the original datasets (Label 0 and Label 1) for both tasks.

4 Results

4.1 Performance Metrics on Test Data

The evaluation metrics for the test data of Tasks 5 and 6 are presented in Table 2. For the primary metric, the F1 score, Task 5 shows a slight improvement of 0.009, while Task 6 shows a deterioration of 0.013.

Table 2: Performance metrics for Tasks 5 and 6 with and without data augmentation using LLM-generated texts (indicated by "Aug"). **Bold** indicates the higher score for each task.

Task	Aug	F1	Precision	Recall
5	No	0.924	**0.966**	0.886
	Yes	**0.933**	0.932	**0.934**
6	No	**0.936**	**0.947**	**0.926**
	Yes	0.923	0.921	0.924

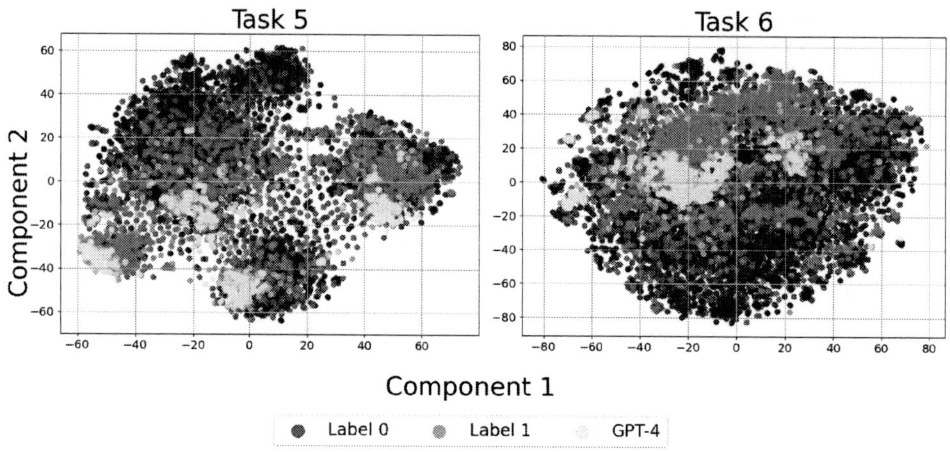

Figure 2: t-SNE visualization of sentence embeddings for Tasks 5 and 6. The plots show the distribution of original and GPT-4 generated sentences in a two-dimensional space after dimensionality reduction using t-SNE. Each point represents a sentence, with colors indicating the label (Label 0, Label 1, or GPT-4).

4.2 *N*-gram Analysis

Based on the results presented in Table 3, it is evident that the texts generated by GPT-4 consistently exhibits a lower number of unique n-grams compared to the original dataset, across both Tasks 5 and 6. This observation holds true for all values of n (1, 2, and 3) considered in the analysis.

Table 3: Number of unique n-grams for each label and task. "Label 0" and "Label 1" represent data from the original dataset, while "GPT-4" represents text generated by GPT-4.

Task	Data	n=1	n=2	n=3
5	Label 0	5,380	22,313	30,577
	Label 1	5,001	21,056	29,102
	GPT-4	2,522	12,922	19,386
6	Label 0	5,471	15,776	18,601
	Label 1	3,337	10,696	13,186
	GPT-4	1,640	6,940	10,331

4.3 Clustering of Text Embeddings

Figure 2 illustrates that GPT-4-generated text embeddings form localized clusters with limited spread compared to the original data, particularly Label 0 sentences. This suggests a lack of diversity in GPT-4 outputs, as the model tends to generate semantically similar sentences, in contrast to the wider distribution and linguistic variety observed in human-generated text.

5 Discussion & Conclusion

The data provided for the Tasks 5 and 6 were both imbalanced datasets with less positive examples than negative ones. To address such imbalances, data augmentation can be an effective approach. Previous methods using deep learning have been proposed as techniques applicable to named entity recoginition and text classification, such as label-wise token replacement, synonym replacement, and entity replacement within sequences (Dai and Adel, 2020; Ding et al., 2021; Zhou et al., 2022). There have also been proposals for approaches combining context and entity levels using LLMs (Ye et al., 2024). In this study, we proposed a method that employs few-shot learning techniques to generate new text using LLMs (Brown et al., 2020). By leveraging LLMs, a vast amount of completely new data can be generated, and by producing high-quality data, improvements in the performance of classification models can be expected. However, in this study, the addition of synthetic data did not significantly improve the performance compared to the baseline model.

As evident from the visualization by clustering, the synthetic data generated by GPT-4 for the Tasks 5 and 6 exhibits a localized distribution in the embeddings extracted from the pre-trained language model, compared to the original data. This suggests that the synthetic data lacks diversity in comparison to the original data, and as a result, the addition of the synthetic data did not improve the overall diversity of the training data, resulting in no explicit improvement in the model's performance.

This study has several limitations. Since we used only GPT-4 for generating synthetic data from sources like Reddit and Twitter, future work should explore new methods to increase diversity. Prior research suggests that diversity can be enhanced by specifying attributes (Yu et al., 2024). For instance, identifying the characteristics of posters or the types of children's diseases could lead to greater diversity. Future efforts could include using multiple language models, presenting more original examples, or adjusting hyperparameters like temperature to improve data diversity.

The Tasks 5 and 6 involved relatively few positive examples, resulting in imbalanced data, which is often the case in real-world settings. While the use of synthetic data generated by language models can be an effective solution for augmenting training data, this study suggests that various efforts are necessary to create diverse data that is effective for training language models.

Acknowledgments

We would like to acknowledge that all ethical aspects of this research were strictly adhered to, in accordance with the regulations set forth by the data providers. Additionally, this study was conducted without the use of any research funding.

References

Tom Brown, Benjamin Mann, Nick Ryder, Melanie Subbiah, Jared D Kaplan, Prafulla Dhariwal, Arvind Neelakantan, Pranav Shyam, Girish Sastry, Amanda Askell, et al. 2020. Language models are few-shot learners. *Advances in neural information processing systems*, 33:1877–1901.

Xiang Dai and Heike Adel. 2020. An analysis of simple data augmentation for named entity recognition. In *Proceedings of the 28th International Conference on Computational Linguistics*, pages 3861–3867, Barcelona, Spain (Online). International Committee on Computational Linguistics.

Ning Ding, Guangwei Xu, Yulin Chen, Xiaobin Wang, Xu Han, Pengjun Xie, Haitao Zheng, and Zhiyuan Liu. 2021. Few-NERD: A few-shot named entity recognition dataset. In *Proceedings of the 59th Annual Meeting of the Association for Computational Linguistics and the 11th International Joint Conference on Natural Language Processing (Volume 1: Long Papers)*, pages 3198–3213, Online. Association for Computational Linguistics.

Pengcheng He, Jianfeng Gao, and Weizhu Chen. 2021a. Debertav3: Improving deberta using electra-style pre-training with gradient-disentangled embedding sharing. *arXiv preprint arXiv:2111.09543*.

Pengcheng He, Xiaodong Liu, Jianfeng Gao, and Weizhu Chen. 2021b. Deberta: Decoding-enhanced bert with disentangled attention. In *International Conference on Learning Representations*.

Nils Reimers and Iryna Gurevych. 2019. Sentence-BERT: Sentence embeddings using Siamese BERT-networks. In *Proceedings of the 2019 Conference on Empirical Methods in Natural Language Processing and the 9th International Joint Conference on Natural Language Processing (EMNLP-IJCNLP)*, pages 3982–3992, Hong Kong, China. Association for Computational Linguistics.

Zahra Shakeri Hossein Abad, Adrienne Kline, Madeena Sultana, Mohammad Noaeen, Elvira Nurmambetova, Filipe Lucini, Majed Al-Jefri, and Joon Lee. 2021. Digital public health surveillance: a systematic scoping review. *NPJ digital medicine*, 4(1):41.

Dongfang Xu, Guillermo Lopez Garcia, Lisa Raithel, Rolland Roller, Philippe Thomas, Eiji Aramaki, Shuntaro Yada, Pierre Zweigenbaum, Sai Tharuni Samineni, Karen O'Connor, Yao Ge, Sudeshna Das, Abeed Sarker, Ari Klein, Lucia Schmidt, Vishakha Sharma, Raul Rodriguez-Esteban, Juan Banda, Ivan Flores Amaro, Davy Weissenbacher, and Graciela Gonzalez-Hernandez. 2024. Overview of the 9th social media mining for health applications (smm4h) shared tasks at acl 2024. In *Proceedings of The 9th Social Media Mining for Health Research and Applications Workshop and Shared Tasks*, Bangkok, Thailand. Association for Computational Linguistics.

Junjie Ye, Nuo Xu, Yikun Wang, Jie Zhou, Qi Zhang, Tao Gui, and Xuanjing Huang. 2024. Llm-da: Data augmentation via large language models for few-shot named entity recognition. *arXiv preprint arXiv:2402.14568*.

Yue Yu, Yuchen Zhuang, Jieyu Zhang, Yu Meng, Alexander J Ratner, Ranjay Krishna, Jiaming Shen, and Chao Zhang. 2024. Large language model as attributed training data generator: A tale of diversity and bias. *Advances in Neural Information Processing Systems*, 36.

Ran Zhou, Xin Li, Ruidan He, Lidong Bing, Erik Cambria, Luo Si, and Chunyan Miao. 2022. MELM: Data augmentation with masked entity language modeling for low-resource NER. In *Proceedings of the 60th Annual Meeting of the Association for Computational Linguistics (Volume 1: Long Papers)*, pages 2251–2262, Dublin, Ireland. Association for Computational Linguistics.

A Appendix

A.1 Full Prompts

The full texts of the prompts are provided below. The variables pos_exs and neg_exs each contain five example texts for Labels 1 and 0, respectively.

A.1.1 Prompt for Task 5

```
The following tweets are from a
   parent with a child.
The label has a definition like '
   This binary classification
   task involves automatically
   distinguishing tweets, posted
   by users who had reported
   their pregnancy on Twitter,
   that report having a child
   with attention-deficit/
   hyperactivity disorder (ADHD),
    autism spectrum disorders (
   ASD), delayed speech, or
   asthma (annotated as "1"),
   from tweets that merely
   mention a disorder (annotated
   as "0").

Examples of label 0:
{neg_exs}

Examples of label 1:
{pos_exs}

Create 10 fictional tweets of '
   label 1'.

You should output following style
   .
1. output1
2. output2
3. output3
4. output4
5. output5
6. output6
7. output7
8. output8
9. output9
10. output10
```

A.1.2 Prompt for Task 6

```
The posts in the following
   dataset come from social media
    platforms Twitter and Reddit.
    These posts are used to
   automatically train models
   focused on identifying
   instances where the user's
   exact age is explicitly
   mentioned. This is
   particularly useful for health
    and demographic research
   applications that require
   precise age data.

 H e r e s  a summary of the
   definitions for labels 0 and 1
    in this dataset:

Label 1: This label is assigned
   to posts where the user's
   exact age can be determined
   directly from the text at the
   time the entry was posted.
   Examples include explicit
   mentions of age, such as "It's
    my 21st birthday today" or
   inferred statements where the
   user indicates they will be a
   certain age, like "tomorrow I'
   ll be 20."
Label 0: This label is used for
   posts where the age of the
   user cannot be determined or
   is ambiguous. Examples include
    unclear references to age,
   mentions of age that may not
   be current (like past or
   future tense without a clear
   indicator of current age), or
   mentions of someone  e l s e s
   age (e.g., a sibling or child)
   .

Examples of label 1:
{pos_exs}

Examples of label 0:
{neg_exs}

Create 10 fictional tweets of '
   label 1'.

You should output following style
   .
1. output1
2. output2
3. output3
4. output4
5. output5
```

```
6. output6
7. output7
8. output8
9. output9
10. output10
```

A.2 Examples of Generated Texts

Here are three actual generated examples from the 1,000 texts produced by GPT-4 for each of the two tasks.

A.2.1 Examples of Task 5

1. Just had a parent-teacher conference about my daughter's ADHD. The teacher recommended some strategies to help her stay focused in class. Feeling hopeful and supported. #AD-HDawareness

2. Navigating ADHD with my child has been a journey of patience, love, and a lot of learning. But seeing his improvements makes it all worth it. #ParentingADHD

3. My toddler with a speech delay said "mama" clear as day. I cried. These moments are everything. (pleading face emoji)(heart with arrow emoji) #speechdelay

A.2.2 Examples of Task 6

1. Just hit the big 25 today, can't believe I'm a quarter of a century old! (party popper emoji)(birthday cake emoji) #birthdayvibes

2. Just signed the lease to my very first apartment, a perfect 27th birthday present to myself. Here's to independence and new beginnings! (house emoji)(birthday cake emoji) #NewHome #27Years

3. Turning 22 in a pandemic means virtual birthday parties and lots of Zoom shots! #QuarantineBirthday

HBUT at # SMM4H 2024 Task1: Extraction and Normalization of Adverse Drug Events with a Large Language Model

Yuanzhi Ke
Hubei University of Technology
keyuanzhi@hbut.edu.cn

Hanbo Jin
Hubei University of Technology
jinhanbo@hbut.edu.cn

Xinyun Wu
Hubei University of Technology
xinyun@hbut.edu.cn

Caiquan Xiong
Hubei University of Technology
xiongcq@hbut.edu.cn

Abstract

In this paper, we describe our proposed systems for the Social Media Mining for Health 2024 shared task 1. We built our system on the basis of GLM, a pre-trained large language model with few-shot learning capabilities, using a two-step prompting strategy to extract adverse drug events (ADEs) and an ensemble method for normalization. In the first step of extraction phase, we extract all the potential ADEs with in-context few-shot learning. In the second step for extraction, we let GLM to filter out false positive outputs in the first step by a tailored prompt. Then we normalize each ADE to its MedDRA preferred term ID (ptID) by an ensemble method using Reciprocal Rank Fusion (RRF). Our method achieved an excellent recall rate. It obtained 41.1%, 42.8%, and 40.6% recall rate for ADE normalization, ADE recognition, and normalization for unseen ADEs, respectively. Compared to the performance of the average and median among all the participants in terms of recall rate, our recall rate scores are generally 10%-20% higher than the other participants' systems.

1 Introduction

Extracting medical entities from social media is a challenging task. BERT (Devlin et al., 2019) is one of the most popular models used for named entity recognition (NER). Among its family, BioBERT (Lee et al., 2019) and clinical-Bert (Huang et al., 2019) are especially reported useful for Medical Named Entity Recognition tasks. But without sufficient resources and data for fine-tuning, BERT tends to underperform. Some researches (Brown et al., 2020; Kojima et al., 2022) have demonstrated that tailored prompts can drive large language models (LLMs) to perform various downstream tasks well with only a small amount of data.

This paper describes our work in the Social Media mining for Health 2024 (SMM4H2024). SMM4H2024 task 1 was a mission for mining adverse drug events (ADE) from Twitter (X) and normalizing to their corresponding preferred terms within the Medical Dictionary for Regulatory Activities (MedDRA) terminology. Inspired by the works on Few-shot (Brown et al., 2020) and Zero-shot (Kojima et al., 2022) learning with LLMs, we introduce a system for ADE identification and normalization.

The identification phase of our system contains two steps: a potential ADE extraction step and a self-improvement step:

1. In the first step, we use a LLM with Few-shot in-context learning to find out all the potential ADEs.

2. In the second step, we use another prompt template that provides the potential ADEs and their original twitter sentences from the first step to let the LLM distinguish whether each ADE is caused by any drug mentioned in the twitter. In this way, false positive outputs in the first step are identified. Then a script drops such outputs to get the final text extraction result for normalization in the next phase.

In the normalization phase, we use an ensemble method, utilizing the Reciprocal Rank Fusion (RRF) (Cormack et al., 2009) method to integrate the relevance scores based on five pre-trained Embedding models and Levenshtein Distance (Yujian and Bo, 2007). The resulted fused score serves as the final measure of relevance. In this phase, each extracted ADE text in the previous phase is labeled with the preferred term ID (ptID) of the most related ADE in MedDRA measured in this way.

Our system scored above mean and median on multiple metrics among all the participants and had the advantage of not requiring fine-tuning on downstream tasks.

Proceedings of The 9th Social Media Mining for Health Applications (#SMM4H) Workshop and Shared Tasks, pages 48–54
August 15, 2024 ©2024 Association for Computational Linguistics

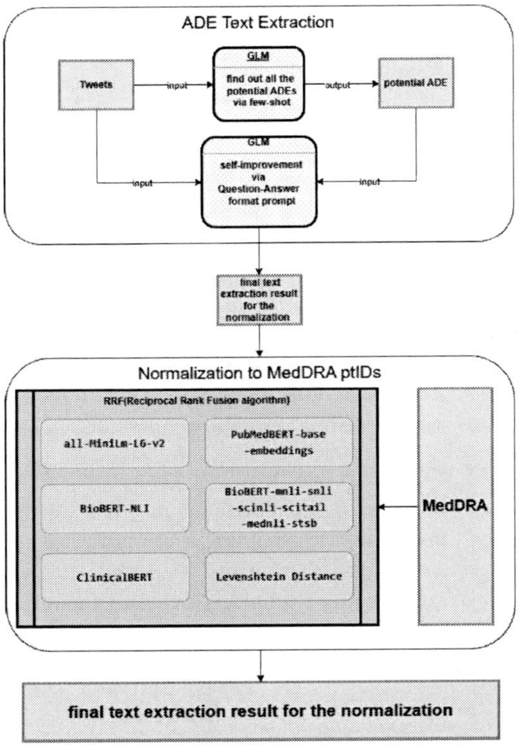

Figure 1: Framework of our system. In ADE Text Extraction, the green rectangle are GLM models with different prompt, and blue are the texts. In Normalization to MedDRA ptIDs, the red rectangle is RRF algorithm, including five language models and Levenshtein Distance.

2 Methodology

The entire framework is illustrated in Figure 1. We have two main steps in our system: a potential ADE Text Extraction phase and a normalization phase to MedDRA ptIDs. In this section, we will introduce the details of these phases.

2.1 ADE Text Extraction Phase

2.1.1 Potential ADE Text Extraction Step

In this section, considering the relatively small dataset size, we employ an approach that utilizes a pre-trained LLM to perform few-shot learning to extract potential ADE mentions. We used GLM-4 (Du et al., 2022) as our base model in this phase. In the first step, we input the tweets into GLM using few-shot prompts to extract potential ADEs.

Some tweets in the datasets do not contain any ADEs. We carefully designed our prompts to avoid outputting ADEs for such tweets. The prompt is shown in Appendix A.3.

2.1.2 Self-improvement Step

In our local experiments, we found that although our system captured many candidate ADE mentions in the first step, most of them were not caused by drugs regarding the context. To improve performance of our system, we designed a second step.

In the second step, inspired by conventional works that proposed a self-consistency check in zero-shot NER (Xie et al., 2023), we propose to utilize a question-and-answer style prompt to let the LLM model improve the outputs by itself.

We input the tweets containing the identified ADEs from the first step, along with their corresponding ADEs, into GLM in a question-and-answer format that ask the LLM to distinguish whether the ADEs identified in the fist step are caused by a drug rather than any other factors. The prompt can be found in Appendix A.3.

By employing this approach, our system filters out the tweets that do not have ADEs caused by drugs regarding the context of the input tweet. Then the remaining candidate ADE mentions are going to be mapped to the MedDRA ptIDs in the normalization phase.

2.2 Normalization Phase

We employ an ensemble approach involving multiple models and methods for normalization. The models and methods used in ensemble are summarized as follows:

- all-MiniLm-L6-v2, a general sentence-transformers (Reimers and Gurevych, 2019) model based on nreimers/MiniLM-L6-H384-uncased[1] and fine-tuned on a corpus including Reddit Comments, WikiAnswers, etc , with a total of 1B tokens [2].

- PubMedBERT-base-embeddings, a PubMedBERT-base model fine-tuned on the titles and abstracts of medical papers from the PubMed dataset [3].

- BioBERT-NLI, a BioBERT (Lee et al., 2019) model further fine-tuned on the SNLI (Bowman et al., 2015) and MultiNLI (Williams et al., 2018) datasets [4].

[1] https://huggingface.co/nreimers/MiniLM-L6-H384-uncased

[2] https://huggingface.co/sentence-transformers/all-MiniLM-L6-v2

[3] https://huggingface.co/NeuML/pubmedbert-base-embeddings

[4] https://huggingface.co/gsarti/biobert-nli

- BioBERT-mnli-snli-scinli-scitail-mednli-stsb (Deka et al., 2022), a sentence-transformer model trained on SNLI, MNLI, SCINLI, SCITAIL, MEDNLI, and STSB datasets [5].

- ClinicalBERT (Wang et al., 2023), a BERT model trained on a 1.2B disease-related dataset [6].

- The Levenshtein Distance between the extracted results and the corresponding text of each ADE in MedDRA.

For Sentence-transformers and BERT models, we calculate the cosine similarity between the embeddings of the extracted results and the embeddings of each ADE as the relevance measure. For Levenshtein Distance, we directly use it as a relevant metric.

Due to the inconsistency in measures among these methods above, we employ Rank-based Reciprocal Rank Fusion (RRF) (Cormack et al., 2009) to fuse the relevance scores since it is a rank-based method.

Denote the embedding from embedding model i as $E_i(\cdot)$, the extracted text piece of a candidate ADE mention as t_{sys}, and any ADE term text in MedDRA as t_{dra}. The similarity based on embedding model i is

$$S_i(t_{sys}, t_{dra}) = \frac{(E_i(t_{sys})) \cdot (E_i(t_{dra}))}{|(E_i(t_{sys}))||(E_i(t_{dra}))|}. \quad (1)$$

We rank the ADEs in MedDRA separately based on the similarities and Levenshtein Distances obtained from the above models, and then fuse the rank results by RRF. Denote the ranking by method i as r_i, the rank of a MedDRA ADE preferred term t_{dra} as $r_i(t_{dra})$, the set of all candidate rankings as $\mathcal{R}$, and the set of all ADE texts in MedDRA as $\mathcal{D}$. The final score is

$$S(t_{dra} \in \mathcal{D}) = \sum_{r_i \in \mathcal{R}} \frac{1}{k + r_i(t_{dra})}. \quad (2)$$

k is a preset constant. We used $k = 5$.

The ptID corresponding to the ADE text with the highest final score is extracted and submitted as the ptID in the submission file.

[5] https://huggingface.co/pritamdeka/BioBERT-mnli-snli-scinli-scitail-mednli-stsb

[6] https://huggingface.co/medicalai/ClinicalBERT

3 Experiments

Besides our method described in Section 2, there are several prompting methods that used in similar tasks, including few-shot prompting (Brown et al., 2020), zero-shot prompting and least-to-most prompting (Zhou et al., 2022) as follows,

1. **Few-shot**: This method provides several examples along in the prompts. Our system uses this method in the first-step in the ADE extraction phase.

2. **Zero-shot**: This method input instructions to a LLM without examples.

3. **Least-to-most**: This prompting strategy proposes to divide a complex problem into subproblems in prompting. To apply it to this task in our experiments, we divided the NER task into three subtasks, including identifying whether the effect is negative, determining whether it is caused by the medication, and determining if it is an ADE caused by the medication based on the results of the previous two tasks.

We compared these methods in our local experiments.

Moreover, GLM-4 are reported as an improved version of GLM-3-Turbo. Thus, we also evaluated it for this task.

In details, we compared five methods that used different prompts and LLM models: few-shot prompt + GLM-3-Turbo WSI(without self-improvement), few-shot prompt + GLM-3-Turbo, few-shot prompt + GLM-4, zero-shot prompt + GLM-3-Turbo, and least-to-most prompt + GLM-3-Turbo. We evaluated these five methods in validation dataset in order to find the best method. The result is shown in Appendix A.1. The few-shot + GLM-4 achieved leading results across all metrics, so we ultimately decided to use the few-shot prompt + GLM-4 method on the test set.

4 Results and Discussions

4.1 Overall Results

The metrics used to evaluate results in the task are as follows,

- **F1-Norm**: The ADE Normalization F1 Score.

- **P-Norm**: The ADE Normalization Precision Score.

	Ours	Mean	Median
F1-Norm	20.5	28.3	29.3
P-Norm	13.7	29.2	33.9
R-Norm	**41.1**	33.4	32.6
F1-NER	21.6	32.7	37.6
P-NER	14.5	35.6	43.7
R-NER	**42.8**	34.0	37.4
F1-Norm-Unseen	10.6	20.9	14.1
P-Norm-Unseen	6.1	20.5	14.4
R-Norm-Unseen	**40.6**	28.7	36.5

Table 1: The final result of our system in the task in comparison with the mean and median scores among all the participants.

- **R-Norm**: ADE Normalization Recall Score.
- **F1-NER**: ADE Extraction F1 Score.
- **P-NER**: ADE Extraction Precision Score.
- **R-NER**: ADE Extraction Recall Score.
- **F1-Norm-Unseen**: ADE Normalization on Unseen MedDRA IDs F1 Score.
- **P-Norm-Unseen**: ADE Normalization on Unseen MedDRA IDs Precision Score.
- **R-Norm-Unseen**: ADE Normalization on Unseen MedDRA IDs Recall Score.

Table 1 presents the results of our method on the test dataset. F1-Norm scored 20.5, P-Norm scored 13.7, and R-Norm scored 41.1. F1-NER scored 21.6, P-NER scored 14.5, and R-NER scored 42.8. F1-norm-unseen scored 10.6, P-Norm-Unseen scored 6.1, and R-Norm-Unseen scored 40.6.

Our method works well without the need for large amounts of training data. At the same time, in terms of recall rate, we have achieved scores higher than both the median and the average scores among the participants.

5 Conclusion

Our method surpasses the median and average results of the SMM4H 2024 Task 1 in some metrics, with improvements of 10%-20% over the median in R-Norm, R-NER, and R-Norm-Unseen. This proves the feasibility of using large language models for social media text mining tasks with a well-designed prompting strategy, especially in the cases that high recall rate is required.

However, the decline in the precision of ADE mining is still an issue that needs to be addressed. We consider the reason for the low accuracy score in ADEs mining is that the large language model we used generalize based on their pre-trained data and prompt engineering, while the text of ADEs in social media comments can be very different from the standard ADE text. This gap leads to the failure of the large language model in identifying ADEs when it mines ADEs from social media.

References

Samuel R. Bowman, Gabor Angeli, Christopher Potts, and Christopher D. Manning. 2015. A large annotated corpus for learning natural language inference. In *Proceedings of the 2015 Conference on Empirical Methods in Natural Language Processing*, pages 632–642, Lisbon, Portugal. Association for Computational Linguistics.

Tom Brown, Benjamin Mann, Nick Ryder, Melanie Subbiah, Jared D Kaplan, Prafulla Dhariwal, Arvind Neelakantan, Pranav Shyam, Girish Sastry, Amanda Askell, Sandhini Agarwal, Ariel Herbert-Voss, Gretchen Krueger, Tom Henighan, Rewon Child, Aditya Ramesh, Daniel Ziegler, Jeffrey Wu, Clemens Winter, Chris Hesse, Mark Chen, Eric Sigler, Mateusz Litwin, Scott Gray, Benjamin Chess, Jack Clark, Christopher Berner, Sam McCandlish, Alec Radford, Ilya Sutskever, and Dario Amodei. 2020. Language models are few-shot learners. In *Proceedings of the 34th International Conference on Neural Information Processing Systems*, NeurIPS '20, Red Hook, NY, USA. Curran Associates Inc.

Gordon V. Cormack, Charles L A Clarke, and Stefan Buettcher. 2009. Reciprocal rank fusion outperforms condorcet and individual rank learning methods. In *Proceedings of the 32nd International ACM SIGIR Conference on Research and Development in Information Retrieval*, SIGIR '09, page 758–759, New York, NY, USA. Association for Computing Machinery.

Pritam Deka, Anna Jurek-Loughrey, and Deepak P. 2022. Evidence extraction to validate medical claims in fake news detection. In *International Conference on Health Information Science*, pages 3–15. Springer.

Jacob Devlin, Ming-Wei Chang, Kenton Lee, and Kristina Toutanova. 2019. Bert: Pre-training of deep bidirectional transformers for language understanding. In *North American Chapter of the Association for Computational Linguistics*.

Zhengxiao Du, Yujie Qian, Xiao Liu, Ming Ding, Jiezhong Qiu, Zhilin Yang, and Jie Tang. 2022. GLM: General language model pretraining with autoregressive blank infilling. In *Proceedings of the 60th Annual Meeting of the Association for Computational Linguistics (Volume 1: Long Papers)*, pages 320–335,

Dublin, Ireland. Association for Computational Linguistics.

Kexin Huang, Jaan Altosaar, and Rajesh Ranganath. 2019. Clinicalbert: Modeling clinical notes and predicting hospital readmission. *ArXiv*, abs/1904.05342.

Takeshi Kojima, Shixiang Shane Gu, Machel Reid, Yutaka Matsuo, and Yusuke Iwasawa. 2022. Large language models are zero-shot reasoners. *ArXiv*, abs/2205.11916.

Jinhyuk Lee, Wonjin Yoon, Sungdong Kim, Donghyeon Kim, Sunkyu Kim, Chan Ho So, and Jaewoo Kang. 2019. Biobert: a pre-trained biomedical language representation model for biomedical text mining. *Bioinformatics*, 36:1234 – 1240.

Nils Reimers and Iryna Gurevych. 2019. Sentence-bert: Sentence embeddings using siamese bert-networks. D19-1:3982–3992.

Guangyu Wang, Xiaohong Liu, Zhen Ying, Guoxing Yang, Zhiwei Chen, Zhiwen Liu, Min Zhang, Hongmei Yan, Yuxing Lu, Yuanxu Gao, Kanmin Xue, Xiaoying Li, and Ying Chen. 2023. Optimized glycemic control of type 2 diabetes with reinforcement learning: a proof-of-concept trial. *Nature Medicine*, 29(10):2633–2642.

Adina Williams, Nikita Nangia, and Samuel Bowman. 2018. A broad-coverage challenge corpus for sentence understanding through inference. In *Proceedings of the 2018 Conference of the North American Chapter of the Association for Computational Linguistics: Human Language Technologies, Volume 1 (Long Papers)*, pages 1112–1122, New Orleans, Louisiana. Association for Computational Linguistics.

Tingyu Xie, Qi Li, Jian Zhang, Yan Zhang, Zuozhu Liu, and Hongwei Wang. 2023. Empirical study of zero-shot NER with ChatGPT. In *Proceedings of the 2023 Conference on Empirical Methods in Natural Language Processing*, pages 7935–7956, Singapore. Association for Computational Linguistics.

Li Yujian and Liu Bo. 2007. A normalized levenshtein distance metric. *IEEE Transactions on Pattern Analysis and Machine Intelligence*, 29(6):1091–1095.

Denny Zhou, Nathanael Schärli, Le Hou, Jason Wei, Nathan Scales, Xuezhi Wang, Dale Schuurmans, Claire Cui, Olivier Bousquet, Quoc Le, and Ed Chi. 2022. Least-to-Most Prompting Enables Complex Reasoning in Large Language Models. *arXiv e-prints*, arXiv:2205.10625.

A Appendix

A.1 Local Experiment Result on Evaluation Set

As shown in Table 2 and Table 3, our system achieved excellent results on the evaluation set (DEV set). Normalization F1: 0.465, Normalization Precision: 0.500, Normalization Recall: 0.435. We also achieved great performance in extraction: Extraction F1: 0.580, Extraction Precision: 0.627, Extraction Recall: 0.540. Other methods performed lower compared to the method we used. Therefore, we believe that Few-shot learning is more suitable for large language models in named entity recognition tasks compared to Zero-shot and least to most approaches. Model improvement is also crucial. Under the same conditions, replacing the model from GLM-3-Turbo to GLM-4 alone can yield a 25% performance improvement.

A.2 Dataset Details

The test dataset contains a total of 11439 tweets. According to the dataset, we removed all sentences which are too short (those with fewer than 5 words). Besides, we deleted all non-English and non-numeric characters.

A.3 Prompts for the Two-step ADE Extraction

Our prompts used for the two-step ADE extraction is shown in Table 4. Our system used prompts were written in Chinese because GLM works better with Chinese prompts. The translated version is provided in Table 5.

Method	F1-Norm	P-Norm	R-Norm	F1-NER	P-NER	R-NER
Few-shot + GLM-3-Turbo WSI	0.168	0.247	0.127	0.332	0.483	0.253
Few-shot + GLM-3-Turbo	0.300	0.320	0.282	0.580	0.627	0.540
Few-shot + GLM-4	0.471	0.514	0.435	0.580	0.627	0.540
Zero-shot + GLM-3-Turbo	0.134	0.101	0.200	0.267	0.202	0.391
Least-to-most + GLM-3-Turbo	0.172	0.197	0.153	0.222	0.258	0.153

Table 2: The overall results achieved by different combinations of prompting strategies and LLMs, evaluated using the DEV set in our local experiments.

Method	F1-Norm-Unseen	P-Norm-Unseen	R-Norm-Unseen
Few-shot + GLM-3-Turbo WSI	0.000	0.000	0.000
Few-shot + GLM-3-Turbo	0.095	0.054	0.400
Few-shot + GLM-4	0.214	0.130	0.600
Zero-shot + GLM-3-Turbo	0.000	0.000	0.000
Least-to-most + GLM-3-Turbo	0.000	0.000	0.000

Table 3: The results for unseen ADEs regarding the training set, achieved by different combinations of prompting strategies and LLMs, evaluated using the DEV set in our local experiments.

Prompts for Coarse-grained ADE Text Extraction
你是一个超级医药化学专家，请你找出我下列英文句子中的化学物品或药物的副作用，句子中不一定包含副作用。若句子没有副作用则不需要输出。如果有副作用，请找到后仅提供最终对应副作用的结果，只输出第一个副作用，无需展示推理过程。副作用可能有多个。我会分多次给你进行提示 有且只有一个副作用的句子请这样输出 例句：avelox has hurt your liver, avoid tylenol always,it further damages liver, eat grapefruit unless taking cardiac drugs 副作用：hurt your liver
Prompts for Self-improvement
我会给你一句话和多个词，你需要使用你的推理能力来判断，这些ADEs是否是由于药物导致的而不是原本就存在的问题。例如：avelox has hurt your liver, avoid tylenol always,it further damages liver, eat grapefruit unless taking cardiac drugs 副作用：hurt your liver 推理：因为例句中avelox has hurt your liver 这句话所以可以推断出hurt your liver是由于avelox导致的。不需要输出原因，只需要输出是或者不是。请你用你的逻辑能力回答：[tweet]中的[ADE]是否是药物导致，还是本身或者其他不良习惯导致，只需要输出是或者不是。

Table 4: Our prompts for the two steps for ADE Text Extraction, which is originally written in Chinese.

Prompts for Coarse-grained ADE Text Extraction in English
You are a super medical chemistry expert, please identify any chemical substances or drugs and their Adverse Drug Event (ADE) in the following English sentences. The sentences may or may not contain ADEs. If the sentence does not contain any ADE, you don't need output anything. If there are ADEs, please find them and provide only the result of corresponding side effects without showing the reasoning process. Side effects may be multiple. I will give you some examples. Example: avelox has hurt your liver, avoid tylenol always, it further damages liver, eat grapefruit unless taking cardiac drugs ADE: hurt your liver

Prompts for Self-improvement in English
I will give you a sentence and multiple words, and you will need to use your reasoning ability to determine whether these ADEs are caused by the medication rather than pre-existing conditions. Example: avelox has hurt your liver, avoid tylenol always, it further damages liver, eat grapefruit unless taking cardiac drugs ADE: hurt your liver reasoning: Because of the sentence "avelox has hurt your liver," it can be inferred that "hurt your liver" is caused by avelox. There is no need to output the reason, just yes or no. Please use your logical abilities to answer: In the [tweet], is the [ADE] caused by the medication, or is it due to the person's own issues or other bad habits? Just output "yes" or "no".

Table 5: Our prompts for the two steps for ADE Text, translated in English.

PheonixTrio918 at SMM4H 2024: 5 Fold Cross Validation for Classification of tweets reporting children's disorders

B Rahul Naik
Indian Institute of Technology Jodhpur
naik.9@iitj.ac.in

Oppangi Poojita
Indian Institute of Technology Jodhpur
poojita.1@iitj.ac.in

PothiReddy Kovidh Reddy
Indian Institute of Technology Jodhpur
reddy.19@iitj.ac.in

Lipika Dey
Ashoka University
lipika.dey@ashoka.edu.in

Abstract

This document describes our system used for the Social Media Mining for Health (SMM4H) 2024 Task 05. The objective of this task was to perform binary classification on the tweets provided in the dataset. The dataset contained two categories of tweets: those reporting medical disorders and those merely mentioning the disease. We tackled this problem using a 5-fold cross-validation approach. Our method utilizes the RoBERTa-Large model with 5-fold cross-validation. The evaluation results yielded an F1-score of 0.886 on the validation dataset and 0.823 on the test dataset.

1 Introduction

Recent advancements in Natural Language Processing have been groundbreaking, simplifying classification systems while making them more resilient to complex challenges. Integrating real-world data with NLP techniques has enhanced system efficiency, and for practical tasks such as Intent Recognition, Sentiment Analysis, or Sentence Classification, the RoBERTa model (Liu et al., 2019) and its variants like RoBERTa-small and RoBERTa-large have demonstrated outstanding performance. These models' comprehension mechanisms have consistently been the preferred choice for these applications.

Social Media platforms, such as Twitter, consistently produce vast quantities of data in various formats, including text. Extensively utilized for sharing and disseminating users' opinions on numerous topics, Twitter data has led to initiatives like the Social Media Mining for Health Application (SMM4H) Shared Task in 2024 (Xu et al., 2024). This paper outlines our approach for shared task 05: Binary classification of tweets that report children's medical disorders (in English).

2 Methodology

2.1 Data

We utilized the dataset made available by the organizers of Task 05. In this task, we were provided with two categories of tweets. Tweets that merely mention Child Disorder are labeled as 0, while tweets that aim to report the disorders are labeled as 1. The dataset is divided into three sections. The training set encompasses 7,398 tweets, with 5,118 tweets labeled as 0 and 2,280 tweets labeled as 1. The validation set includes 389 tweets, with 254 tweets labeled as 0 and 135 tweets labeled as 1. The test set consists of approximately 10,000 tweets without labels.

Label	Training	Validation
0	5118 (69.2%)	254 (65.3%)
1	2280 (30.8%)	135 (34.7%)

Table 1: Data distribution across Training, Validation sets for each label.

2.2 Preprocessing

Neattext is a simple NLP package designed for cleaning textual data and preprocessing text by efficiently removing various types of unnecessary text elements, such as usernames, hashtags, links, emojis, and dates.[1] In this dataset, Neattext played a crucial role in enhancing the preprocessing phase by using its library of functions to systematically remove these elements, thereby enriching the model's understanding. Initially, we utilized Neattext's functions to remove user handles, hashtags, emojis, and URLs, ensuring a basic level of preprocessing. However, we observed that Neattext's function for removing emojis was not exhaustive. Consequently, we created a custom Python function to

[1]https://pypi.org/project/neattext/

Proceedings of The 9th Social Media Mining for Health Applications (#SMM4H) Workshop and Shared Tasks, pages 55–57
August 15, 2024 ©2024 Association for Computational Linguistics

ensure the removal of all types of emojis, further refining the preprocessing.

The detailed preprocessing steps, including the removal of user handles, hashtags, emojis, and URLs, significantly improved the model's performance. By eliminating these extraneous elements, the model focused more on the core textual content, which enhanced its ability to detect relevant patterns and reduced noise. This preprocessing resulted in better feature extraction and improved the accuracy and robustness of the model's predictions.

After preprocessing, we merged the training and validation datasets to augment the data available to the model. This strategy aimed to enhance its ability to generalize and mitigate overfitting. The combined dataset totaled 7,787 tweets, with 5,372 labeled as 0 and 2,415 as 1.

2.3 Resampling

We noticed a clear imbalance in the dataset before and after concatenating, where tweets labeled as mere mentions labeled as 0 outnumbered those labeled with medical disorders labeled as 1. To address this, we employed the RandomUnderSampler technique (Lemaitre et al., 2017) to balance the dataset by reducing the number of examples in the Majority class. This approach equalized both classes to 2,415 samples each, resulting in a total dataset of 4,830 samples. This resampling technique helps mitigate the bias in the dataset and ensures that the model is trained more effectively on both classes, improving its ability to generalize across different categories.

	Label 0	Label 1
Before Resampling	5372	2415
After Resampling	2415	2415

Table 2: Data distribution of labels before and after applying the resampling.

2.4 Model

RoBERTa(Liu et al., 2019), an optimized variant of BERT, demonstrates superior performance in natural language understanding tasks. It utilizes larger training datasets, longer training times, and dynamic masking patterns during pretraining to effectively capture intricate linguistic nuances. RoBERTa was chosen for its proven efficacy in various NLP benchmarks, consistently outperforming baseline models like BERT due to refined training methodologies and architectural enhancements.

2.5 Cross validation

Cross-validation is a data resampling technique employed to evaluate models and is computationally efficient. The fundamental approach to cross-validation is k-fold cross-validation (Brownlee, 2023). In this study, we implemented 5-fold cross-validation to robustly assess the model's performance across different subsets of the dataset.

2.6 Model Parameters

In implementing the 5-fold cross-validation method, we trained our model using parameters set as follows: 15 epochs, a batch size of 6, and a maximum tokenization length of 280 characters. The criterion employed was Cross Entropy Loss, with optimization performed using the Adam Optimizer at a learning rate of 1.5e-8.

3 Results

After applying the model parameters and conducting 5-fold cross-validation on both the validation and test datasets, we achieved an f1 score of 0.886 with precision of 0.831 and recall of 0.948 on the validation dataset. On the test dataset, the model yielded an f1 score of 0.823, with precision and recall values of 0.721 and 0.959, respectively.

	F1 Score	Precision	Recall
Valid dataset	0.886	0.831	0.948
Test dataset	0.823	0.721	0.959

Table 3: Results of the model on Validation and Test Datasets.

4 Discussion

Despite achieving an average f1 score of 0.823, our methodology falls short of matching the baseline model performance. Several factors may contribute to this disparity. One significant factor is the suboptimal choice of the learning rate. In our methodology, the learning rate was not effectively optimized, potentially impeding the model's ability to converge to the global minimum of the loss function efficiently. Consequently, the model may have struggled to achieve optimal parameter settings.Furthermore, the decision not to implement text conversion to lowercase could have limited the standardization of the model's vocabulary and text, potentially hindering the training process. Additionally, not setting the tokenization length to the maximum of 512 tokens, as recommended for RoBERTa, may have reduced the model's ability to capture the complete contextual meaning of tweets, thus possibly leading to suboptimal performance.Our methodology consistently achieves an impressive recall score across the training, validation, and test datasets. This can be attributed to the RoBERTa model's dynamic masking, which adeptly captures intricate linguistic nuances. Consequently, while our model excels in identifying relevant cases, it also tends to exhibit higher false positive rates, thereby impacting precision.

Acknowledgments

We would like to thank our supervisor, Lipika Dey, for her invaluable guidance, support and mentorship throughout the competition. In addition, we extend our gratitude to the reviewers for their insightful suggestions.

References

Jason Brownlee. 2023. A gentle introduction to k-fold cross-validation. *Machine Learning Mastery*.

Guillaume Lemaitre, Fernando Nogueira, and Christos K. Aridas. 2017. imbalanced-learn: A python toolbox to tackle the curse of imbalanced datasets in machine learning. *Journal of Machine Learning Research*, 18(17):1–5.

Yinhan Liu, Myle Ott, Naman Goyal, Jingfei Du, Mandar Joshi, Danqi Chen, Omer Levy, Mike Lewis, Luke Zettlemoyer, and Veselin Stoyanov. 2019. Roberta: A robustly optimized BERT pretraining approach. *CoRR*, abs/1907.11692.

Dongfang Xu, Guillermo Lopez Garcia, Lisa Raithel, Rolland Roller, Philippe Thomas, Eiji Aramaki, Shuntaro Yada, Pierre Zweigenbaum, Karen O'Connor, Yao Ge, Sudeshna Das, Abeed Sarker, Ari Klein, Lucia Schmidt, Vishakha Sharma, Raul Rodriguez-Esteban, Juan Banda, Ivan Flores Amaro, Sai Tharuni Samineni, and Graciela Gonzalez-Hernandez. 2024. Overview of the 9th social media mining for health applications (#SMM4H) shared tasks at ACL 2024. In *Proceedings of the 9th Social Media Mining for Health Applications Workshop and Shared Task*, Bangkok, Thailand. Association for Computational Linguistics.

HBUT at #SMM4H 2024 Task2: Cross-lingual Few-shot Medical Entity Extraction using a Large Language Model

Yuanzhi Ke
Hubei University of Technology
keyuanzhi@hbut.edu.cn

Zhangju Yin
Hubei University of Technology
yinzhangju@hbut.edu.cn

Xinyun Wu
Hubei University of Technology
xinyun@hbut.edu.cn

Caiquan Xiong
Hubei University of Technology
xiongcq@hbut.edu.cn

Abstract

Named entity recognition (NER) of drug and disorder/body function mentions in web text is challenging in the face of multilingualism, limited data, and poor data quality. Traditional small-scale models struggle to cope with the task. Large language models with conventional prompts also yield poor results. In this paper, we introduce our system, which employs a large language model (LLM) with a novel two-step prompting strategy. Instead of directly extracting the target medical entities, our system firstly extract all entities and then prompt the LLM to extract drug and disorder entities given the all-entity list and original input text as the context. The experimental and test results indicate that this strategy successfully enhanced our system performance, especially for German language. Our code is available on Github [1].

1 Introduction

Discussions on drugs and their adverse drug reactions shared by users in social media, including the efficacy, side effects, and personal treatment journeys serve as valuable references for pharmacovigilance.

We participated in the 9th Social Media Mining for Health Research and Applications Workshop and Shared Tasks (Xu et al., 2024). The task targets both the extraction of drug and disorder/body function mentions (Subtask 2a) and the extraction of relations between those entities (joint Named Entity Detection and Relation Extraction, Subtask 2b). The paper is about the task 2a. Our task was to identify drug and disorder mentions in German, French, and Japanese datasets from X(Twitter) and a German patient forum.

In previous works on similar tasks, people commonly use BERT models (Kenton and Toutanova, 2019). For multilingual tasks, the m-BERT (multilingual BERT) model is often employed (Papadimitriou et al., 2021). However, in this particular open task, the provided dataset is relatively small, especially in the training set. Therefore, it is challenging to perform pre-training and fine-tuning on the traditional BERT model due to the limited amount of data available. Thus, we opted to use a large language model (LLM) for the task.

Because the training data is also not enough to fine-tune a LLM well, we opted for the prompt-based approach. Our system utilizes a two-step prompting strategy. A first step to get a list of all entities, and then a second to further extract drug and disorder entities from the list.

2 Methodology

Due to the small size of the provided dataset, we placed our focus primarily on using large language models (LLMs). We build our system based on a GLM (Du et al., 2022; Zeng et al., 2022) through its online API. It is trained on multilingual dataset and performs almost on a par with ChatGPT and worked well for multilingual tasks in our preliminary experiments.

We utilize few-shot prompt engineering technologies to adapt GLM to the drug and disorder mention extraction task as shown in Figure 1.

2.1 Prompting Method

The task is challenging for the numerous abbreviations and colloquial vocabulary in the dataset, which requires our system to be able to recognize such unofficial representations. In our preliminary experiments, we found that our system failed to work well with conventional prompts that guide the LLM to extract drugs and disorder mentions from the input text.

Thus, we address this issue by a novel strategy. Firstly, instead of instructing the LLM to extract drug/disorder mentions, our system prompts the model to extract all entities, just with additional

[1] https://github.com/YinZhangJu/SMM4H24code

Proceedings of The 9th Social Media Mining for Health Applications (#SMM4H) Workshop and Shared Tasks, pages 58–62
August 15, 2024 ©2024 Association for Computational Linguistics

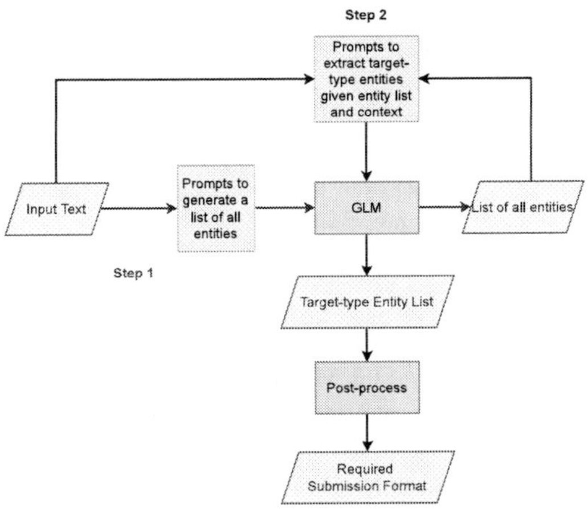

Figure 1: Overall architecture of our system. In the first step, the prompting guides the model to identify all entity lists from the dataset. In the second step, the prompting guides the model to combine the original data content and extract the target entities from the entity list.

attention on medicine-related entities. Then our system prompts the model to extract drugs and disorders given both the entity list and the original input as the context. In this way, our system is more robust for informal representations and ill-formed sentences. The detailed settings of our two-step prompting are described in Tables 1 and Tables 2.

2.2 Preprocessing for Sensitive Words

The GLM API raised errors for some samples in the test dataset, with error messages indicating that the input texts contained some sensitive words.

When our system detects such failures, it collects and writes the sample IDs, sample contents, and error messages in an error log. We manually check the error records, identify and remove the sensitive words in the corresponding samples, and then redo generation for them.

This method may produce incomplete and ill-formed sentences. However, because our two-step design is robust against ill-formed text, the adverse effects on the NER accuracy are relatively small. Replacing sensitive words with alternative words may result in better NER performance. However, we did not have enough time to work out a replacement table for the sensitive words in the shared task. We plan to complete it in future work.

2.3 The Post-Processing Step

We save the output results to a CSV file. Then, to match the submission format, a Python script is used to remove punctuation and symbols, and convert the results into the BRAT format.

3 Experiments and Results

3.1 Experiments on LLM Model Selection

Conventional works reported that Qwen-14B (Bai et al., 2023) performs well for factual tasks. Thus, we also conducted experiments with Qwen-14B in comparison to GLM.

We randomly sampled 100 samples from the

Our Prompt (Original in Chinese)
"role": "user", "content": "作为一名精通日法德三语的语言学家，请你找出下列日语、法语或者德语语句中的实体，尤其注意医学领域的实体，比如药物名称以及副作用等，注意简称，只提供结果，不需要推理过程" [Examples of finding entity lists for German, French, and Japanese] "role": "assistant", "content": "好的，我为您筛选出了语句中的各种实体" "role": "user", "content": "请找出下列[list]中的实体名称"
Our Prompt (Translated into English)
"role": "user", "content": "You are a linguist proficient in German, French, and Japanese. Please help me identify the entities in the following Japanese, French, or German sentences. Please pay attention to entities in the medical field, such as drugs and disorder mentions. Pleas also be aware of abbreviations. Provide only the results without the need for the reasoning process" [Examples of finding entity lists for German, French, and Japanese] "role": "assistant", "content": "I have filtered out various entities from the sentences for you." "role": "user", "content": "Please identify the entity names in the following [list]."

Table 1: The prompt template used in the first step. In this step, the model is tasked with identifying the entity list from the dataset. "[list]" represents the examples randomly sampled from the dataset. We provide the English translation of our prompt together with the original prompt used in the system (which is written in Chinese).

Our Prompt (Original in Chinese)
"role": "user", "content": "作为一名日法德三语药剂师，请你根据提供的日语、法语或者德语句子语境，按照下列格式找出实体列表中的药物以及副作用实体，注意简称，只提供结果，不需要推理过程" [Examples of target entities within the entity lists for German, French, and Japanese:] "role": "assistant", "content": "好的，我为您筛选出了实体列表中的药品名称和副作用" "role": "user", "content": "请根据[list]中的语境，仿照上面格式，找出下列[str1]中药物名称以及副作用"
Our Prompt (Translated into English)
"role": "user", "content": "You are a pharmacist proficient in German, French, and Japanese. Please identify the drug and disorder mentions from the list, according to the context in Japanese, French, or German. Pleas also be aware of abbreviations. Provide only the results without the need for the reasoning process." [Examples of target entities within the entity lists for German, French, and Japanese:] "role": "assistant", "content;": "I have filtered out the drugs and disorder mentions from the entity list for you." "role": "user", "content": "Please, based on the context in [list], follow the format above to identify the drugs and disorder mentions in [str1]."

Table 2: The prompt template used in the second step. In this step, the model is fed with the original text with the generated entity list from the previous step to identify drugs and their corresponding disorder mentions. "[list]" represents the original text of the dataset, and "[str1]" denotes the generated entity list. We provide the English translation of our prompt together with the original prompt used in the system (which is written in Chinese).

outputs generated by Qwen-14B and GLM respectively, and manually checked the correctness. The accuracy scores achieved by the two models in this local experiment is as shown in Table 3. Qwen-14B failed to get satisfying results. Qwen-14B exhibited tokenization issues with German, French, and Japanese data, resulting in significant problems

Model	Accuracy by Human
GLM-3-Turbo	0.4358
Qwen-14B	0.1538

Table 3: Human evaluation results of GLM-3-Turbo and Qwen-14B in our local experiments to choose the candidate base model in our system.

with fabricated entities and incorrect scope.

In case of GLM, there are several available versions. We tried GLM-3-turbo and GLM-4. However, we encountered some encoding issues with GLM-4. When answering questions in German and French, we observed that GLM-4 firstly translated the text into English and then provided an English response. This led to inaccuracies in the output of German and French words. In contrast, GLM-3-turbo exhibited minimal issues in this regard. Therefore, we used GLM-3-turbo in our system. We used the default generation parameters. The temperature is set to 0.95, top p is set to 0.7, and max tokens is set to 1024.

3.2 Experiments on Prompting Methods

We conducted experiments on three prompt strategies as follows,

- Single-word Prompting (Single Prompt): In this approach, we directly guided the model to perform named entity recognition on German, French, and Japanese data.

- Detached Step-by-step Prompting (DSBS Prompt): In this approach, we utilized a step-by-step prompting method. In the first step, we initially prompted the LLM model to generate a list of all entities in the input text, instead of just medical-related entities like that in conventional system. In the second step, we let the LLM model to extract drug and disorder entities from the entity list.

- Integrated Step-by-step Prompting (ISBS Prompt): This approach is also a step-by-step prompting method. The first step is as the same as DSBS prompt. But in the second step, we combined the entity list and the original input text in the prompts.

The results are shown in Table 4.

Single Prompt uses fewer tokens. However, it exhibited lower precision in capturing phrase scope, with larger discrepancies compared to the gold annotation.

DSBS Prompt resulted in more accurate identification of entity words and achieved higher precision. However, as it detached from the original text, it performed poorly in extracting drugs and disorder mentions from the entity list.

ISBS Prompt led to more accurate identification of entity words, higher precision, and effective extraction of drugs and disorder mentions from the entity list.

Prompting Method	F1
Single Prompt	0.3420
DSBS Prompt	0.3023
ISBS Prompt	**0.3533**

Table 4: Comparison of F1 scores achieved by Single Prompt, DSBS Prompt and ISBS Prompt, evaluated using the online scores in validation phase.

4 Final Results

Our final system employed GLM-3-turbo model with ISBS Prompt. The overall results are shown in Table 5. Our system achieved a better precision than the mean precision among all teams (including baseline).

Especially, our method demonstrated excellent performance for the German part. The results for the German part are shown in Table 6. The overall precision, recall and F1 scores are higher than the mean scores among all the participants. However, our system failed to address mentions of function.

Team	Precision	Recall	F1
Ours	0.6052	0.2654	0.3690
Mean	0.5942	0.3434	0.4291
Std	0.0156	0.1103	0.0850

Table 5: Overall results.

Entity Type	Precision	Recall	F1
DISORDER	0.5463	0.2744	0.3653
DRUG	0.7627	0.3629	0.4918
FUNCTION	0.0000	0.0000	0.0000
All	0.6228	0.2751	0.3817
Mean	0.4404	0.1945	0.2699

Table 6: Final result of our system for German data. The first three rows show the results of our system for different entity types. The last row shows the mean result among all participants for all entity types.

5 Conclusion

Due to the small size of our dataset, we found that the performance with conventional small language models (e.g. BERTs) is not able to give satisfying precision. Consequently, we build our system based on GLM-3-Turbo large language model. Besides, we found that conventional straight-forward prompting strategy encountered low precision for this task. To address this issue, we proposed a step-by-step prompting strategy. We firstly prompt to extract a list of all entities. Then we prompt to guide the model to choose drugs and disorder mentions from the list within the context of the corresponding original input text. This prompting approach significantly improved the effectiveness of our system.

References

Jinze Bai, Shuai Bai, Yunfei Chu, Zeyu Cui, Kai Dang, Xiaodong Deng, Yang Fan, Wenbin Ge, Yu Han, Fei Huang, Binyuan Hui, Luo Ji, Mei Li, Junyang Lin, Runji Lin, Dayiheng Liu, Gao Liu, Chengqiang Lu, Keming Lu, Jianxin Ma, Rui Men, Xingzhang Ren, Xuancheng Ren, Chuanqi Tan, Sinan Tan, Jianhong Tu, Peng Wang, Shijie Wang, Wei Wang, Shengguang Wu, Benfeng Xu, Jin Xu, An Yang, Hao Yang, Jian Yang, Shusheng Yang, Yang Yao, Bowen Yu, Hongyi Yuan, Zheng Yuan, Jianwei Zhang, Xingxuan Zhang, Yichang Zhang, Zhenru Zhang, Chang Zhou, Jingren Zhou, Xiaohuan Zhou, and Tianhang Zhu. 2023. Qwen technical report. *arXiv preprint arXiv:2309.16609*.

Zhengxiao Du, Yujie Qian, Xiao Liu, Ming Ding, Jiezhong Qiu, Zhilin Yang, and Jie Tang. 2022. Glm: General language model pretraining with autoregressive blank infilling. In *Proceedings of the 60th Annual Meeting of the Association for Computational Linguistics (Volume 1: Long Papers)*, pages 320–335.

Jacob Devlin Ming-Wei Chang Kenton and Lee Kristina Toutanova. 2019. Bert: Pre-training of deep bidirectional transformers for language understanding. In *Proceedings of NAACL-HLT*, pages 4171–4186.

Isabel Papadimitriou, Ethan A Chi, Richard Futrell, and Kyle Mahowald. 2021. Deep subjecthood: Higher-order grammatical features in multilingual bert. In *Proceedings of the 16th Conference of the European Chapter of the Association for Computational Linguistics: Main Volume*, pages 2522–2532.

Dongfang Xu, Guillermo Lopez Garcia, Lisa Raithel, et al. 2024. Overview of the 9th social media mining for health applications (#SMM4H) shared tasks at ACL 2024. In *Proceedings of The 9th Social Media Mining for Health Research and Applications Workshop and Shared Tasks*, Bangkok, Thailand. Association for Computational Linguistics.

Aohan Zeng, Xiao Liu, Zhengxiao Du, Zihan Wang, Hanyu Lai, Ming Ding, Zhuoyi Yang, Yifan Xu, Wendi Zheng, Xiao Xia, et al. 2022. Glm-130b: An open bilingual pre-trained model. *arXiv preprint arXiv:2210.02414*.

A Appendix

A.1 Dataset Details

The detailed information of the dataset is shown in Table 7.

	Training	**Validation**	**Test**
German	70	23	25
Japanese	392	168	118
French	4	0	96

Table 7: The Japanese data is sourced from X (Twitter), the German data is obtained from a German patient forum, and the French data is translated from the German data (distinct from the German data).

PCIC at SMM4H 2024: Enhancing Reddit Post Classification on Social Anxiety Using Transformer Models and Advanced Loss Functions

Leon Hecht, **Victor Martinez Pozos**, **Helena Gomez Adorno**,
Gibran Fuentes Pineda, **Gerardo Eugenio Sierra Martínez**, **Gemma Bel Enguix**

leon.hecht@comunidad.unam.mx
National Autonomous University of Mexico

Abstract

We present our approach to solving the task of identifying the effect of outdoor activities on social anxiety based on reddit posts. We employed state-of-the-art transformer models enhanced with a combination of advanced loss functions. Data augmentation techniques were also used to address class imbalance within the training set. Our method achieved a macro-averaged F1 score of 0.655 in the test data, exceeding the mean F1 score of the shared task of 0.519. These findings suggest that integrating weighted loss functions improves the performance of transformer models in classifying unbalanced text data, while data augmentation can improve the model's ability to generalize.

1 Introduction

This paper addresses Task 3 of the 9th Social Media Mining for Health (SMM4H) (Xu et al., 2024) workshop at ACL 2024, which focuses on analyzing the impact of outdoor activities on social anxiety through the lens of Reddit posts. Despite advances in natural language processing, accurately classifying such nuanced data poses significant challenges due to linguistic variability and data imbalances.

Previous studies have shown that there exist various possibilities to address data imbalance in classification tasks. For example, in (Shaikh et al., 2021), additional samples for underrepresented classes were generated. In (Hasib et al., 2023), Random Under-Sampling and Synthetic Minority Oversampling Techniques were employed. Other authors proposed to tackle the class imbalance problem by introducing loss functions that focus on the underrepresented classes (Lin et al., 2017).

Our study builds on these works by incorporating a novel combination of weighted loss functions within a transformer model to address the challenge of imbalanced data.

2 Task Description

The aim of this task is to classify into four classes if an outdoor activity mentioned in a post had a positive, neutral, negative, or unrelated impact on the person's social anxiety symptoms. The outdoor activity mentioned in every text was given as a keyword in an extra column, so that the dataset consisted of the columns `'id'`, `'text'`, `'keyword'` and `'label'`. The training dataset includes 1800 posts, and the validation and test set includes 600 posts each.

The keyword is highly important in the classification task since a post can be highly negative, but still, in only one sentence, the user mentions the positive effect of an outdoor activity. In this case, even though the entire text itself was negative, it should be classified as positive. So, it was important to somehow link the classification task to the keyword.

Another challenge was to tackle the highly imbalanced classes. The training dataset containing 1800 samples has 1131 texts for the class `'unrelated'`, 395 for class `'neutral'`, 160 for class `'positive'` and only 114 for class `'negative'`.

3 Methodology

For the experiments run to solve task 3, different transformer models of the huggingface library were used. A small model (DistilBert), two medium size models (RoBERTa and XLNet-base), and a larger model (XLNet-large) were employed (Sanh et al., 2019; Liu et al., 2019; Yang et al., 2019). These models were modified in some experiments to use a combined loss function. This combined loss function consists of the `Focal-Loss`, designed to address class imbalance by increasing the importance of hard-to-classify examples, `Weighted-Cross-Entropy` Loss, which assigns different weights to classes based on

Proceedings of The 9th Social Media Mining for Health Applications (#SMM4H) Workshop and Shared Tasks, pages 63–66
August 15, 2024 ©2024 Association for Computational Linguistics

Run	Class distribution	LR	DistilBert	RoBERTa	XLNet-base	XLNet-large
0	(400, 97, 231, 71)*	5e-6	0.47	0.55	0.49	0.49
1	(400, 97, 231, 71)*	5e-6	0.54	0.55	0.54	0.55
2	(400, 97, 231, 71)*	5e-6	0.47	0.56	0.55	**0.60**
3	(796, 464, 530, 415)*	5e-6	0.49	0.52	0.51	0.57
4	(1000, 580, 1000, 415)*	5e-6	0.51	0.56	0.56	0.53
5	(2000, 928, 1590, 747)*	5e-6	0.52	0.53	0.54	**0.54**

Table 1: Results of the different experiments run to monitor the influence of each modification on the model performance. *Class distribution has the order (unrelated, negative, neutral, positive). Note: LR stands for Learning Rate.

their representation in the training data, and `Weighted-Smooth-Cross-Entropy` Loss. This variation adds smoothing to the class labels to improve generalization (Lin et al., 2017). To combine these three loss functions, the mean of the three values is computed after each batch.

The preprocessing of the text data included cleaning the text of any URLs, extra whitespaces and performing a conversion of the emojis used to text using the python library `'emoji'` (Teahoon and Wurster, 2024). Apart from that, only the sentence(s) in each text containing the keyword and their previous and next sentence were used. This adaptation was made to reduce the text's length to prevent a substantial part of the text from being truncated in the tokenization process. The previous and next sentences were included to provide more context for the keyword phrases. The keyword was appended to the input text by using the model-specific separator token.

For training and validation, the training dataset (1800 instances) was divided into 70% for training and 30% for validation with a seed of 42. The macro-averaged F1 score was evaluated on the validation dataset (600 instances) to test the model performance. As a termination criterion, the F1 score was used for early stopping with a patience of 6 epochs. For the learning rate, the value of 5e-6 was the best result of a hyperparameter optimization and was therefore chosen for the experiments. The batch size was set to either 16 or 32 depending on the max-length parameter of the tokenizer due to hardware restrictions.

For some experiments, data augmentation was performed to address the class imbalance. On one hand, an augmentation by paraphrasing was used. On the other hand, augmentation was performed by punctuation insertion, random deletion, random insertion, and random swapping (from now on referred to as 'traditional augmentation').

For the paraphrasing task, a finetuned model from the huggingface repository was used, which is based on the T5-base model and finetuned on paraphrases generated by ChatGPT (Vorobev and Kuznetsov, 2023). For the punctuation insertion augmentation, the punctuation marks [',', '.', '!', '?', ';'] were inserted at a random position in the text with a frequency of 10% in relation to the number of words in the text. In the random deletion augmentation, each word in the text is deleted with a probability of 20%. In the random-insertion augmentation, four random words in the text were chosen, and then, using the wordnet of the library `'nltk'`, synonyms for these words were searched (Bird et al., 2009). One of the synonyms found was randomly inserted in the text for each of the four words. For the random swapping task, two random words were chosen in the text and then swapped.

Table 2 shows the hardware and software environment with which all experiments were run.

4 Results and Discussion

First, a baseline classification (Run 0) was run with the four different models. In this experiment, the original loss function of the model was used, and no data augmentation was performed.

For the following experiments, we aimed to evaluate the effect of the combined loss function (Run 1), the max-length parameter of the tokenizer (Run 2), data augmentation using paraphrasing (Run 3), and data augmentation using the aforementioned traditional augmentation (Run 4). Finally, an experiment was run where all these methods were combined, using the combined loss function, increased value for the max-length parameter, and both data augmentation methods (Run 5). The results of these experiments are shown in Table 1.

Operating System	Linux
GPU	Nvidia RTX A5000 24GB (1)
CUDA Version	12.1
Deep Learning Framework	PyTorch 2.2.0
Transformers Library Version	4.39.0
Python Version	3.10.12

Table 2: Specifications of the hardware and software environment used in the experiments.

Analyzing the experiment results, we can observe that, with exception of the RoBERTa model, all modifications outperformed the baseline or at least had the same performance. Using the combined loss function, improved the F1 score in the experiments for all models except of RoBERTa, with a mean absolute improvement of 0.06. This combined loss function is used throughout the following experiments.

Changing the max-length parameter of the tokenizer from 128 to 256, yielded the best result in all experiments for XLNet-large, lifting the F1 score up from 0.55 to 0.60. The motivation of this experiment surged from an analysis that we conducted, showing that 334 texts (18.5%) of the tokenized text lengths in the training dataset surpass the previously set length of 128 for the XLNet tokenizer (see figure 1). For comparability, the same max-length parameter was set for the experiments with DistilBert and RoBERTa, even though these models work with a different tokenizer and tokenized text length might differ.

Augmentation by paraphrasing only showed a slight improvement for one model, XLNet-large. At the same time, traditional augmentation showed a mixed effect on the F1 score across the different models. In the experiment with XLNet-large, traditional augmentation seems to have a negative impact on the F1 score, decreasing it from 0.55 to 0.53.

In the following experiment, all of the previous modifications were combined. Hence, the combined loss function was used, the max-length parameter of the tokenizer was set to 256, and the data were augmented using both augmentation methods. Despite the slightly decreased performance on the validation data in this experiment, it is possible that due to the augmentation, the model's ability to generalize improves. Therefore, this set-up with the model XLNet-large was used to obtain the predictions on the test set (0.655 F1 score) and was sent to the task organizers.

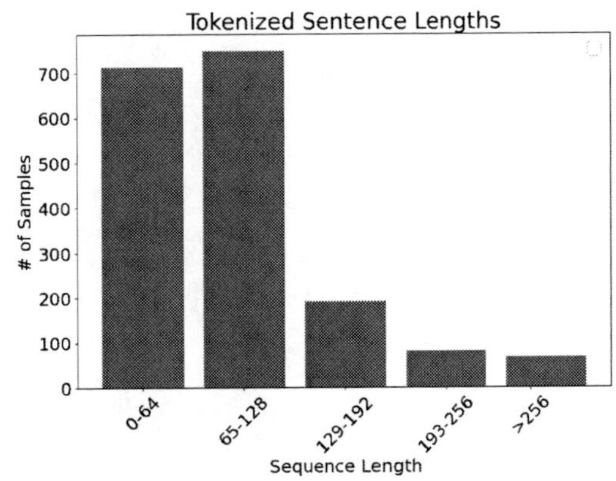

Figure 1: Sequence length analysis of tokenized texts (XLNet tokenizer)

5 Conclusion

This paper presents our approach to solving task 3 of the SMM4H 2024 workshop, which consists of a unique pre-processing to avoid truncating important information and providing only the key phrases of the text to the model. Furthermore, the XLNet model was adapted to use a combined loss function, and data augmentation was performed to address the class imbalance and improve generalizability. The keyword related to the outdoor activity was appended to the input text using the separator token. This setup has resulted in a macro-averaged F1 score of 0.655 on the test data, outperforming the mean of 0.519.

6 Acknowledgments

This work was carried out as part of PAPIIT project IT100822, IN104424 and the project CF2023-G64, which is funded by '*Consejo Nacional de Humanidades, Ciencias y Tecnologías* (CONAHCYT)'. We also want to thank Ricardo Villareal and Rita Rodríguez for supporting us with the computing resources.

References

S Bird, E Klein, and E Loper. 2009. *Natural Language Processing with Python – Analyzing Text with the Natural Language Toolkit.* O'Reilly Media, Inc.

Khan Md Hasib, Nurul Akter Towhid, Kazi Omar Faruk, Jubayer Al Mahmud, and M.F. Mridha. 2023. Strategies for enhancing the performance of news article classification in bangla: Handling imbalance and interpretation. *Engineering Applications of Artificial Intelligence*, 125:106688.

Tsung-Yi Lin, Priya Goyal, Ross Girshick, Kaiming He, and Piotr Dollar. 2017. Focal loss for dense object detection. In *Proceedings of the IEEE International Conference on Computer Vision (ICCV)*.

Yinhan Liu, Myle Ott, Naman Goyal, Jingfei Du, Mandar Joshi, Danqi Chen, Omer Levy, Mike Lewis, Luke Zettlemoyer, and Veselin Stoyanov. 2019. Roberta: A robustly optimized BERT pretraining approach. *CoRR*, abs/1907.11692.

Victor Sanh, Lysandre Debut, Julien Chaumond, and Thomas Wolf. 2019. Distilbert, a distilled version of bert: smaller, faster, cheaper and lighter. *ArXiv*, abs/1910.01108.

Sarang Shaikh, Sher Muhammad Daudpota, Ali Shariq Imran, and Zenun Kastrati. 2021. Towards improved classification accuracy on highly imbalanced text dataset using deep neural language models. *Applied Sciences*, 11(2).

K. Teahoon and K. Wurster. 2024. emoji: A python library to perform operations on emojis in text data. `https://carpedm20.github.io/emoji/docs/index.html`. Accessed: 2024-05-13.

V Vorobev and M Kuznetsov. 2023. A paraphrasing model based on chatgpt paraphrases.

Dongfang Xu, Guillermo Lopez Garcia, Lisa Raithel, Rolland Roller, Philippe Thomas, Eiji Aramaki, Shuntaro Yada, Pierre Zweigenbaum, Sai Tharuni Samineni, Karen O'Connor, Yao Ge, Sudeshna Das, Abeed Sarker, Ari Klein, Lucia Schmidt, Vishakha Sharma, Raul Rodriguez-Esteban, Juan Banda, Ivan Flores Amaro, Davy Weissenbacher, and Graciela Gonzalez-Hernandez. 2024. Overview of the 9th social media mining for health applications (#SMM4H) shared tasks at ACL 2024. In *Proceedings of The 9th Social Media Mining for Health Research and Applications Workshop and Shared Tasks*, Bangkok, Thailand. Association for Computational Linguistics.

Zhilin Yang, Zihang Dai, Yiming Yang, Jaime G. Carbonell, Ruslan Salakhutdinov, and Quoc V. Le. 2019. Xlnet: Generalized autoregressive pretraining for language understanding. *CoRR*, abs/1906.08237.

Transformers at #SMM4H 2024: Identification of Tweets Reporting Children's Medical Disorders And Effects of Outdoor Spaces on Social Anxiety Symptoms on Reddit Using RoBERTa

Kriti Singhal, Jatin Bedi
Computer Science and Engineering Department
Thapar Institute of Engineering and Technology
kritisinghal711@gmail.com, jatin.bedi@thapar.edu

Abstract

With the widespread increase in the use of social media platforms such as Twitter, Instagram, and Reddit, people are sharing their views on various topics. They have become more vocal on these platforms about their views and opinions on the medical challenges they are facing. This data is a valuable asset of medical insights in the study and research of healthcare. This paper describes our adoption of transformer-based approaches for tasks 3 and 5. For both tasks, we fine-tuned large RoBERTa, a BERT-based architecture, and achieved an F1 score of 0.413 and 0.900 in tasks 3 and 5, respectively.

1 Introduction

The past few years have witnessed an exponential rise in the use of social media. People can voice their views and opinions on social media platforms such as Facebook, Reddit, and Twitter. Twitter and Reddit have become major platforms for people seeking help and sharing their medical problems. They also take to these platforms to share their health regimen and medical concerns. Thus, Reddit and Twitter are indispensable resources that aid in better comprehension of health services and exploration of avenues for improvement in health services.

The recent developments in the field of Natural Language Processing (NLP) have garnered great interest from the healthcare research community. Some of the major breakthroughs include Long Short Term Memory (Hochreiter and Schmidhuber, 1997), and Gated Recurrent Units (Chung et al., 2014). However, the advent of transformers (Vaswani et al., 2017) led to a significant improvement in the performance.

The Social Media Mining for Health Applications (SMM4H) (Xu et al., 2024) unites researchers from across the globe with the objective of developing and sharing NLP methods for mining, representation, and analysis of health-related data. This year, SMM4H hosted seven shared tasks involving extraction, classification, and Large Language Model (LLM) identification. Our team participated in two of the classification tasks, namely, Task 3 and Task 5.

Task 3 is a multi-class classification task aimed to qualitatively evaluate the impact of outdoor spaces on Social Anxiety Disorder (SAD). Around one-third of the people with SAD report showing symptoms for ten years before seeking medical help. However, people do share their symptoms on platforms such as Reddit to discuss and share their symptoms and seek help to alleviate those symptoms. Task 3 focuses on classifying such posts on Reddit into one of the four categories, 'positive effect', 'neutral or no effect', 'negative effect', and 'unrelated'.

Task 5 aims at distinguishing tweets that mention a disorder such as delayed speech from the tweets whose users reported their pregnancy on Twitter and also reported having a child with attention-deficit/hyperactivity disorder (ADHD), autism spectrum disorders (ASD), delayed speech, or asthma. This task facilitates the use of Twitter as a medium not just for epidemiological studies but also to investigate the parents' experiences and directly target support interventions.

2 Methodology

Transformers have shown great potential in various NLP classification tasks (Khatri et al., 2022). For the purpose of performing classification in both the tasks, various transformers were tested. However, RoBERTa showed the best performance for both the multi-class and binary classification tasks. The RoBERTa transformer model was first introduced by Facebook in 2019 (Liu et al., 2019). RoBERTa has been trained on 160GB of uncompressed text leading to improved performance on classification tasks. In this work, we present our approach to fine-tuning RoBERTa for the SMM4H classification

Proceedings of The 9th Social Media Mining for Health Applications (#SMM4H) Workshop and Shared Tasks, pages 67–70
August 15, 2024 ©2024 Association for Computational Linguistics

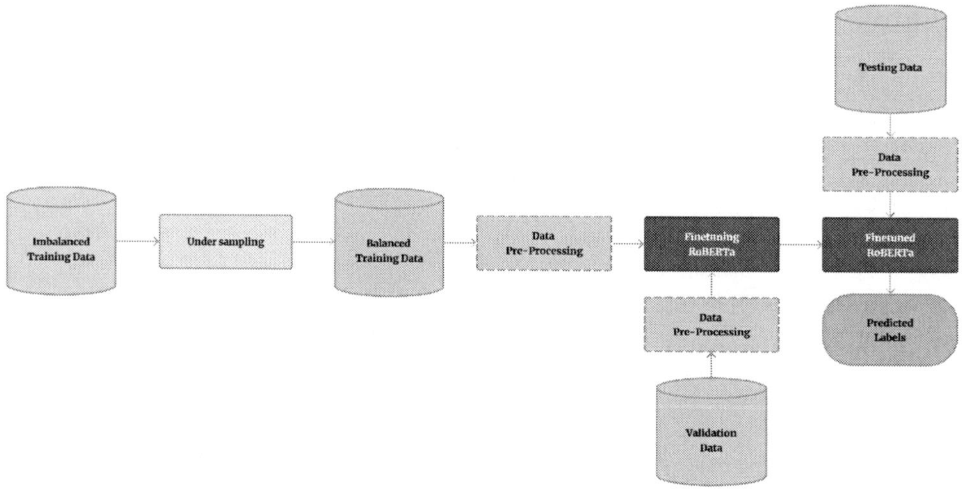

Figure 1: Proposed Methodology

Table 1: Data Distribution for Task 3

Dataset	Label		Total
	0	1	
Training	5118	2280	7398
Validation	254	135	389

Table 2: Data Distribution for Task 5

Dataset	Label				Total
	0	1	2	3	
Training	1131	395	160	114	1800
Validation	377	131	54	38	600

tasks. A pre-trained model, RoBERTa, was trained in a self-supervised manner, i.e., only raw texts were used to train it without the involvement of human intervention for labeling.

2.1 Data Pre-processing

For the tasks, the data was imbalanced in nature, as can be seen from the data distribution in Table 1 and Table 2 for Task 3 and Task 5, respectively. Due to this, there is a possibility that the results can be skewed with a preference towards the majority class. To address this issue, undersampling was performed. In this, random sampling was performed on all classes such that the number of instances for all the classes become equal to the number of instances in the minority class.

The data provided for both the tasks has been sourced from social media platforms. As a consequence of this, there was extensive use of various emojis, numbers and other special characters. In the pre-processing step, all the characters are first converted to lowercase. Then, emojis, numbers, and special characters are removed from the text.

2.2 Transformer Fine-tuning

The procedure adopted to fine-tune the model has been shown in the Figure 1. The RoBERTa transformer was fine-tuned in two different ways. The difference between the two approaches was that pre-processing was performed in one and it was not performed in the other. The results for both these approaches have been detailed in Table 3 and 4 for Task 3 and Task 5, respectively.

In Task 3, to train the transformer, the learning rate was set to 1e-6, and the weighted Adam optimizer was used. The Cross-Entropy loss function was utilized to penalize the mistakes made by the model during the training process. The model was fine-tuned till 45 epochs when data pre-processing was performed. And when data pre-processing was not performed the model was trained for 42 epochs.

In Task 5, to fine-tune the RoBERTa, the learning rate used was 1e-5. The cross-entropy function and weighted Adam optimizer were used as the loss function and optimizer, respectively. The model was fine-tuned for 10 epochs when pre-processing

Table 3: Model Performance for Task 3

Description	F1 Score	Precision	Recall	Accuracy
RoBERTa with Pre-Processing	0.383	0.413	0.482	0.378
RoBERTa without Pre-Processing	0.413	0.431	0.52	0.411
Mean	0.5186	0.5649	0.5379	0.5746
Median	0.5795	0.63	0.5885	0.627

Table 4: Model Performance for Task 5

Description	F1 Score	Precision	Recall
RoBERTa with Pre-Processing	0.900	0.854	0.950
RoBERTa without Pre-Processing	0.870	0.807	0.944
Mean	0.822	0.818	0.838
Median	0.901	0.885	0.917

was performed, and when no pre-processing was performed, the model was fine-tuned for 11 epochs.

To determine the number of iterations for which the model should be trained, the early stopping was used. If there was no significant improvement in the performance of the validation data for 5 consecutive epochs during training, then the process was not carried further and was halted at that point.

3 Results and Discussion

An in-depth analysis was performed on the performance of the large RoBERTa transformer model in the work. The performance was analyzed in different scenarios, both with and without pre-processing. The results obtained on the test data, along with the mean and median of the overall performance of all teams, have been summarised in Table 3 and Table 4 for Task 3 and Task 5, respectively.

To fine-tune the transformers, we first perform under-sampling on the data to avoid bias in the model. Then, we use two approaches for training using the balanced data. In the first approach, we use the text as is, without any pre-processing, and in the second approach, we perform pre-processing as described in Section 2.1. The early stopping approach was used to determine the number of epochs.

In Task 3, large RoBERTa without pre-processing performed better and achieved an F1-score of 0.413 than large RoBERTa with pre-processing, which achieved an F1-score of 0.383.

However, in Task 5, large RoBERTa with pre-processing achieved an F1-score of 0.900, whereas large RoBERTa without pre-processing achieved only 0.870.

4 Conclusion and Future Work

In this work, we present our adoption of the large RoBERTa transformer model to perform classification on social media data sourced from Reddit and Twitter. In Task 3, we use the model to perform multi-class classification on the effects of outdoor spaces on social anxiety using Reddit posts. In Task 5, we use the model to perform binary classification of English tweets to classify whether or not they report medical disorders in children. We also analyze the performance of the transformer, both with and without pre-processing.

In the future, an ensembling approach can be implemented. This approach can help amalgamate the results of different transformers, which may lead to improved results (Dima et al., 2020; Montañés-Salas et al., 2022; Lin et al., 2022).

References

Junyoung Chung, Caglar Gulcehre, KyungHyun Cho, and Yoshua Bengio. 2014. Empirical evaluation of gated recurrent neural networks on sequence modeling. *arXiv preprint arXiv:1412.3555.*

George-Andrei Dima, Andrei-Marius Avram, and Dumitru-Clementin Cercel. 2020. Approaching SMM4H 2020 with ensembles of BERT flavours. In *Proceedings of the Fifth Social Media Mining for Health Applications Workshop & Shared Task*, pages 153–157, Barcelona, Spain (Online). Association for Computational Linguistics.

Sepp Hochreiter and Jürgen Schmidhuber. 1997. Long short-term memory. *Neural computation*, 9(8):1735–1780.

Roshan Khatri, Sougata Saha, Souvik Das, and Rohini Srihari. 2022. UB health miners@SMM4H'22: Exploring pre-processing techniques to classify tweets using transformer based pipelines. In *Proceedings of The Seventh Workshop on Social Media Mining for Health Applications, Workshop & Shared Task*, pages 114–117, Gyeongju, Republic of Korea. Association for Computational Linguistics.

Tzu-Mi Lin, Chao-Yi Chen, Yu-Wen Tzeng, and Lung-Hao Lee. 2022. NCUEE-NLP@SMM4H'22: Classification of self-reported chronic stress on Twitter using ensemble pre-trained transformer models. In *Proceedings of The Seventh Workshop on Social Media Mining for Health Applications, Workshop & Shared Task*, pages 62–64, Gyeongju, Republic of Korea. Association for Computational Linguistics.

Yinhan Liu, Myle Ott, Naman Goyal, Jingfei Du, Mandar Joshi, Danqi Chen, Omer Levy, Mike Lewis, Luke Zettlemoyer, and Veselin Stoyanov. 2019. Roberta: A robustly optimized bert pretraining approach.

Rosa Montañés-Salas, Irene López-Bosque, Luis García-Garcés, and Rafael del Hoyo-Alonso. 2022. ITAINNOVA at SocialDisNER: A transformers cocktail for disease identification in social media in Spanish. In *Proceedings of The Seventh Workshop on Social Media Mining for Health Applications, Workshop & Shared Task*, pages 71–74, Gyeongju, Republic of Korea. Association for Computational Linguistics.

Ashish Vaswani, Noam Shazeer, Niki Parmar, Jakob Uszkoreit, Llion Jones, Aidan N Gomez, Łukasz Kaiser, and Illia Polosukhin. 2017. Attention is all you need. *Advances in neural information processing systems*, 30.

Dongfang Xu, Guillermo Lopez Garcia, Lisa Raithel, Rolland Roller, Philippe Thomas, Eiji Aramaki, Shuntaro Yada, Pierre Zweigenbaum, Karen O'Connor, Yao Ge, Sudeshna Das, Abeed Sarker, Ari Klein, Lucia Schmidt, Vishakha Sharma, Raul Rodriguez-Esteban, Juan Banda, Ivan Flores Amaro, Sai Tharuni Samineni, and Graciela Gonzalez-Hernandez. 2024. Overview of the 9th social media mining for health applications (#SMM4H) shared tasks at ACL 2024. In *Proceedings of the 9th Social Media Mining for Health Applications Workshop and Shared Task*, Bangkok, Thailand. Association for Computational Linguistics.

CHAAI@SMM4H'24: Enhancing Social Media Health Prediction Certainty by Integrating Large Language Models with Transformer Classifiers

Sedigh Khademi, Christopher Palmer, Muhammad Javed,
Jim Buttery, Gerardo Luis Dimaguila
Murdoch Children's Research Institute
{sedigh.khademi, chris.palmer, muhammad.javed,
jim.buttery, gerardoluis.dimaguil}@mcri.edu.au

Abstract

This paper presents our approach for SMM4H 2024 Task 5, focusing on identifying tweets where users discuss their child's health conditions of ADHD, ASD, delayed speech, or asthma. Our approach uses a pipeline that combines transformer-based classifiers and GPT-4 large language models (LLMs). We first address data imbalance in the training set using topic modelling and under-sampling. Next, we train RoBERTa-based classifiers on the adjusted data. Finally, GPT-4 refines the classifier's predictions for uncertain cases (confidence below 0.9). This strategy achieved significant improvement over the baseline RoBERTa models. Our work demonstrates the effectiveness of combining transformer classifiers and LLMs for extracting health insights from social media conversations.

1 Introduction

The SMM4H Shared Tasks focus on using machine learning and natural language processing to solve the challenges associated with extracting health insight from social media (Xu et al., 2024). This paper presents our work in SMM4H 2024 Task 5. The focus of task 5 is to identify tweets where users talk about their own child having attention-deficit/hyperactivity disorder (ADHD), autism spectrum disorders (ASD), delayed speech, or asthma, and differentiate them from tweets that simply discuss these conditions in general. For a description of this dataset, see (Klein et al., 2024).

One of the main challenges of processing social media text is dealing with the lay language that often lacks proper grammar and structure, containing misspellings, abbreviations, and unconventional phrasing (Gonzalez-Hernandez et al., 2017). Additionally, texts from advertising, news, social bots, and other non-personal posting modalities need to be excluded when identifying personal conversations (Javed et al., 2023). This requires robust algorithms capable of understanding colloquial language while maintaining accuracy in analysis and interpretation (Khademi Habibabadi et al., 2022). In this paper, we present a pipeline that combines transformer-based classifiers with the GPT-4 LLM to classify the tweets containing the health conditions of interest.

1.1 Task description

The dataset of Task 5 contained texts from Twitter. The training data consisted of an imbalanced set of 7,398 tweets (2,280 positive and 5,118 negative). A separate validation set of 389 tweets (135 positive and 254 negative) was also provided. An unlabelled test dataset of 10,000 records was supplied after the model building and validation stages of the competition, predictions over this test data were the competition submission.

To prepare the text data, we used the clean-text and html python libraries to fix Unicode errors and convert character entities, transliterate to ASCII, replace user mentions and URLs with generic placeholders, and remove line breaks. We observed that in the validation data emojis were replaced with double question marks, so we adopted the same strategy.

2 Method

Our approach consisted of three main steps. First, for data preparation we used a topic-modelling based under-sampling method to address data imbalance issues. Secondly, we trained a RoBERTa-based classifier on this adjusted training data. Finally, we employed GPT-based LLMs to refine the classifiers' predictions where the classifier's confidence in its predictions dropped below a threshold of 0.9.

2.1 Data preparation

To select a more balanced set our strategy was to retain all the positive labels and under-sample neg-

Proceedings of The 9th Social Media Mining for Health Applications (#SMM4H) Workshop and Shared Tasks, pages 71–73
August 15, 2024

ative labels. Using BERTopic, we performed topic modelling on all the data. We discovered one topic with a repetitive subject and content, for which we retained only a few records that represented the subject. For the other topics, we sampled 60% of the negative labels per topic, selecting them based on decreasing proximity from the topic's cluster centroid, as determined by text similarity. This method (Khademi et al., 2023) ensured a diverse selection of negative examples. We identified 10 records in the negative training data that appeared more likely positive, so these were relabeled. The dataset then consisted of 2,290 positive and 3,345 negative records. Subsequently, we divided this into training and validation datasets of 5,345 and 500 respectively, preserving the competition's supplied 389 records validation dataset as a hold-out test set.

2.2 Classifier training

We trained BERTweet-Large (Nguyen et al., 2020) and Twitter-RoBERTa (Loureiro et al., 2023) classifiers for 6 epochs with a batch size of 16 for training and validation. The AdamW optimizer was used; the evaluation strategy was "steps", the learning rate was 2e-5, weight decay was 0.01, with no warmup steps.

Checkpoints were saved every 10 steps and the best checkpoints were subsequently identified based on a balance of loss, ROC-AUC, and F1-score on validation data. Their F1 scores were evaluated against the hold-out test set. The top-performing BERTweet models achieved a rounded F1 score of 0.912, while the best Twitter-RoBERTa model reached 0.938.

2.3 LLM prompt engineering

We used two state-of-the-art language models: GPT-4-Turbo-2024-04-09 ("GPT4") with a 128K token context window and GPT-3.5-Turbo-16K-0613 ("GPT3") with a 16K token context window. To ensure consistent predictions, we set the model temperature to zero for all experiments. Default context length settings for each model were used. Our evaluation involved a comparison of zero-shot (where no training data is provided) and few-shot (with limited training data) prompting strategies combined with standard and chain of thoughts (CoT) prompting approaches. Based on its superior performance in our evaluations, GPT-4's zero-shot CoT prompting was chosen. Appendix A provides the prompt used for this task.

Model	Precision	Recall	F1
LLM+Clf	0.949	0.963	0.956
Clf	0.921	0.956	0.938
LLM	0.777	0.956	0.857

Table 1: Validation scores

Model	Precision	Recall	F1
LLM+Clf	0.907	0.949	0.927
Median	0.885	0.917	0.901
Mean	0.818	0.838	0.822

Table 2: Test scores

2.4 Predictions

On validation data, we compared the LLM's predictions against those of the Twitter-RoBERTa classifier's predictions, when the classifier's probabilities fell below 0.9, which we took as a threshold for a degree of certainty. The LLM's accuracy in correcting predictions in this cohort outweighed its errors, so by using its predictions for these records instead of the classifier's predictions, the F1 score increased by two percentage points. We employed this strategy for our entry with the competition's test data.

3 Results

The Twitter-RoBERTa classifier performed well independently, while the LLM showed comparatively lower performance when assessed against the validation dataset. However, by using the LLM for those texts where the classifier was less certain, we gained 2 percentage points over using the classifier only – this model is depicted as LLM+Clf in Table 1, with the other models below it in order of F1 score. Using the same strategy on the supplied test dataset gave us an F1 score of 0.927, which exceeded the median of all the competitors' results by almost 3 percentage points, as shown in Table 2.

4 Conclusions

While standard Transformer models excel at the classification task, using LLMs as arbiters on less interpretable texts can be beneficial. With good prompt design an LLM can be tuned to discern subtleties in texts that are difficult to teach a classifier without extensive training data. However, LLMs on their own may not be ideal for all classification tasks due to limitations in overall accuracy, efficiency, and resource requirements.

References

Graciela Gonzalez-Hernandez, Abeed Sarker, Karen O'Connor, and Guergana Savova. 2017. Capturing the patient's perspective: a review of advances in natural language processing of health-related text. *Yearbook of medical informatics*, 26(01):214–227.

Muhammad Javed, Gerardo Luis Dimaguila, Sedigh Khademi Habibabadi, Chris Palmer, and Jim Buttery. 2023. Learning from machines? social bots influence on covid-19 vaccination-related discussions: 2021 in review. In *Proceedings of the 2023 Australasian Computer Science Week*, pages 190–197.

Sedigh Khademi, Christopher Palmer, Muhammad Javed, Gerardo Luis Dimaguila, Jim P Buttery, and Jim Black. 2023. Detecting asthma presentations from emergency department notes: An active learning approach. In *Australasian Conference on Data Science and Machine Learning*, pages 284–298. Springer.

Sedigheh Khademi Habibabadi, Pari Delir Haghighi, Frada Burstein, and Jim Buttery. 2022. Vaccine adverse event mining of twitter conversations: 2-phase classification study. *JMIR Medical Informatics*, 10(6):e34305.

Ari Z Klein, José Agustín Gutiérrez Gómez, Lisa D Levine, and Graciela Gonzalez-Hernandez. 2024. Using longitudinal twitter data for digital epidemiology of childhood health outcomes: an annotated data set and deep neural network classifiers. *Journal of Medical Internet Research*, 26:e50652.

Daniel Loureiro, Kiamehr Rezaee, Talayeh Riahi, Francesco Barbieri, Leonardo Neves, Luis Espinosa Anke, and Jose Camacho-Collados. 2023. Tweet insights: a visualization platform to extract temporal insights from twitter. *arXiv preprint arXiv:2308.02142*.

Dat Quoc Nguyen, Thanh Vu, and Anh Tuan Nguyen. 2020. Bertweet: A pre-trained language model for english tweets. *arXiv preprint arXiv:2005.10200*.

Dongfang Xu, Guillermo Lopez Garcia, Lisa Raithel, Rolland Roller, Philippe Thomas, Eiji Aramaki, Shuntaro Yada, Pierre Zweigenbaum, Karen O'Connor, Yao Ge, Sudeshna Das, Abeed Sarker, Ari Klein, Lucia Schmidt, Vishakha Sharma, Raul Rodriguez-Esteban, Juan Banda, Ivan Flores Amaro, Sai Tharuni Samineni, and Graciela Gonzalez-Hernandez. 2024. Overview of the 9th Social Media Mining for Health Applications (#SMM4H) Shared Tasks at ACL 2024.

A Prompts

Here is the zero-shot chain of thought prompt used in this task.

You are a highly intelligent and accurate tweet classifier with reasoning capabilities.

You will receive a tweet enclosed within triple quotes. Set the final Answer to Yes, and answer the following questions with Yes or No: Questions:

Q1: Does the tweet report a child having ADHD, ASD, autism, speech delay, asthma, or being nonverbal, or mention medications used for these conditions? Descriptions of assessment or testing of the child for these health conditions do not count as definite evidence of having them. Enquiries about the possibility of the child having these health conditions also do not count. If the answer to Q1 is yes, describe the health condition of the child and go to Q2. Otherwise, set the final answer to No and go to Instruction1.

Q2: Does the tweet describe a child with one of the health conditions—ADHD, ASD, autism, delayed speech, being nonverbal, or asthma—as the child of the author of the tweet, irrespective of any other relationships discussed? State the relationship of the writer of the tweet with the child who has one of these health conditions. If the answer to Q2 is yes, go to Q3; otherwise, set the final answer to No and go to Instruction1.

Q3: Is the text containing the health conditions of ADHD, ASD, autism, delayed speech, being nonverbal, or asthma an original statement, not a quote from someone else? If the answer to Q3 is yes, go to Instruction1; otherwise, set the final answer to No and go to Instruction1.

Intruction1: The final answer should be Yes if Q1, Q2, and Q3 are Yes; otherwise, the final answer is No. Your output should include answers to Q1, Q2, and Q3, followed by the final answer and reasoning sentences that show the proof step by step within 30 words.

PolyuCBS at SMM4H 2024: LLM-based Medical Disorder and Adverse Drug Event Detection with Low-rank Adaptation

Yu Zhai[1], Xiaoyi Bao[1], Emmanuele Chersoni[1], Beatrice Portelli[2], Sophia Yat Mei Lee[1], Jinghang Gu[1] and Chu-Ren Huang[1]

[1]The Hong Kong Polytechnic University, Hong Kong, China

[2]University of Udine & University of Naples, Italy

{tonyayu.zhai,xiaoyi.bao}@connect.polyu.hk,
portelli.beatrice@spes.uniud.it
{emmanuele.chersoni,ym.lee,jinghang.gu,churen.huang}@polyu.edu.hk

Abstract

This is the demonstration of systems and results of our team's participation in the Social Medical Mining for Health (SMM4H) 2024 Shared Task. Our team participated in two tasks: Task 1 and Task 5. Task 5 requires the detection of tweet sentences that claim children's medical disorders from certain users. Task 1 needs teams to extract and normalize Adverse Drug Event terms in the tweet sentence. The team selected several Pre-trained Language Models and generative Large Language Models to meet the requirements. Strategies to improve the performance include cloze test, prompt engineering, Low Rank Adaptation etc. The test result of our system has an F1 score of 0.935, Precision of 0.954 and Recall of 0.917 in Task 5 and an overall F1 score of 0.08 in Task 1.

1 Introduction

The rise in people using social media for health information has resulted in a significant increase in health-related data, which allows researchers to harness the information, along with NLP and Machine Learning methods, to contribute to public health (Klein et al., 2024). The 9th Social Media Mining for Health Applications (SMM4H) Shared Tasks, aiming to advance methods that utilise social media data for health research, have a special focus on Large Language Models (LLMs) and Generalizability for Social Media NLP.

There are 7 tasks given in the 9th SMM4H workshop (Xu et al., 2024). Our team focused on Task 1 and Task 5. Task 1 has two sub-tasks, which are (1) detecting Adverse Drug Event terms in tweets and (2) normalizing these colloquial mentions to their standard concept to Preferred Term IDs according to MedDRA. Task 5 focuses on classifying tweets reporting children's medical disorders. The challenge of this task is differentiating between tweets from pregnancy users who declared that their child has specific conditions with attention-deficit/hyperactivity disorder (ADHD), autism spectrum disorders (ASD), delayed speech, or asthma, and tweets that just mention the disorder. The main challenges of both tasks are as follows: (1) medical terms conveyed in colloquial language, a common issue in social media data, which might misguide the trained model; (2) How to activate, transfer and utilize the knowledge learnt from the pre-training in Pre-trained Language Models(PLM) and Large Language Models (LLM). To address these issues, we use both PLMs and LLMs as basis and implement parameter-efficient tuning during training. To move a step further, we also conduct prompt engineering along with task decomposition to fully activate the knowledge of the LLM and ensure its understanding of the task, which could bridge the gap between the LLM and the specific downstream task.

2 Methodology

2.1 Task 1

For Task 1, BERT (Devlin et al., 2019) was utilized as the baseline model to conduct domain continual pre-training (Gu et al., 2021; Peng et al., 2021) and fine-tuning on the training dataset, during which BIO tags were utilized. In the extraction task, the models chosen are BERT and LLaMA-2(Touvron et al., 2023). The normalization task was regarded as a classification problem, the models chosen are SapBERT (Liu et al., 2021) and CODER (Yuan et al., 2022).

Fine-tune: For ADE extraction in Task 1, two fine-tuning modes are being experimented with: the first is by extracting the [CLS] in the last layer of the BERT model and utilizing the data tagged by BIO tag set to train the BERT classification model; the second is by treating the extraction task as a cloze test task. Given a label sets $Y = \{y_1, y_2, \dots y_k\}$, for each tweet sentence $\mathbf{S} = \{t_1, t_2, \dots t_n\}$ where t is the token and n

Proceedings of The 9th Social Media Mining for Health Applications (#SMM4H) Workshop and Shared Tasks, pages 74–78
August 15, 2024

is the number of tokens in S. For a token t_q, we create template $\mathbf{P} = \{p_1, p_2, \ldots p_m\}$ where $P = S + t_q + cloze$, eg. *A tweet sentence, a token in this sentence, this is __.* We mask p_i, the ith token in P, into [mask] and construct token-label converter v_y to ensure the label y_1 has a token set $C_{y_1} = \{c_1, c_2 ... c_n\}$ that mapping token c into label y_1. We extract the last layer's representation in the [mask] position to get the fixed-length labels. For a token t_n in a tweet, the possibility of its label equals to y_1 is as follows:

$$\Pr\left(y_1 \mid t_q\right) = \frac{\exp M\left(v_1 \,\middle|\, \frac{P}{\{p_i\}i}\right)}{\sum_{j=1}^{k} \exp\left(v_{y_j} \,\middle|\, \frac{P}{\{p_i\}i}\right)} \quad (1)$$

$\frac{P}{\{p_i\}i}$ means p_i in template P were replaced with [mask]. $M\left(v_1 \,\middle|\, \frac{P}{\{p_i\}i}\right)$ represents the probability that a masked token predicted by the model is mapped to the label y_1.

2.2 Task 5

In task 5, we first designed the prompt by several prompt engineering methods, then converted the original data into instructions by the designed template. These instructions were fed into LoRa adapters to finetune the LLaMA-2 model cost-effectively.

Low-Rank Adaptation (LoRA) (Hu et al., 2021) aims to improve the efficiency of fine-tuning large language models by training much smaller low-rank decomposition matrices of certain weights. Consider a weight matrix $\mathbf{W}_0 \in \mathbf{R}^{d \times k}$ from the pre-trained model. During training, the original weight matrix W_0 remains frozen and does not receive gradient updates. The trainable parameters are the matrices $\mathbf{B} \in \mathbf{R}^{d \times r}$ and $\mathbf{A} \in \mathbf{R}^{r \times k}$ ($r \ll \min(d, k)$), which represent the low-rank decomposition. The forward pass with LoRA is as follows:

$$W_0 x + \Delta W x = W_0 x + \mathbf{BA}x \quad (2)$$

Instruction Tuning. The importance of prompt templates has been demonstrated in various information extraction studies (Lu et al., 2021; Bao et al., 2022, 2023), particularly in the context of LLM. Task 5 presents a challenge due to its complex requirements. For example, Task 5 involves four types of disorders and requires that the report be from parents regarding their child. If we provide all the requirements in a single pass, LLMs may not

Prompt

```
Clarify the tweets whether meet the following two
requirements:

### The first requirement ###:
Mentioning the following four types of disorder :
[a] attention-deficit/hyperactivity disorder (ADHD),
[b] autism spectrum disorders (ASD),
[c] delayed speech,
[d] asthma,

### The second requirement ###:
If the tweet mentions one of the four types, the
patient with that disorder must be their child.

If the tweets do not meet the two requirements
above, the tweet is just merely mention another
disorder or the patient is not their child.

the tweets: <Input Text>
```

Target Sequence

```
Yes, it reports having a child with one of the
four types.

No, just merely mention a disorder or the patient
with one of the four types is not their child
```

Figure 1: Illustration of prompt and target sequence.

accurately capture the semantic information. Based on the analysis above, we propose a decomposed prompt approach. As illustrated in Figure 1, we break down the task into independent units representing each requirement and fine-tune the LLM to address them individually. The first requirement guides the LLM to focus on determining the presence of the four specific diseases, while the second requirement ensures that the patients reported are children and that the reports come from their parents.

As shown in the penultimate paragraph of the prompt in Figure 1, the LLM is asked to make the final decision about whether the tweet fulfills the two requirements at the same time and if not, a No answer should be given. The tag <Input Text> will be replaced by the specific tweets before fed into the model. When it comes to the target sequence, they are proposed under a similar motivation to the prompt, intending to have the LLM aware of the reason for giving the choice. If the tweet cannot fulfill the requirements, then the second line of the target sequence should be given; otherwise, the first one.

3 Experiment Results

3.1 Task 1

In Task 1, tweets in training data were treated as continual pre-training data for the masked language modelling process. For both extraction and normalization tasks with PLMs, we run 30 epochs with a

	Validation(F/P/R)		Test(F/P/R)	
Model	Extraction	Norm-single	Extraction	Norm
BERT-cloze+SapBERT	0.32/0.32/0.32	0.713/0.709/0.718	0.024/0.013/0.136	0.039/0.021/0.224
BERT+SapBERT	0.51/0.49/0.54	0.713/0.709/0.718	0.010/0.005/0.053	0.044/0.024/0.243
LLaMA-2+CODER	**0.659**/0.687/0.632	**0.741**/0.741/0.741	**0.112**/0.062/0.633	**0.080**/0.044/0.450

Table 1: Task 1 overall results on extraction and normalization of the validation and test sets

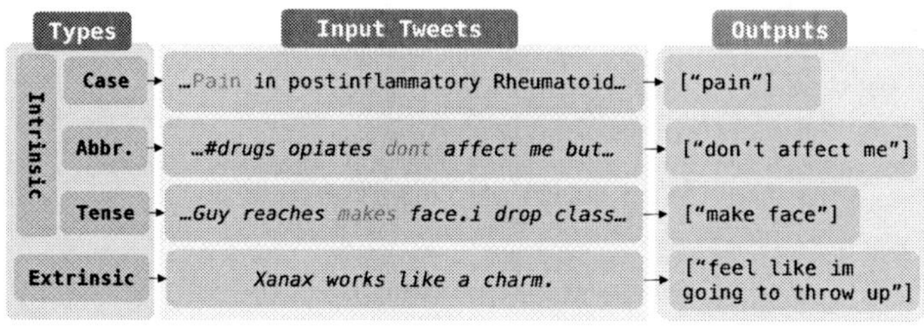

Figure 2: Hallucinations from LLaMA-2 in Task 1.

batch size of 4.

In Table 1, Norm task scores on the validation set were calculated based on the assumption that there is no error in the extraction task. For the extraction task, the basic BERT [1] classification method proposed in section 2.1 reached 0.51 in F score, while the cloze test method didn't get the best performance because of the weaker generation ability of encoder-only models. Therefore, LLaMA-2 was introduced.

On the validation set, the F1 score of the extraction saw an around 0.15 rise with LLaMA-2, but the test result still saw a drastic fall for both models (the Task1 test score with LLaMA-2 was uploaded during post-evaluation).

For LLaMA-2 predictions, except for around 1.4% results are empty, 25.9% of the answer in the test set suffers from hallucination problems. This section adopts hallucination categories of "Intrinsic Hallucination" and "Extrinsic Hallucination" (Zhou et al., 2021). Most hallucinations in Task 1 are intrinsic, taking up about 77.2% of the total hallucination errors.

In Figure 2, we demonstrated three common types of intrinsic hallucinations: Case, Abbreviation, and Tense. Some of the intrinsic hallucination happened due to the colloquial nature of the tweet sentence. There are 22.8% extrinsic hallucinations generated without clear grounding in the input. As shown in Figure 2, the model generated *"feel like im going to throw up"* which is irrelevant to the original input tweet. Hallucination issues might cause severe consequences when facing health-sensitive data. We used heuristic rules to correct part of the intrinsic hallucinations in this task and got a 0.015 improvement on F1. Another solution could be Retrieval Augmented Generation (Chen et al., 2024) which we will implement in future work.

3.2 Task 5

For our LLM, we employ LLaMA-2-7B[2] and LoRA fine-tune the adapter parameters. We tune the parameters of our models by grid searching on the validation dataset. We fine-tune the model with 20 epochs and save the model parameters for inference. The LoRA alpha is set to 128 and the LoRA rank is set to 64.

Model	Validation (F/P/R)	Test(F/P/R)
BERTweet	0.85/0.84/0.85	-
GPT-2	0.81/0.79/0.83	-
LLaMA-2	**0.932**/0.947/0.919	**0.935**/0.954/0.917
SMM4H Mean	-	0.822/0.818/0.838

Table 2: Task 5 results on validation and test sets

The model parameters are optimized by Adam (Kingma and Ba, 2015), and the learning rate of fine-tuning is 5e-5. The batch size is set to 4 with a cut-off length of 1024. The LoRA adapter would be merged with the original LLaMA-2-7B parameters and frozen during the inference process. During inference, we do the greedy search. Our experiments are carried out with an Nvidia RTX 4090 GPU. The model finally got an F1 score of 0.935 on the test set.

[1] Bert-base-uncased, `https://huggingface.co/google-bert/bert-base-uncased`

[2] LLaMA-2-7B-Chat, `https://huggingface.co/meta-LLaMA/LLaMA-2-7b-chat-hf`

4 Conclusion

In conclusion, this paper presents PLM and LLM-based methods for Task 1 and Task 5 in SMM4H 2024. Our team balanced efficiency and efficacy by employing strategies like cloze test shifting, instruction tuning, and low-rank adaptation. Through empirical evaluations, this paper proved that the LLM did surpass certain PLMs in the two tasks, though it suffered from the hallucination issue. More methods need to be explored to ensure solid and reliable results from LLM in tasks of the public health domain.

Acknowledgments

This research work is supported by a General Research Fund (GRF) project sponsored by the Research Grants Council, Hong Kong (Project No. 15611021).

References

Xiaoyi Bao, Zhongqing Wang, Xiaotong Jiang, Rong Xiao, and Shoushan Li. 2022. Aspect-based sentiment analysis with opinion tree generation. In *Proceedings of the Thirty-First International Joint Conference on Artificial Intelligence, IJCAI 2022, Vienna, Austria, 23-29 July 2022*, pages 4044–4050. ijcai.org.

Xiaoyi Bao, Zhongqing Wang, and Guodong Zhou. 2023. Exploring graph pre-training for aspect-based sentiment analysis. In *Findings of the Association for Computational Linguistics: EMNLP 2023*, pages 3623–3634, Singapore. Association for Computational Linguistics.

Jiawei Chen, Hongyu Lin, Xianpei Han, and Le Sun. 2024. Benchmarking large language models in retrieval-augmented generation. In *Proceedings of the AAAI Conference on Artificial Intelligence*, volume 38, pages 17754–17762.

Jacob Devlin, Ming-Wei Chang, Kenton Lee, and Kristina Toutanova. 2019. BERT: Pre-training of deep bidirectional transformers for language understanding. In *Proceedings of the 2019 Conference of the North American Chapter of the Association for Computational Linguistics: Human Language Technologies, Volume 1 (Long and Short Papers)*, pages 4171–4186, Minneapolis, Minnesota. Association for Computational Linguistics.

Yu Gu, Robert Tinn, Hao Cheng, Michael Lucas, Naoto Usuyama, Xiaodong Liu, Tristan Naumann, Jianfeng Gao, and Hoifung Poon. 2021. Domain-specific language model pretraining for biomedical natural language processing. *ACM Transactions on Computing for Healthcare (HEALTH)*, 3(1):1–23.

Edward J Hu, Yelong Shen, Phillip Wallis, Zeyuan Allen-Zhu, Yuanzhi Li, Shean Wang, Lu Wang, and Weizhu Chen. 2021. Lora: Low-rank adaptation of large language models. *arXiv preprint arXiv:2106.09685*.

Diederik P. Kingma and Jimmy Ba. 2015. Adam: A method for stochastic optimization. In *3rd International Conference on Learning Representations, ICLR 2015, San Diego, CA, USA, May 7-9, 2015, Conference Track Proceedings*.

Ari Z Klein, Juan M Banda, Yuting Guo, Ana Lucia Schmidt, Dongfang Xu, Ivan Flores Amaro, Raul Rodriguez-Esteban, Abeed Sarker, and Graciela Gonzalez-Hernandez. 2024. Overview of the 8th social media mining for health applications (smm4h) shared tasks at the amia 2023 annual symposium. *Journal of the American Medical Informatics Association*, 31(4):991–996.

Fangyu Liu, Ehsan Shareghi, Zaiqiao Meng, Marco Basaldella, and Nigel Collier. 2021. Self-alignment pretraining for biomedical entity representations. In *Proceedings of the 2021 Conference of the North American Chapter of the Association for Computational Linguistics: Human Language Technologies*, pages 4228–4238, Online. Association for Computational Linguistics.

Yaojie Lu, Hongyu Lin, Jin Xu, Xianpei Han, Jialong Tang, Annan Li, Le Sun, Meng Liao, and Shaoyi Chen. 2021. Text2Event: Controllable sequence-to-structure generation for end-to-end event extraction. In *Proceedings of the 59th Annual Meeting of the Association for Computational Linguistics and the 11th International Joint Conference on Natural Language Processing (Volume 1: Long Papers)*, pages 2795–2806, Online. Association for Computational Linguistics.

Bo Peng, Emmanuele Chersoni, Yu-Yin Hsu, and Chu-Ren Huang. 2021. Is domain adaptation worth your investment? comparing BERT and FinBERT on financial tasks. In *Proceedings of the Third Workshop on Economics and Natural Language Processing*, pages 37–44, Punta Cana, Dominican Republic. Association for Computational Linguistics.

Hugo Touvron, Louis Martin, Kevin Stone, Peter Albert, Amjad Almahairi, Yasmine Babaei, Nikolay Bashlykov, Soumya Batra, Prajjwal Bhargava, Shruti Bhosale, Dan Bikel, Lukas Blecher, Cristian Canton Ferrer, Moya Chen, Guillem Cucurull, David Esiobu, Jude Fernandes, Jeremy Fu, Wenyin Fu, Brian Fuller, Cynthia Gao, Vedanuj Goswami, Naman Goyal, Anthony Hartshorn, Saghar Hosseini, Rui Hou, Hakan Inan, Marcin Kardas, Viktor Kerkez, Madian Khabsa, Isabel Kloumann, Artem Korenev, Punit Singh Koura, Marie-Anne Lachaux, Thibaut Lavril, Jenya Lee, Diana Liskovich, Yinghai Lu, Yuning Mao, Xavier Martinet, Todor Mihaylov, Pushkar Mishra, Igor Molybog, Yixin Nie, Andrew Poulton, Jeremy Reizenstein, Rashi Rungta, Kalyan Saladi, Alan Schelten,

Ruan Silva, Eric Michael Smith, Ranjan Subramanian, Xiaoqing Ellen Tan, Binh Tang, Ross Taylor, Adina Williams, Jian Xiang Kuan, Puxin Xu, Zheng Yan, Iliyan Zarov, Yuchen Zhang, Angela Fan, Melanie Kambadur, Sharan Narang, Aurelien Rodriguez, Robert Stojnic, Sergey Edunov, and Thomas Scialom. 2023. Llama 2: Open foundation and fine-tuned chat models.

Dongfang Xu, Guillermo Lopez Garcia, Lisa Raithel, Rolland Roller, Philippe Thomas, Eiji Aramaki, Shuntaro Yada, Pierre Zweigenbaum, Karen O'Connor, Yao Ge, Sudeshna Das, Abeed Sarker, Ari Klein, Lucia Schmidt, Vishakha Sharma, Raul Rodriguez-Esteban, Juan Banda, Ivan Flores Amaro, Sai Tharuni Samineni, and Graciela Gonzalez-Hernandez. 2024. Overview of the 9th social media mining for health applications (#SMM4H) shared tasks at ACL 2024. In *Proceedings of the 9th Social Media Mining for Health Applications Workshop and Shared Task*, Bangkok, Thailand. Association for Computational Linguistics.

Zheng Yuan, Zhengyun Zhao, Haixia Sun, Jiao Li, Fei Wang, and Sheng Yu. 2022. Coder: Knowledge-infused cross-lingual medical term embedding for term normalization. *Journal of Biomedical Informatics*, 126:103983.

Chunting Zhou, Graham Neubig, Jiatao Gu, Mona Diab, Francisco Guzmán, Luke Zettlemoyer, and Marjan Ghazvininejad. 2021. Detecting hallucinated content in conditional neural sequence generation. In *Findings of the Association for Computational Linguistics: ACL-IJCNLP 2021*, pages 1393–1404, Online. Association for Computational Linguistics.

Deloitte at #*SMM4H* 2024: Can GPT-4 Detect COVID-19 Tweets Annotated by Itself?

Harika Abburi[1], Nirmala Pudota[1], Balaji Veeramani[2], Edward Bowen[2], Sanmitra Bhattacharya[2]

[1]Deloitte & Touche Assurance & Enterprise Risk Services India Private Limited India
[2]Deloitte & Touche LLP, USA
{abharika, npudota, bveeramani, edbowen, sanmbhattacharya}@deloitte.com

Abstract

The advent of Large Language Models (LLMs) such as Generative Pre-trained Transformers (GPT-4) mark a transformative era in Natural Language Generation (NLG). These models demonstrate the ability to generate coherent text that closely resembles human-authored content. They are easily accessible and have become invaluable tools in handling various text-based tasks, such as data annotation, report generation, and question answering. In this paper, we investigate GPT-4's ability to discern between data it has annotated and data annotated by humans, specifically within the context of tweets in the medical domain. Through experimental analysis, we observe GPT-4 outperform other state-of-the-art models. The dataset used in this study was provided by the SMM4H (Social Media Mining for Health Research and Applications) shared task. Our model achieved an accuracy of 0.51, securing a second rank in the shared task.

1 Introduction

The field of Natural Language Generation (NLG) is undergoing a significant transformation driven by the emergence of LLMs like GPT-4 (OpenAI, 2023), and many other Large Language Models. These models are capable of generating text of human-level quality for a wide range of applications, such as data annotation (Tan et al., 2024), medical question answering (Kung et al., 2023), conversation response generation (Mousavi et al., 2023), and code auto-completion (Tang et al., 2023). Notably, the ability of these models to learn without extensive training data (zero-shot learning) or with just a few examples (few-shot learning), simplifies their integration into various language generation applications.

While the LLMs demonstrate the ability to understand the context and generate coherent human-like responses, they do not have a true understanding of what they are producing (Li et al., 2023; Turpin et al., 2023). This could potentially lead to adverse consequences when used in downstream applications. For example, consider an application of a LLM tasked with summarizing a medicinal drug datasheet inadvertently produces wrong dosage information. This generation of plausible but false content (referred as *hallucination* (Bang et al., 2023; Ji et al., 2023)), can unintentionally spread misinformation, false narratives, fake news, and spam. Similarly, the use of LLMs in data annotation has sparked a debate within the research community due to potential issues like inherent biases and hallucinations associated with these models (Yao et al., 2024; Bogdanov et al., 2024). Motivated by these challenges, the automatic detection of AI-generated outputs has emerged as an active area of research.

The detection of data annotations generated by LLMs closely resembles the process of identifying AI-generated text, which aims to distinguish between human-authored and machine-generated content (Abburi et al., 2023b,a). In this paper, we explore methodologies utilized for AI-generated text detection, with a particular emphasis on zero-shot detection techniques, and examine the synergies between these two areas. The AI-generated text detection methods predominantly involve the analysis of outputs from LLMs utilizing features such as entropy, log-probability scores, and perplexity (Wu et al., 2023; Yang et al., 2023a; Bao et al., 2023; Hans et al., 2024) to distinguish between human-written and machine-generated content. Building upon this foundation, DetectGPT (Mitchell et al., 2023) introduced the concept of analyzing negative log probability curvature, identifying a distinct pattern in AI-generated text. Subsequent advancements, such as DNA-GPT (Yang et al., 2023b), improved performance by analyzing the divergence of n-grams between the original text and LLM-prompted versions.

While zero-shot detection methods are effective,

Proceedings of The 9th Social Media Mining for Health Applications (#SMM4H) Workshop and Shared Tasks, pages 79–82
August 15, 2024 ©2024 Association for Computational Linguistics

their success often rely on direct access to the internal mechanisms of the specific LLM that generated the text. However, the underlying architecture and weights of many LLMs, including OpenAI's GPT-4, are not publicly available. As a result, these techniques often depend on a substitute LLM, presumed to have mechanisms similar to the proprietary model. The reliance on a proxy LLM may limit the robustness and generalizability of zero-shot detection in various scenarios.

2 GPT-4 as AI-annotated detector

While GPT-4 has exhibited proficiency in various Natural Language Processing (NLP) tasks, its potential for distinguishing between human-annotated and AI-annotated data remains largely untapped. In this study, we explore the effectiveness of GPT-4 in distinguishing between tweets it annotated and those annotated by humans in the medical domain. Our experimental analysis indicates that GPT-4 outperform other state-of-the-art models in detecting AI-annotated data.

2.1 Dataset

We use a dataset provided by the SMM4H shared task. The dataset consists of both human annotated (human) and GPT-4 annotated (generated) tweets detailing COVID-19 symptoms written in Latin American Spanish. In total 3,682 tweets were available for training and 2,110 tweets were available for testing. More details about the dataset can be found in the SMM4H overview paper (Xu et al., 2024).

2.2 Choice of prompt and experimental settings

We experimented with various prompts to identify the most effective one for label prediction. Our findings indicated that complex prompts often caused GPT-4 to generate incorrect labels. Therefore, we chose a simpler prompt, which resulted in better performance. The chosen prompt is as follows:

"Imagine you're a data annotation expert. You're presented with a tweet containing a description of potential COVID-19 symptoms. This tweet has already been annotated as containing COVID-19 symptoms, but you don't know if that is annotated by a human expert or AI model.
Your task is to analyze the tweet to determine whether the initial annotation of 'contains COVID-19 symptoms' was made by a human expert or by an AI model. Can you determine the source of the label (human or generated)? Answer in a single word, predicted label should be one of 'human' or 'generated'.
text : {text}
prediction:{}"

Evaluation sets	Number of samples annotated by human	Number of samples annotated by AI
Set1	368	322
Set2	350	340
Set3	357	333

Table 1: Statistics of evaluation sets

Models	Set 1		Set 2		Set 3	
	Acc	F_{mac}	Acc	F_{mac}	Acc	F_{mac}
(Hans et al., 2024)	**0.47**	0.36	0.49	0.38	0.48	0.37
(Bao et al., 2023)	**0.47**	0.33	0.49	0.33	0.48	0.33
Our approach	0.46	**0.40**	**0.49**	**0.42**	**0.49**	**0.41**

Table 2: Performance comparison of zero-shot detection models on 3 evaluation sets. Acc: Accuracy, F_{mac}: F1-macro

Each tweet is passed to the prompt as *{text}*. Since this task involves classifying whether the given text is annotated by human or AI in a zero-shot setting, we do not fine-tune the model. Instead, we directly prompt GPT-4 to classify the test samples annotated by 'human' or 'generated'. To limit randomness in the model's output, which is crucial for classification accuracy, we set the *temperature* parameter to 0.

3 Results

To assess the performance of our approach in the zero-shot setting, we curated three distinct evaluation sets – Set 1, Set 2, and Set 3 as shown in Table 1. Each set consisted of 690 samples, randomly selected from the training data, to avoid overlap.

We conducted a comparative evaluation of our GPT-4-based approach against multiple zero-shot detection models. The two most effective baselines methods we identified were: 1) Binoculars (Hans et al., 2024), which compares the perplexity and cross-perplexity of two closely related language models to identify AI-generated text; and 2) Fast-DetectGPT (Bao et al., 2023), which utilizes conditional probability curvature to efficiently detect AI content, particularly from models like GPT.

Table 2 provides a comparative performance analysis of the baselines and our approach across the three distinct evaluation sets. Despite our GPT-4 prompting-based approach consistently achieving the highest Acc and F_{mac} across each of the sets, the top scores did not exceed 0.5. This highlights the significant challenge posed by AI annotation detection, especially for tweets in the medical domain. Nevertheless, we used GPT-4 prompting to generate the predictions on the test set (predictions are uploaded to codalab), resulting in accuracy 0.51, which placed our team in second place in the shared task.

4 Conclusion

In this paper, we investigated the potential of GPT-4 to detect tweets annotated by itself in zero-shot setting. Our experiments highlighted the complexity of this task, particularly for short texts in the medical domain, achieving an accuracy of 0.51. For future work, we aim to improve the model's performance by incorporating the reasoning behind its predictions. By using this reasoning as additional input, we aim to enhance the model's ability to differentiate between data annotations predicted by LLM and those annotated by human experts.

References

Harika Abburi, Kalyani Roy, Michael Suesserman, Nirmala Pudota, Balaji Veeramani, Edward Bowen, and Sanmitra Bhattacharya. 2023a. A simple yet efficient ensemble approach for AI-generated text detection. In *Proceedings of the Third Workshop on Natural Language Generation, Evaluation, and Metrics (GEM)*, pages 413–421, Singapore. Association for Computational Linguistics.

Harika Abburi, Michael Suesserman, Nirmala Pudota, Balaji Veeramani, Edward Bowen, and Sanmitra Bhattacharya. 2023b. Generative ai text classification using ensemble llm approaches. In *IberLEF@ SEPLN*.

Yejin Bang, Samuel Cahyawijaya, Nayeon Lee, Wenliang Dai, Dan Su, Bryan Wilie, Holy Lovenia, Ziwei Ji, Tiezheng Yu, Willy Chung, et al. 2023. A multitask, multilingual, multimodal evaluation of chatgpt on reasoning, hallucination, and interactivity. *arXiv preprint arXiv:2302.04023*.

Guangsheng Bao, Yanbin Zhao, Zhiyang Teng, Linyi Yang, and Yue Zhang. 2023. Fast-detectgpt: Efficient zero-shot detection of machine-generated text via conditional probability curvature. *arXiv preprint arXiv:2310.05130*.

Sergei Bogdanov, Alexandre Constantin, Timothée Bernard, Benoit Crabbé, and Etienne Bernard. 2024. Nuner: Entity recognition encoder pre-training via llm-annotated data. *arXiv preprint arXiv:2402.15343*.

Abhimanyu Hans, Avi Schwarzschild, Valeriia Cherepanova, Hamid Kazemi, Aniruddha Saha, Micah Goldblum, Jonas Geiping, and Tom Goldstein. 2024. Spotting llms with binoculars: Zero-shot detection of machine-generated text. *arXiv preprint arXiv:2401.12070*.

Ziwei Ji, Nayeon Lee, Rita Frieske, Tiezheng Yu, Dan Su, Yan Xu, Etsuko Ishii, Ye Jin Bang, Andrea Madotto, and Pascale Fung. 2023. Survey of hallucination in natural language generation. *ACM Computing Surveys*, 55(12):1–38.

Tiffany H Kung, Morgan Cheatham, Arielle Medenilla, Czarina Sillos, Lorie De Leon, Camille Elepaño, Maria Madriaga, Rimel Aggabao, Giezel Diaz-Candido, James Maningo, et al. 2023. Performance of chatgpt on usmle: Potential for ai-assisted medical education using large language models. *PLoS digital health*, 2(2):e0000198.

Hanzhou Li, John T Moon, Saptarshi Purkayastha, Leo Anthony Celi, Hari Trivedi, and Judy W Gichoya. 2023. Ethics of large language models in medicine and medical research. *The Lancet Digital Health*.

Eric Mitchell, Yoonho Lee, Alexander Khazatsky, Christopher D Manning, and Chelsea Finn. 2023. Detectgpt: Zero-shot machine-generated text detection using probability curvature. *arXiv preprint arXiv:2301.11305*.

Seyed Mahed Mousavi, Simone Caldarella, and Giuseppe Riccardi. 2023. Response generation in longitudinal dialogues: Which knowledge representation helps?

OpenAI. 2023. Gpt-4 technical report.

Zhen Tan, Alimohammad Beigi, Song Wang, Ruocheng Guo, Amrita Bhattacharjee, Bohan Jiang, Mansooreh Karami, Jundong Li, Lu Cheng, and Huan Liu. 2024. Large language models for data annotation: A survey.

Ze Tang, Jidong Ge, Shangqing Liu, Tingwei Zhu, Tongtong Xu, Liguo Huang, and Bin Luo. 2023. Domain adaptive code completion via language models and decoupled domain databases. *arXiv preprint arXiv:2308.09313*.

Miles Turpin, Julian Michael, Ethan Perez, and Samuel R Bowman. 2023. Language models don't always say what they think: Unfaithful explanations in chain-of-thought prompting. *arXiv preprint arXiv:2305.04388*.

Kangxi Wu, Liang Pang, Huawei Shen, Xueqi Cheng, and Tat-Seng Chua. 2023. Llmdet: A large language models detection tool.

Dongfang Xu, Guillermo Lopez Garcia, Lisa Raithel, Rolland Roller, Philippe Thomas, Eiji Aramaki, Shuntaro Yada, Pierre Zweigenbaum, Sai Tharuni Samineni, Karen O'Connor, Yao Ge, Sudeshna Das, Abeed Sarker, Ari Klein, Lucia Schmidt, Vishakha Sharma, Raul Rodriguez-Esteban, Juan Banda, Ivan Flores Amaro, Davy Weissenbacher, and Graciela Gonzalez-Hernandez. 2024. Overview of the 9th social media mining for health (#SMM4H) research and applications workshop and shared tasks at ACL 2024. In *Proceedings of The 9th Social Media Mining for Health Research and Applications Workshop and Shared Tasks*.

Xianjun Yang, Wei Cheng, Linda Petzold, William Yang Wang, and Haifeng Chen. 2023a. Dna-gpt: Divergent n-gram analysis for training-free detection of gpt-generated text.

Xianjun Yang, Wei Cheng, Linda Petzold, William Yang Wang, and Haifeng Chen. 2023b. Dna-gpt: Divergent n-gram analysis for training-free detection of gpt-generated text. *arXiv preprint arXiv:2305.17359*.

Yuxuan Yao, Sichun Luo, Haohan Zhao, Guanzhi Deng, and Linqi Song. 2024. Can llm substitute human labeling? a case study of fine-grained chinese address entity recognition dataset for uav delivery. In *Companion Proceedings of the ACM on Web Conference 2024*, pages 1099–1102.

A List of prompts explored

1. *"Imagine you're a data annotation expert specializing in identifying the origin of annotations on social media posts. You are presented with a tweet that has been annotated as "contains COVID-19 symptoms. Your task is to analyze the tweet and determine whether this annotation was made by a human expert or generated by an AI model.*
   Please follow these steps to make your determination:
   1.Content Analysis:
   Examine the language and structure of the tweet. Consider the complexity, coherence, and nuance in describing COVID-19 symptoms.
   Evaluate the specificity and accuracy of the symptoms mentioned. Human experts tend to provide precise and medically accurate descriptions, whereas AI models might be more general or formulaic.
   2.Annotation Style:
   Assess the style and quality of the annotation itself. Human annotations often reflect domain expertise and may include subtle contextual understanding.
   Look for patterns typical of AI-generated annotations, such as repetitive phrasing, lack of deep contextual insight, or overly broad categories.
   3.Consistency and Commonality:
   Compare the tweet with common patterns and characteristics known from human annotations versus AI-generated annotations. Humans may show more variability and adaptability in their descriptions.
   Based on your analysis, provide your prediction on whether the annotation was made by a human expert or generated by an AI model. Answer in a single word, predicted label should be one of 'human' or 'generated'.
   text : {text}
   prediction:{}"

2. *"Imagine you're a data annotation expert specializing in identifying the origin of annotations on social media posts. You are presented with a tweet that has been annotated as 'contains COVID-19 symptoms.' Your task is to analyze the tweet and determine whether this annotation was made by a human expert or generated by an AI model. Answer in a single word, predicted label should be one of 'human' or 'generated'.*
   text : {text}
   prediction:{}"

3. "You are a GPT4 data annotation expert. Given a text, predict who annotated the text: human or AI. Predicted label should be one of 'human' or 'generated'.
   text : {text}
   prediction:{}"

IMS_medicALY at #SMM4H 2024: Detecting Impacts of Outdoor Spaces on Social Anxiety with Data Augmented Ensembling

Amelie Wührl[1,2,✐], **Lynn Greschner**[2,✐],
Yarik Menchaca Resendiz[1,2,✐] and **Roman Klinger**[2]
[1]University of Stuttgart, Germany, [2]University of Bamberg, Germany
firstname.lastname@ims.uni-stuttgart.de, firstname.lastname@uni-bamberg.de

Abstract

Many individuals affected by Social Anxiety Disorder turn to social media platforms to share their experiences and seek advice. This includes discussing the potential benefits of engaging with outdoor environments. As part of #SMM4H 2024, Shared Task 3 focuses on classifying the effects of outdoor spaces on social anxiety symptoms in Reddit posts. In our contribution to the task, we explore the effectiveness of domain-specific models (trained on social media data – SocBERT) against general domain models (trained on diverse datasets – BERT, RoBERTa, GPT-3.5) in predicting the sentiment related to outdoor spaces. Further, we assess the benefits of augmenting sparse human-labeled data with synthetic training instances and evaluate the complementary strengths of domain-specific and general classifiers using an ensemble model. Our results show that (1) fine-tuning small, domain-specific models generally outperforms large general language models in most cases. Only one large language model (GPT-4) exhibits performance comparable to the fine-tuned models (52% F_1). Further, we find that (2) synthetic data does improve the performance of fine-tuned models in some cases, and (3) models do not appear to complement each other in our ensemble setup.

1 Introduction

Social Anxiety Disorder is a medical condition that can significantly impact an individual's life (Vilaplana-Pérez et al., 2021). Social media platforms have emerged as spaces where affected individuals can communicate their experiences and seek support. These platforms are rich with biomedical information, providing an opportunity for medical practitioners to gain novel insights about medical conditions. However, this data is highly diverse and annotation is expensive, especially for the medical domain. Access to a broad variety of classification and generative models has the potential to close this gap as it (1) allows to explore the capability of domain-specific and general models to solve these types of tasks, and (2) opens the possibility to generate synthetic training instances to complement sparse, human-labeled data. As part of #SMM4H 2024, Shared Task 3 focuses on classifying the effects of outdoor spaces on social anxiety symptoms in Reddit posts. Given the post, the goal is to predict the user's sentiment towards the effect of outdoor space in a multi-class classification setup with the target labels POSITIVE, NEGATIVE, NEUTRAL and UNRELATED.

With our contribution, we investigate three research questions (RQs):

RQ1 Given the sparsity of the data, do fine-tuned, domain-specific models outperform general models?

RQ2 Does incorporating synthetic data complement human-labeled data and enhance model robustness?

RQ3 Is Reddit's text diversity better captured by a set of different models in an ensemble setup?

2 System Description

We hypothesize that the diversity of texts shared on Reddit benefits from aggregating multiple approaches. Therefore, we design an ensemble model that takes as input the predictions of individual models varying in architecture and training procedure. The individual models are:

Fine-tuned language models. We fine-tune `BERT` (Devlin et al., 2019), `RoBERTa` (Liu et al., 2019), `DistilBERT` (Sanh et al., 2019) and `SocBERT` (Guo and Sarker, 2023) on the training split of the task data to obtain customized models. We set truncation and padding to True, batch size to 4, and train

✐The first three authors contributed equally.

Proceedings of The 9th Social Media Mining for Health Applications (#SMM4H) Workshop and Shared Tasks, pages 83–87
August 15, 2024

for 3 epochs, with a learning rate of $5 \cdot 10^{-5}$ using the AdamW optimizer.

Few-shot prompting. To explore the capacity of general large language models (LLMs), we prompt `Mistral-7B-v0.1` (Jiang et al., 2023), `Llama-2-7b-chat-hf` (Touvron et al., 2023), `GPT-3.5` (OpenAI, 2022) and `GPT-4` (OpenAI, 2023) to generate labels. We provide them with task instructions in a one-shot setup, where the example instance is randomly chosen from the training data. Table 3 shows the prompt templates.

Ensemble. Our neural `Ensemble` consists of an input layer, a one dimensional convolutional layer and max-pooling layer, a dense layer with 128 neurons, and a classification layer (Dense) with 4 neurons. The model inputs include 8 predictions and probabilities from fine-tuned models (4 models trained with and 4 trained without synthetic data) and 4 predictions from LLMs, 12 in total. We train the ensemble using 400 instances from the validation data over 20 epochs and evaluate it using the remaining 200 instances from the same subset.

Synthetic data augmentation. We further hypothesize that additional training data for the minority classes POSITIVE, NEGATIVE, and NEUTRAL benefits model performance in the fine-tuning setup. To investigate this, we generate synthetic instances using `Mistral-7B` and `GPT-3.5`. We generate as many instances as needed to match the size of the largest class (UNRELATED, 1,131). Table 4 shows prompt templates and examples.

3 Results

Table 1 shows the performance ofe all classifiers on the shared task's validation set. Table 2 reports the results of three of our systems on the test set.

RQ1: How do domain-specific models compare to general models within this task? Overall, the models show a mixed performance. `GPT-4` achieves the best results (.52 F_1). The general-domain models are slightly more robust than `SocBERT` which is specialized for the social media domain (Δ .07 F_1). Compared to prompting, fine-tuning leads to more consistent performances across models.

RQ2: How do synthetic training instances affect classification performance? Table 1 reports the results for models fine-tuned with (+) and without additional synthetic training instances. We observe that for two out of four models (`DistilBERT`, `RoBERTa`), fine-tuning on the additional synthetic data leads to a more robust performance, compared to only training on gold data. This indicates that to a certain degree, the synthetic data is complementary to the human annotations.

	Model	P	R	F_1
Gold	BERT	0.58	0.47	**0.50**
	DistilBERT	0.70	0.37	0.39
	RoBERTa	0.58	0.43	0.43
	SocBERT	0.54	0.43	0.45
Gold+Syn	BERT+	0.58	0.47	**0.50**
	DistilBERT+	0.47	0.45	0.45
	RoBERTa+	0.47	0.48	0.47
	SocBERT+	0.54	0.43	0.45
Prompt	GPT-3.5	0.32	0.46	0.28
	GPT-4	0.56	0.55	**0.52**
	Llama-2	0.19	0.30	0.15
	Mistral	0.28	0.27	0.11
	Ensemble	0.56	0.49	0.51

Table 1: Macro F_1 of individual classifiers on the validation set. + indicates that the models are fine-tuned with additional synthetic data. We evaluate the ensemble on 200 unseen instances from the validation set.

Model	Acc	P	R	F_1
Ensemble	0.48	0.35	0.40	0.34
GPT-4	0.56	0.60	0.52	0.50
SocBert	0.62	0.63	0.53	0.56

Table 2: Performance of three classifiers – Domain-specific (SocBert), General-Domain (GPT-4), and the Ensemble model – on the Task 3 test set of #SMM4H 2024.

RQ3: Do domain-specific & general models complement each other? Table 2 reports the performance of our models on the test set. We submit the predictions from the best domain-specific (`SocBERT`), general-domain (`GPT-4`), and the `Ensemble` model. The best model is `SocBERT` (.56 F_1), followed by `GPT-4` (.50 F_1). The predictions from the `Ensemble` obtain an F_1-score of .34, indicating that (1) models do not complement each other or (2) the ensemble might benefit from additional features that go beyond prediction probabilities. The result may also be attributed to the limited amount of data available for training the `Ensemble`.

4 Conclusion

We present our contribution to the #SMM4H 2024 Shared Task 3 which focuses on classifying the effects of outdoor spaces on social anxiety symptoms in Reddit posts. We find that fine-tuning models overall show a more robust performance compared to LLM prompting. Synthetic data may increase the robustness of the models. We can not show a superior performance of an ensemble. This leads to important future work: For the prompting approaches, we have to evaluate the impact of the prompt design for the task. For fine-tuned models, a thorough analysis of the synthetic data is key to gauging the impact of generated instances in more detail. For all models, an in-depth error analysis is crucial to understand model capabilities and the impact the individual predictions may have on the `Ensemble`. Further, testing alternative ensemble designs (e.g., Gradient-boosted Decision Trees) is key to understanding the interaction between the probability-based, class predictions we obtain from the fine-tuned models, and the class-only predictions from LLMs.

Acknowledgements

Yarik Menchaca Resendiz is funded by a CONACYT scholarship (2020-000009-01EXTF-00195). Lynn Greschner is funded by the EMCONA project (DFG, project number KL 2869/5-1). Amelie Wührl is funded by the FIBISS project (DFG, KL 2869/12-1) as well as the CEAT project (DFG, KL 2869/1-2.). We thank the reviewers for their valuable feedback.

References

Jacob Devlin, Ming-Wei Chang, Kenton Lee, and Kristina Toutanova. 2019. BERT: Pre-training of deep bidirectional transformers for language understanding. In *Proceedings of the 2019 Conference of the North American Chapter of the Association for Computational Linguistics: Human Language Technologies, Volume 1 (Long and Short Papers)*, pages 4171–4186, Minneapolis, Minnesota. Association for Computational Linguistics.

Yuting Guo and Abeed Sarker. 2023. SocBERT: A pretrained model for social media text. In *Proceedings of the Fourth Workshop on Insights from Negative Results in NLP*, pages 45–52, Dubrovnik, Croatia. Association for Computational Linguistics.

Albert Q. Jiang, Alexandre Sablayrolles, Arthur Mensch, Chris Bamford, Devendra Singh Chaplot, Diego de las Casas, Florian Bressand, Gianna Lengyel, Guillaume Lample, Lucile Saulnier, Lélio Renard Lavaud, Marie-Anne Lachaux, Pierre Stock, Teven Le Scao, Thibaut Lavril, Thomas Wang, Timothée Lacroix, and William El Sayed. 2023. Mistral 7b. *Preprint*, arXiv:2310.06825.

Yinhan Liu, Myle Ott, Naman Goyal, Jingfei Du, Mandar Joshi, Danqi Chen, Omer Levy, Mike Lewis, Luke Zettlemoyer, and Veselin Stoyanov. 2019. Roberta: A robustly optimized BERT pretraining approach. *CoRR*, abs/1907.11692.

OpenAI. 2022. Gpt-3.5 turbo. https://www.openai.com. Accessed: [18.04.2024].

OpenAI. 2023. Gpt-4 turbo. https://www.openai.com. Accessed: [18.04.2024].

Victor Sanh, Lysandre Debut, Julien Chaumond, and Thomas Wolf. 2019. Distilbert, a distilled version of bert: smaller, faster, cheaper and lighter. *ArXiv*, abs/1910.01108.

Hugo Touvron, Louis Martin, Kevin Stone, Peter Albert, Amjad Almahairi, Yasmine Babaei, Nikolay Bashlykov, Soumya Batra, Prajjwal Bhargava, Shruti Bhosale, Dan Bikel, Lukas Blecher, Cristian Canton Ferrer, Moya Chen, Guillem Cucurull, David Esiobu, Jude Fernandes, Jeremy Fu, Wenyin Fu, Brian Fuller, Cynthia Gao, Vedanuj Goswami, Naman Goyal, Anthony Hartshorn, Saghar Hosseini, Rui Hou, Hakan Inan, Marcin Kardas, Viktor Kerkez, Madian Khabsa, Isabel Kloumann, Artem Korenev, Punit Singh Koura, Marie-Anne Lachaux, Thibaut Lavril, Jenya Lee, Diana Liskovich, Yinghai Lu, Yuning Mao, Xavier Martinet, Todor Mihaylov, Pushkar Mishra, Igor Molybog, Yixin Nie, Andrew Poulton, Jeremy Reizenstein, Rashi Rungta, Kalyan Saladi, Alan Schelten, Ruan Silva, Eric Michael Smith, Ranjan Subramanian, Xiaoqing Ellen Tan, Binh Tang, Ross Taylor, Adina Williams, Jian Xiang Kuan, Puxin Xu, Zheng Yan, Iliyan Zarov, Yuchen Zhang, Angela Fan, Melanie Kambadur, Sharan Narang, Aurelien Rodriguez, Robert Stojnic, Sergey Edunov, and Thomas Scialom. 2023. Llama 2: Open foundation and fine-tuned chat models. *Preprint*, arXiv:2307.09288.

Alba Vilaplana-Pérez, Ana Pérez-Vigil, Anna Sidorchuk, Gustaf Brander, Kayoko Isomura, Eva Hesselmark, Ralf Kuja-Halkola, Henrik Larsson, David Mataix-Cols, and Lorena Fernández de la Cruz. 2021. Much more than just shyness: the impact of social anxiety disorder on educational performance across the lifespan. *Psychological Medicine*, 51(5):861–869.

A Appendix

A.1 One-shot Prompting

We use the template provided in Table 3 to prompt the four LLMs.

A.2 Synthetic Data Generation

Prompt Design. Using `Mistral-7B`, we simulate a one turn user-assistant conversation. We instruct the model to produce a post by a Reddit user that describes how an outdoor space or activity affects their social anxiety symptoms POSITIVELY, NEGATIVELY or has a NEUTRAL effect. We randomly pick an instance from the training data as a one-shot example in the prompt. Similarly, we randomly choose an outdoor space/activity from the spaces mentioned in the training data as well as a persona for the Reddit user to ensure more variability. We provide the prompt template in Table 4. We follow a similar process for `GPT-3.5`, we generate new instances by randomly selecting a humman generated Reddit post and using it as a few-shot example for `GPT-3.5` to create another Reddit post, the newly generated post uses the same keywords (e.g., *beach*, *forest*), see Table 4 for examples.

Personas. 'teenager', 'young adult', 'middle-aged adult', 'senior citizen', 'child', 'adolescent', 'adult', 'elderly person', 'teacher', 'doctor', 'nurse', 'computer scientist', 'engineer', 'scientist', 'researcher', 'professor', 'academic', 'student', 'florist', 'farmer', 'chef', 'cook', 'baker', 'waiter', 'waitress', 'cashier', 'bank teller', 'receptionist', 'librarian', 'archivist', 'historian', 'writer', 'author', 'PhD student', 'graduate student', 'undergraduate student'.

Activities. 'ocean', 'swim', 'outdoors', 'running', 'go for a run', 'soccer', 'pond', 'golf', 'playground', 'rowing', 'coast', 'climb', 'bonfire', 'basketball', 'horses', 'snowboard', 'forest', 'hills', 'lawn', 'tennis', 'hill', 'bicycle', 'cabin', 'mountain', 'snowboards', 'surfing', 'backyard', 'fresh air', 'outside', 'bikes', 'waterfall', 'runners', 'backpacking', 'go for a walk', 'riding', 'runs', 'park', 'jog', 'river', 'nature', 'mountains', 'baseball', 'hiking', 'bench', 'outdoor', 'sea', 'skating', 'hike', 'skate', 'breeze', 'barbecue', 'bay', 'waves', 'runner', 'stream', 'swimming', 'shore', 'horse', 'biking', 'open space', 'sand', 'run', 'canoe', 'camping', 'benches', 'surf', 'jogging', 'boat', 'cruise', 'wildlife', 'parks', 'hikes', 'tree', 'camps', 'bleachers', 'walk', 'pool', 'bike', 'camp', 'sun', 'grass', 'garden', 'beach', 'lake', 'landscape', 'trees'.

Examples of generated instances. Using the prompt structures from Table 4.

Mistral I am a middle aged man who suffers from depression and anxiety. Running has always been a great way for me to clear my mind and relieve stress. However, lately I've noticed that running has also been triggering my anxiety. Every time I step outside and see other runners, I start to compare myself to them and feel like I'll never be able to catch up. This leads to feelings of shame and self-doubt, making it harder for me to enjoy the exercise. I've tried different types of runs, such as solo runs and group runs, but nothing seems to help. Has anyone else experienced this before? How do you cope with these feelings?

Being outside in nature has been a game-changer for my social anxiety. The vast openness and fresh air provide a sense of freedom and calm that I can't find anywhere else. Just the simple act of stepping outside can instantly ease my racing thoughts and tension.

GPT3.5 If you're feeling up to it, I highly recommend giving backpacking a try. Start with a short trip or even just a day hike to see how it resonates with you. You might be surprised at how much it can help alleviate some of your social anxiety symptoms. Remember, you're not alone in this journey. Feel free to reach out if you need any advice or support. Take care of yourself and keep pushing through! You got this.

Reddit post: "I struggle with social anxiety, and finding ways to cope with it has been a journey. One thing that has surprisingly helped me is spending time outside. There's something about nature that calms my anxieties and helps me feel more at ease. Whether it's taking a walk in the park, going for a hike, or simply sitting in my backyard, being outside helps me clear my mind and feel less overwhelmed by social situations.

Hey everyone, I just wanted to share my experience with being outside and how it has helped me with my social anxiety. For the longest time, I struggled with being around people and entering social situations made me extremely anxious. However, I found that spending time outside in nature has been incredibly beneficial for my mental health.

Prompt
We analyze effects of outdoor spaces on social anxiety symptoms in Reddit posts. You will be presented with a user-written post. The posts were filtered based on a list of nature-related keywords related to outdoor spaces and activities. Your task is to categorize posts into one of four categories: 0) unrelated: the nature-related keyword does not reference nature (e.g., it is used in a metaphor or idiomatic expression), the user is/has not personally experienced the nature-related keyword, or it. Note, that each post has only one classification. 1) positive effect: the nature-related space/activity helps the user's mental well-being. 2) neutral or no effect: the nature keyword is referencing nature, however, the user makes no mention of it having a positive or negative effect on the user's mental well-being. 3) negative: the nature-related space/activity has a negative effect on the user's mental well-being. Provide the output in a json format with the key being 'label' and the value being the category number as an integer. For example: if you believe the post should be categorized as 1) positive, your json output should be: **{'label': 1}** Now consider the following example: *I'm supposed to go on a hike with friends, but I'm feeling tense about it. they still haven't made any proper plans yet. I was kind of hoping they would forgot about it, so I wouldn't have to go through the hassle of getting ready and dealing with the crowds. Now I'll have to wake up early and be prepared, just in case.* What is the correct category for this post? Here is the correct category formatted as json: **{'label': 3}**

Table 3: Template for few-shot classification with four LLMs: The task/instruction description is in monospace, class descriptions (0, 1, 2, and 3) are in normal font, few-shot examples are in *italics*, and the expected LLM output is in **bold**.

M.	Role	Prompt
Mistral 7-B	user	Imagine you are a person who is suffering from social anxiety. Write a Reddit post in which you describe the effects of an nature-related space or outdoor activity on your social anxiety symptoms. The outdoor space or activity could be something like 'surfing', 'backyard', 'fresh air' or 'basketball'. Your post should describe how the outdoor space or activity has a **<target sentiment>** effect on your symptoms, so the nature-related space/activity **<helps your mental well-being./does not help your mental well-being/has no effect on your symptoms.>**. Provide the output in json format with the key being 'post' and the value being the text of your post. Write a post for the outdoor space/activity * **<activity>***.
	assistant	Here is the post I came up with formatted as json: {'post': '**<random training instance for target sentiment>**'}
	user	Perfect! Let's try another one. Imagine you are a **<persona>**. Write a post for the outdoor space/activity ***<activity>***. Your post should describe how the outdoor space or activity has a **<target sentiment>** effect on your symptoms, so the nature-related space/activity **<helps your mental well-being./does not help your mental well-being/has no effect on your symptoms.>**. Only output the json, no additional text or explanation.
GPT 3.5	system	Imagine you are a person who is suffering from social anxiety. Write a Reddit post in which you describe the effects of nature-related space or outdoor activity on your social anxiety symptoms. Use the following example: **<keywords>** Reddit post: **<Reddit post example>**. Do not write more than 350 words and only write the post itself.
	user	Keywords: **<keywords>**

Table 4: Prompt template to generate additional training instances with Mistral-7B-v0.1 and GPT 3.5. We randomly sample an example instance from the training instances with the target sentiment and instruct the model to write from the perspective of a randomly sampled persona to increase variety in the synthetic data.

1024m at SMM4H 2024: Tasks 3, 5 & 6 - Self Reported Health Text Classification through Ensembles

Ram Mohan Rao Kadiyala
University of Maryland, College Park
rkadiyal@terpmail.umd.edu

M.V.P. Chandra Sekhara Rao
RVR&JC College of Engineering
mvpcs@rvrjc.ac.in

Abstract

Social media is a great source of data for users reporting information regarding their health and how various things have had an effect on them. This paper presents various approaches using Transformers and Large Language Models and their ensembles, their performance along with advantages and drawbacks for various tasks of SMM4H'24 - Classifying texts on impact of nature and outdoor spaces on the author's mental health (Task 3), Binary classification of tweets reporting their children's health disorders like Asthma, Autism, ADHD and Speech disorder (task 5), Binary classification of users self-reporting their age (task 6).

1 Introduction

Social media has become a key way for people to share their experiences and feelings. This has opened up new opportunities for researchers to understand how different aspects of life affect our well-being. The paper explores three tasks of SMM4H 2024(Xu et al., 2024) - 4-way classification of texts based on effect of nature, outdoor spaces and activities on author's mental health (Task 3), Binary classification of texts reporting health disorders in author's child including ADHD, Autism, Asthma and Speech disorder (Task 5)(Klein et al., 2024), Binary classification of texts self-reporting author's exact age directly / indirectly (Task 6).The paper explores usage of transformer models like RoBERTa(Liu et al., 2019), DeBERTa(He et al., 2021), Longformer(Beltagy et al., 2020) and LLs including both proprietary and open-source like GPT-4(OpenAI, 2024), Claude-Opus(Anthropic, 2024), Llama-3 8B(Touvron et al., 2023), Mistral 7B(Jiang et al., 2023), Gemma 7B(GemmaTeam, 2024), and ensembles along with advantages and drawbacks of each approach using the models. Similar previous works can be found in (Weissenbacher et al., 2022), (Magge et al., 2021) and (Klein et al., 2020).

2 Datasets

The dataset for Task 3 consists of 3000 reddit posts from r/socialanxiety belonging to four classes based on self reported impact of outdoor spaces and activities on the author's mental health - 0: unrelated to the task, 1: had a positive impact, 2: is neutral or had no effect, 3: had a negative effect. The dataset for Task 5 consists of 9734 tweets belonging to two classes - 1: users reporting having a child having ADHD, Asthma, Autism or Speech disorder and the rest as class 0. Similarly for Task 6, the dataset of 21200 texts consists of both tweets and reddit posts from r/AskDocs for two classes - Class 1 being texts through which the author's current age in years may be determined and rest as Class 0. The distribution of labels for the three tasks can be seen in Table 1, Table 2 and Table 3.

	Training	Development	Testing
Class 0	1131	377	?
Class 1	160	54	?
Class 2	395	131	?
Class 3	114	38	?
Total	1800	600	600

Table 1: Dataset split and class distribution : Task 3

	Training	Development	Testing
Class 0	5118	254	?
Class 1	2280	135	?
Total	7398	389	1947

Table 2: Dataset split and class distribution : Task 5

	Training	Development	Testing
Class 0	5966	2435	?
Class 1	2834	1765	?
Total	8800	4200	8200

Table 3: Dataset split and class distribution : Task 6

Proceedings of The 9th Social Media Mining for Health Applications (#SMM4H) Workshop and Shared Tasks, pages 88–94
August 15, 2024 ©2024 Association for Computational Linguistics

	F1	P	R
Bart-Large* (2-stage)	0.673	0.666	**0.687**
Bart-Large (direct)	0.654	0.676	0.643
Bart-Large (2-stage)	**0.679**	**0.677**	0.682
Mean	0.519	0.565	0.538
Median	0.580	0.630	0.589

Table 4: Precision, Recall and F1 on Test set compared to other participants : Task 3

* indicates model is trained without using Dev set

	F1	P	R
Bart-Large* (direct)	0.912	0.896	**0.929**
Bart-Large (direct)	**0.918**	**0.923**	0.912
Mean	0.822	0.818	0.838
Median	0.901	0.885	0.917

Table 5: Precision, Recall and F1 on Test set compared to other participants : Task 5

* indicates model is trained without using Dev set

3 Systems Description

For Task 3, two approaches were tested. One where classification was done directly in a 4-way and the other where classification was done is two stages, this involved first classifying the text whether it is related to the task or not i.e class 0 or not and then classifying the effect on the user in the second stage. For Task 5 and 6 it was done directly as a binary classification task[1][2]. In LLM approaches, The proprietary versions were used as zero-shot and the rest of the LLMs were tested in a zero-shot and fine-tuned manner. Additionally they were tested in a two stage classification for Task 3. In the case of ensembles, It was through majority voting in a set of models, through and-rule for high precision requirement and through or-rule for high recall requirements. For Task 5 and 6, while using LLMs, classification was done by dividing the criteria into parts and aggregating the individual results. i.e In the case of Task 5, individual prompts test for each condition that needed to be satisfied to classify as positive and AND-rule is used for generating final label. Similarly OR-rule was used for Task 6. The performance of different approaches can be seen in Table 7, Table 8 and Table 9.The data during training was shuffled after every epoch and also internally in each mini-batch.

[1]Code available at: https://github.com/1024-m/SMM4H-ACL-2024

[2]Models available at: https://huggingface.co/1024m

	F1	P	R
Bart-Large (direct)	**0.959**	**0.953**	**0.965**
GPT-4 (and-rule)	0.922	0.895	0.951
Mean	0.924	0.924	0.926
Median	0.936	0.934	0.949

Table 6: Precision, Recall and F1 on Test set compared to other participants : Task 6

4 Error Analysis

The LLMs performed equally good on all kinds of data while transformers models performed less effectively when the kind of language used is off from rest of the data or when criteria for classification was mentioned in one sentence and referred to the conditions indirectly later on. It was observed that positively labelled samples were predicted correctly by either the LLM approach or transformers, hence ensembles of both had recall over 0.99 with just 1 percent drop in F1 scores in Task 5 and 6. Many of the positively misclassified samples were in the format of advertisements where the title appears to match the criteria for positive classification. This is one area where LLMs were still able to distinguish effectively while other models did not.

5 Conclusion

the performance of some of the models compared to others on the test set can be seen in Table 4, Table 5 and Table 6. The LLM approach did yield comparatively good results despite using in a 4bit precision due to lack of computational resources. It is likely the performance would be better that the current models in full precision. Many of the positive label texts have been filtered out during the data collection process. For example, texts self-reporting age in text format instead of numerical. Due to this, a higher focus on recall is necessary. A custom metric with higher importance to recall is better suited for Task 5 and 6 compared to F1 scores. Ensemble approaches like majority voting and filtering guaranteed positive label texts using LLM predictions could improve performance without a significant drop in the F1 scores. Finally, the performance improved on all the tasks while using dev set as additional training data compared to just the training data, hinting at the possibility of improving the performance by adding more training data. Augmentation through paraphrasing existing data however did not improve the results.

	Direct Classification			2-Stage Classification		
Model	**Macro-F1**	**Precision**	**Recall**	**Macro-F1**	**Precision**	**Recall**
			Transformers (fine-tuned)			
longformer-large	0.603	0.610	0.596	0.667	**0.671**	0.660
RoBERTa-large	0.595	0.601	0.585	0.664	0.669	0.652
BART-large	0.603	0.597	0.611	**0.670**	0.652	**0.687**
DeBERTa-large	0.601	0.598	0.606	0.661	0.657	0.669
			Proprietary LLMs (zero-shot)			
GPT-4	0.536	0.545	0.546	0.584	0.592	0.571
Claude-Opus	0.504	0.492	0.605	0.579	0.565	0.594
			Open-source LLMs (fine-tuned)			
LLaMa-3-8B	0.643	0.622	0.653	-	-	-
Mistral-7B	0.637	0.621	0.646	-	-	-
Gemma-7B	0.639	0.624	0.644	-	-	-

Table 7: performance of different approaches on Dev set : Task 3

	Direct Classification			And-rule Classification		
Model	**Class1-F1**	**Precision**	**Recall**	**Class1-F1**	**Precision**	**Recall**
			Transformers (fine-tuned)			
longformer-large	0.937	**0.940**	0.933	-	-	-
RoBERTa-large	0.926	0.926	0.926	-	-	-
BART-large	**0.940**	0.933	0.947	-	-	-
DeBERTa-large	0.927	0.914	0.941	-	-	-
			Proprietary LLMs (zero-shot)			
GPT-4	0.786	0.862	0.956	0.859	0.785	0.948
Claude-Opus	0.689	0.809	**0.985**	0.851	0.782	0.943
			Open-source LLMs (fine-tuned)			
LLaMa-3-8B	0.925	0.939	0.911	-	-	-
Mistral-7B	0.921	0.921	0.921	-	-	-
Gemma-7B	0.920	0.934	0.907	-	-	-

Table 8: performance of different approaches on Dev set : Task 5

	Direct Classification			Or-rule Classification		
Model	**Class1-F1**	**Precision**	**Recall**	**Class1-F1**	**Precision**	**Recall**
			Transformers (fine-tuned)			
longformer-large	0.898	0.884	0.914	-	-	-
RoBERTa-large	0.891	0.862	0.920	-	-	-
BART-large	**0.901**	0.878	0.926	-	-	-
DeBERTa-large	0.894	0.869	0.923	-	-	-
			Proprietary LLMs (zero-shot)			
GPT-4	0.861	0.791	**0.960**	0.897	0.870	0.925
Claude-Opus	0.858	0.794	0.952	0.893	0.873	0.937
			Open-source LLMs (fine-tuned)			
LLaMa-3-8B	0.898	**0.912**	0.886	-	-	-
Mistral-7B	0.894	0.908	0.883	-	-	-
Gemma-7B	0.894	0.901	0.889	-	-	-

Table 9: performance of different approaches on Dev set : Task 6

References

Anthropic. 2024. Proprietary documentation of Company Name. [link].

Iz Beltagy, Matthew E. Peters, and Arman Cohan. 2020. Longformer: The long-document transformer.

GemmaTeam. 2024. Gemma: Open models based on gemini research and technology.

Pengcheng He, Xiaodong Liu, Jianfeng Gao, and Weizhu Chen. 2021. Deberta: Decoding-enhanced bert with disentangled attention.

Albert Q. Jiang, Alexandre Sablayrolles, Arthur Mensch, Chris Bamford, Devendra Singh Chaplot, Diego de las Casas, Florian Bressand, Gianna Lengyel, Guillaume Lample, Lucile Saulnier, Lélio Renard Lavaud, Marie-Anne Lachaux, Pierre Stock, Teven Le Scao, Thibaut Lavril, Thomas Wang, Timothée Lacroix, and William El Sayed. 2023. Mistral 7b.

Diederik P. Kingma and Jimmy Ba. 2017. Adam: A method for stochastic optimization.

Ari Klein, Ilseyar Alimova, Ivan Flores, Arjun Magge, Zulfat Miftahutdinov, Anne-Lyse Minard, Karen O'Connor, Abeed Sarker, Elena Tutubalina, Davy Weissenbacher, and Graciela Gonzalez-Hernandez. 2020. Overview of the fifth social media mining for health applications (#SMM4H) shared tasks at COLING 2020. In *Proceedings of the Fifth Social Media Mining for Health Applications Workshop & Shared Task*, pages 27–36, Barcelona, Spain (Online). Association for Computational Linguistics.

Ari Z Klein, Juan M Banda, Yuting Guo, Ana Lucia Schmidt, Dongfang Xu, Jesus Ivan Flores Amaro, Raul Rodriguez-Esteban, Abeed Sarker, and Graciela Gonzalez-Hernandez. 2023. Overview of the 8lt;supgt;thlt;/supgt; social media mining for health applications (smm4h) shared tasks at the amia 2023 annual symposium. *medRxiv : the preprint server for health sciences*, page 2023.11.06.23298168.

Ari Z Klein, José Agustín Gutiérrez Gómez, Lisa D Levine, and Graciela Gonzalez-Hernandez. 2024. Using longitudinal twitter data for digital epidemiology of childhood health outcomes: An annotated data set and deep neural network classifiers. *J Med Internet Res*, 26:e50652.

Yinhan Liu, Myle Ott, Naman Goyal, Jingfei Du, Mandar Joshi, Danqi Chen, Omer Levy, Mike Lewis, Luke Zettlemoyer, and Veselin Stoyanov. 2019. Roberta: A robustly optimized bert pretraining approach.

Arjun Magge, Ari Klein, Antonio Miranda-Escalada, Mohammed Ali Al-Garadi, Ilseyar Alimova, Zulfat Miftahutdinov, Eulalia Farre, Salvador Lima López, Ivan Flores, Karen O'Connor, Davy Weissenbacher, Elena Tutubalina, Abeed Sarker, Juan Banda, Martin Krallinger, and Graciela Gonzalez-Hernandez. 2021. Overview of the sixth social media mining for health applications (#SMM4H) shared tasks at NAACL 2021. In *Proceedings of the Sixth Social Media Mining for Health (#SMM4H) Workshop and Shared Task*, pages 21–32, Mexico City, Mexico. Association for Computational Linguistics.

OpenAI. 2024. Gpt-4 technical report.

Hugo Touvron, Thibaut Lavril, Gautier Izacard, Xavier Martinet, Marie-Anne Lachaux, Timothée Lacroix, Baptiste Rozière, Naman Goyal, Eric Hambro, Faisal Azhar, Aurelien Rodriguez, Armand Joulin, Edouard Grave, and Guillaume Lample. 2023. Llama: Open and efficient foundation language models.

Davy Weissenbacher, Juan Banda, Vera Davydova, Darryl Estrada Zavala, Luis Gasco Sánchez, Yao Ge, Yuting Guo, Ari Klein, Martin Krallinger, Mathias Leddin, Arjun Magge, Raul Rodriguez-Esteban, Abeed Sarker, Lucia Schmidt, Elena Tutubalina, and Graciela Gonzalez-Hernandez. 2022. Overview of the seventh social media mining for health applications (#SMM4H) shared tasks at COLING 2022. In *Proceedings of The Seventh Workshop on Social Media Mining for Health Applications, Workshop & Shared Task*, pages 221–241, Gyeongju, Republic of Korea. Association for Computational Linguistics.

Dongfang Xu, Guillermo Lopez Garcia, Lisa Raithel, Rolland Roller, Philippe Thomas, Eiji Aramaki, Shuntaro Yada, Pierre Zweigenbaum, Sai Tharuni Samineni, Karen O'Connor, Yao Ge, Sudeshna Das, Abeed Sarker, Ari Klein, Lucia Schmidt, Vishakha Sharma, Raul Rodriguez-Esteban, Juan Banda, Ivan Flores Amaro, Davy Weissenbacher, and Graciela Gonzalez-Hernandez. 2024. Overview of the 9th social media mining for health applications (#SMM4H) shared tasks at ACL 2024. In *Proceedings of The 9th Social Media Mining for Health Research and Applications Workshop and Shared Tasks*, Bangkok, Thailand. Association for Computational Linguistics.

A Task 3 System Overview

Classifying the class of unrelated texts (class 0) from the other 3 separately had improved the performance by reducing mis-classification between Class 0 and others. The overview of the process can be seen in Figure 1. The fine-tuned transformer models used had the best results with a learning rate of 0.00002 and weight decay of 0.01 over 30 epochs for 2-stage classification and 50 epochs for direct classification. In case of the fine-tuned LLMs, the base models were loaded in 4-bit configuration due to computational limitations, later fine-tuned and used in 16-bit precision for inference. During training, RoPE scaling was used for texts longer than 2048 tokens. They were fine-tuned over 3 epochs with a learning rate of 0.0002 and weight decay of 0.01 using Alpaca prompts.

The prompts ussed over the LLMs were as follows :

- **2-stage 1st prompt** : "Did outdoor spaces or activities get mentioned? Respond only with a 1 for yes or 0 for no. Only one character (0/1) nothing else."
- **2-stage 1st prompt** : "What impact did outdoor spaces or activities have on the user's mental health ? Respond only with a 1 for positive or 2 for neutral or 3 for negative. Only one character (1/2/3) nothing else."
- **Direct classification prompt** : "What impact did outdoor spaces or activities have on the user's mental health ? Respond only with a 1 for positive or 2 for neutral or 3 for negative or 0 for no mention. Only one character (1/2/3/0) nothing else."
- **Fine-tuned LLMs prompt** : "What impact did outdoor spaces or activities have on the user's mental health ? Respond only with a 1 for positive or 2 for neutral or 3 for negative or 0 for no mention. Only one character (1/2/3/0) nothing else"

The models that resulted in the best performance on the test set are available at :

- https://huggingface.co/1024m/SMM4H-Task3-BartL-1A30
- https://huggingface.co/1024m/SMM4H-Task3-BartL-1B30

B Task 5 System Overview

The overview of the process can be seen in Figure 2. The fine-tuned transformer models used had the same hyper-parameters as used in Task 3, and were fine-tuned over 20 epochs. In case of the fine-tuned LLMs, the process is same as what was used in task 3. The proprietary systems were tested additionally using multiple separate prompts for each sub-condition that is to be true to be classified as a positive class text. In case of And-rule approach, the texts were marked as positive (class 1) if all of the conditions were met to achieve higher F1 with a lower recall trade-off.

The prompts used over the LLMs were as follows :

- **Direct classification prompt** : "The tweets already mention at least one of the following: attention-deficit/hyperactivity disorder (ADHD), autism spectrum disorders (ASD), delayed speech (speech disorder), or asthma. In some cases, the tweets discuss hypothetical cases or the possibility of having the condition. It might be about someone else's child or an adult son/daughter. Respond with '1' if the tweet explicitly mentions an existing formal diagnosis of one of those conditions AND it concerns a child/baby AND the child is the user's own. In all other cases, respond with a '0'. Respond with only one character ('0'/'1') and nothing else."
- **AND-rule prompt 1** : "The tweets already mention at least one of the following: attention-deficit/hyperactivity disorder (ADHD), autism spectrum disorders (ASD), delayed speech (speech disorder), or asthma. In some cases, the tweets discuss hypothetical cases or the possibility of having the condition. Respond with '1' if the tweet explicitly mentions an existing formal diagnosis of one of those conditions. In all other cases, respond with a '0'. Respond with only one character ('0'/'1') and nothing else."
- **AND-rule prompt 2** : "The tweets already mention... ...Respond with '1' if the tweet explicitly mentions it concerns a child/baby having one of those conditions. In all other cases, respond with a '0'. Respond with only one character ('0'/'1') and nothing else."

- **AND-rule prompt 3** : "The tweets already mention... ...Respond with '1' if the tweet explicitly mentions the child is the user's own having diagnosed with one of those conditions. In all other cases, respond with a '0'. Respond with only one character ('0'/'1') and nothing else."

The model that resulted in the best performance on the test set is available at :

- https://huggingface.co/1024m/SMM4H-Task5-BartL-2A

C Task 6 System Overview

The overview of the process can be seen in Figure 3. The fine-tuned transformer models used had the same hyper-parameters as used in Task 3, and were fine-tuned over 20 epochs. In case of the fine-tuned LLMs, the process is same as what was used in task 3. The proprietary systems were tested additionally using multiple separate prompts for each sub-condition that can be true to be classified as a positive class text. In case of OR-rule approach, the texts were marked as positive (class 1) if at least one of the conditions were met to achieve higher F1 with a lower recall trade-off. The classification was done separately fro twitter and reddit posts with separate models i.e one for each platform's posts.

The prompts used over the LLMs were as follows :

- **Direct classification prompt** : "Respond only with 0 or 1 and nothing else : based on whether current age of the AUTHOR in years can be known from the texts. The texts have a two digit number which is likely an age if not clear. The age needed to know in context is current age of THE author and not someone else. In some cases formats like 25m , 24f are used where m refers to Male and f refers to Female."

- **OR-rule prompt 1** : "Respond only with 0 or 1 and nothing else based on whether the current age of the author was reported in the given text."

- **OR-rule prompt 2** : "Respond only with 0 or 1 and nothing else based on whether the current age of the author can be determined from the given text."

- **OR-rule prompt 3** : "Respond only with 0 or 1 and nothing else based on whether the current age of the author was expressed using formats like 25m , 24f are used where 'm' refers to Male and 'f' refers to Female."

The models that resulted in the best performance on the test set are available at :

- https://huggingface.co/1024m/SMM4H-Task6-BartL-A20 For Reddit texts

- https://huggingface.co/1024m/SMM4H-Task6-BartL-B20 For Twitter texts

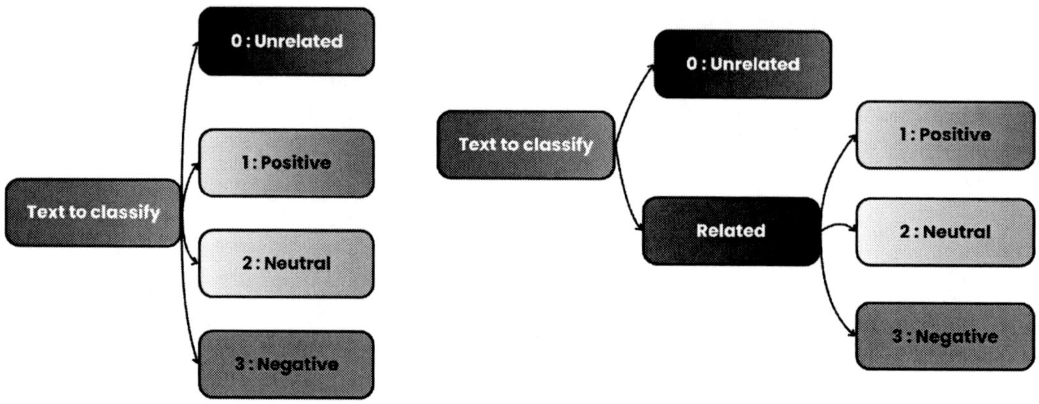

Figure 1: Overview of approaches used for Task 3 : Direct (left) and 2-Stage (right)

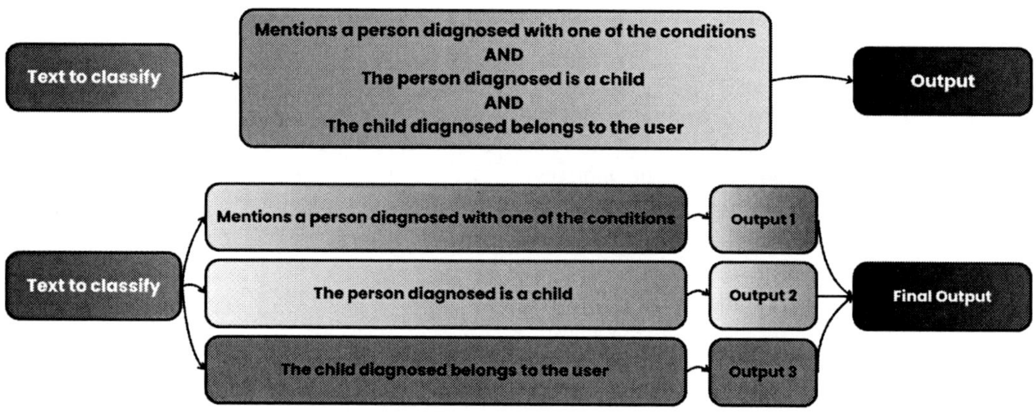

Figure 2: Overview of approaches used for Task 5 : Direct (top) and AND-rule (bottom)

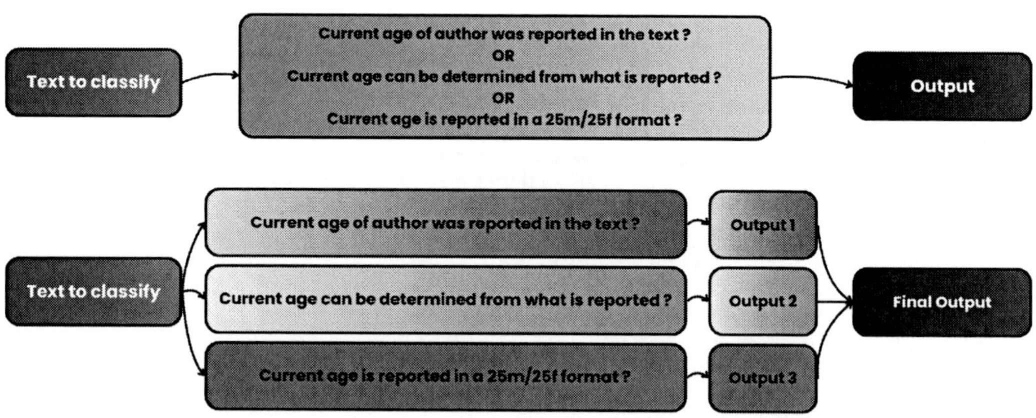

Figure 3: Overview of approaches used for Task 6 : Direct (top) and OR-rule (bottom)

LAMA at SMM4H 2024: Experimenting with Transformer-based and Large Language Models for Classifying Effects of Outdoor Spaces on Social Anxiety in Social Media Data

Falwah AlHamed
Department of Computing,
Imperial College London
London, UK
f.alhamed20@imperial.ac.uk

Julia Ive
Queen Mary
University of London
London, UK
j.ive@qmul.ac.uk

Lucia Specia
Department of Computing,
Imperial College London
London, UK
l.specia@imperial.ac.uk

Abstract

Social Anxiety Disorder (SAD) is a common condition, affecting a significant portion of the population. While research suggests spending time in nature can alleviate anxiety, the specific impact on SAD remains unclear. This study explores the relationship between discussions of outdoor spaces and social anxiety on social media. We leverage transformer-based and large language models (LLMs) to analyze a social media dataset focused on SAD. We developed three methods for the task of predicting the effects of outdoor spaces on SAD in social media. A two-stage pipeline classifier achieved the best performance of our submissions with results exceeding baseline performance.

1 Introduction

Social anxiety disorder (SAD) is a prevalent anxiety disorder that affects up to 12% of the population at some point in their lives (Kessler et al., 2005). Interestingly, social media platforms like Reddit have become a space for people with SAD to connect, share their experiences, and seek advice on managing symptoms. While research suggests that spending time outdoors in natural environments can be beneficial for alleviating anxiety in general (Barton and Pretty, 2010; Berman et al., 2008), little is known about the specific impact of such environments on SAD. This study investigates the effect of outdoor spaces on social anxiety using social media posts as a source of data.

2 Data and Task Description

This work leverages the dataset provided by the organizers of SMM4H (Xu et al., 2024). The data originates from the r/socialanxiety subreddit on Reddit and consists of 3,000 annotated social media posts. The training set comprises 1,800 posts, each labelled with a code based on the user's sentiment towards the mentioned nature-related spaces or activities. A four-class labelling scheme was employed: **positive**: the space/activity benefits the user's well-being; **neutral**: mention of nature without a clear impact; **negative**: the space/activity has a negative impact; and **unrelated**: posts where the nature keyword is metaphorical or unclear in meaning. The task is four-class classification on these labels.

3 Methods

3.1 Models Exploration

We address this multi-class classification task by leveraging several transformer-based and LLMs on the provided training dataset. While the task definition specifies four classes, there is natural division into two broader categories: Related and Unrelated. The Related category encompasses posts where the mentioned outdoor space is relevant to the user's experience, and the effect can be positive, neutral, or negative. Conversely, the Unrelated category consists solely of posts where the outdoor space reference is metaphorical and has no bearing on the user's experience. We performed experiments utilizing five models: BERT, RoBERTa, MentalBERT, GPT3.5, and LlamA. For each model, we conducted three distinct experiments:

Experiment 1: Binary Classification: The model classifies posts into: Related and Unrelated classes.

Experiment 2: Multi-Class Classification within Related class: The model focuses solely on "related" posts and further categorizes them into three classes: positive, neutral, and negative.

Experiment 3: Four-Class Classification: This model classifies posts into all classes (positive, neutral, negative, and unrelated).

The primary objective of this exploration was to identify the optimal model for each experiment, which would then be used in subsequent stages of our submission models design. For transformer-based models, we utilized the Hugging Face transformers library (Wolf et al., 2019) with

Proceedings of The 9th Social Media Mining for Health Applications (#SMM4H) Workshop and Shared Tasks, pages 95–97
August 15, 2024 ©2024 Association for Computational Linguistics

Experiment	BERT			RoBERTa			Mental BERT			Llama			GPT		
	P	R	F1	P	R	F1	P	R	F1	P	R	F1	P	R	F1
Experiment 1	0.97	0.97	0.97	0.98	0.98	0.98	**0.99**	**0.98**	**0.98**	Random Labels			0.70	0.60	0.58
Experiment 2	0.66	0.67	0.62	**0.71**	**0.66**	**0.68**	0.83	0.54	0.53	N/A					
Experiment 3	0.73	0.77	0.73	**0.82**	**0.71**	**0.73**	0.63	0.65	0.64	N/A					

Table 1: Results of best model exploration experiments (bold fonts signify the best results)

Submission	F1 Score	Precision	Recall	Accuracy
Our RoBERTa 4-classes	0.543	0.607	0.534	**0.651**
Our 2-Stages Pipeline	**0.545**	**0.633**	**0.536**	0.636
Our Multi-task Learning	0.321	0.346	0.326	0.448
All Teams Mean	0.519	0.565	0.538	0.575

Table 2: Results of submitted systems (bold fonts signify the best results)

model cards (`bert-base-uncased`, `roberta-base`, and `mental/mental-bert-base-uncased`. These models received only the post text as input and were trained to predict a pre-defined class label depending on the experiment. LLMs employed a zero-shot learning approach, where prompts incorporating both the post text and relevant keywords were fed to the models. We experimented with various prompts and the best performing prompt is " `Do you think the person is really [keyword] in this paragraph or just a metaphor? return 1 if you think they are doing this otherwise, return 0 Text:[post_text]`.

3.2 Methods Description

Based on experiment results, we selected the top-performing models for each task and employed them to develop three submission methods:

Four-Class Classification: This method utilizes the chosen model from Experiment 3 to classify posts into all four original classes (positive, neutral, negative, and unrelated).

Two-Stage Pipeline: This approach leverages two models in a sequential pipeline. The first stage employs the model selected from Experiment 1 for the binary classification of posts into Related and Unrelated. Subsequently, posts classified as "Related" are passed to the model chosen from Experiment 2, which performs a three-class classification into positive, neutral, and negative.

Multi-Task Learning: This method employs a multi-task learning model trained on two tasks simultaneously. The first task is a binary classification of all (posts) into Related and Unrelated categories. The second task is a multi-class classification focusing solely on posts classified as "Related" in the first task, categorizing them into positive, neutral, and negative.

4 Results

The evaluation metric for this task is the macro-averaged F1-score over all 4 classes. Table 1 summarizes the performance of the models across various experiments, evaluated using precision, recall, and F1-score metrics. MentalBERT emerged as the most effective model for classifying posts into "Related" and "Unrelated" categories. On the other hand, RoBERTa achieved the best performance in both classifying "Related" posts into positive, neutral, and negative classes and in the four-class classification task. LLMs performance in classifying posts as related/unrelated was disappointing, thus it was discarded from the remaining experiments. Based on these results, we selected RoBERTa for the final submission employing the four-class classification approach. Additionally, the pipeline approach for the submission utilized MentalBERT for the initial binary classification of posts into "Related" and "Unrelated" categories and RoBERTa model for the following three-class classification into positive, neutral, and negative. RoBERTa was also employed in the multi-task approach. Results of our submission provided by task organizers is shown in Table 2.

5 Conclusion

This study investigated the connection between social anxiety and outdoor spaces on social media. We employed transformer-based and LLMs classifiers on the provided dataset. Our results exceeded the baseline, and the pipeline approach achieved the highest performance as evaluated by the task organizers.

References

J Barton and J Pretty. 2010. Tenacious nature: A review of the evidence for greater well-being within natural environments. *Journal of Environmental Psychology*, 30(4):380–390.

Marc G Berman, John Jonides, and Stephen Kaplan. 2008. The cognitive benefits of interacting with nature. *Proceedings of the National Academy of Sciences*, 105(48):19127–19132.

Ronald C Kessler, Pamela Berglund, Olga Demler, and Elizabeth E Walters. 2005. Prevalence, severity, and comorbidity of twelve-month dsm-iv disorders in the national comorbidity survey replication. *Archives of General Psychiatry*, 62(6):617–627.

Thomas Wolf, Lysandre Debut, Victor Sanh, Julien Chaumond, Clement Delangue, Anthony Moi, Pierric Cistac, Tim Rault, Rémi Louf, Morgan Funtowicz, et al. 2019. Huggingface's transformers: State-of-the-art natural language processing. *arXiv preprint arXiv:1910.03771*.

Dongfang Xu, Guillermo Lopez Garcia, Lisa Raithel, Rolland Roller, Philippe Thomas, Eiji Aramaki, Shuntaro Yada, Pierre Zweigenbaum, Sai Tharuni Samineni, Karen O'Connor, Yao Ge, Sudeshna Das, Abeed Sarker, Ari Klein, Lucia Schmidt, Vishakha Sharma, Raul Rodriguez-Esteban, Juan Banda, Ivan Flores Amaro, Davy Weissenbacher, and Graciela Gonzalez-Hernandez. 2024. Overview of the 9th social media mining for health applications (#SMM4H) shared tasks at ACL 2024. In *Proceedings of The 9th Social Media Mining for Health Research and Applications Workshop and Shared Tasks*, Bangkok, Thailand. Association for Computational Linguistics.

interrupt-driven@SMM4H'24: Relevance-weighted Sentiment Analysis of Reddit Posts

Jessica Elliott and **Roland Elliott**
Independent researchers
send-interrupt@jessicajeanne.co.za
and elliott.roland@gmail.com

Abstract

This paper describes our approach to Task 3 of the Social Media Mining for Health 2024 (SMM4H'24) shared tasks. The objective of the task was to classify the sentiment of social media posts, taken from the social anxiety subreddit, with reference to the outdoors, as positive, negative, neutral, or unrelated. We classified posts using a relevance-weighted sentiment analysis, which scored poorly, at 0.45 accuracy on the test set and 0.396 accuracy on the evaluation set. We consider what factors contributed to these low scores, and what alternatives could yield improvements, namely: improved data cleaning, a sentiment analyzer trained on a more suitable data set, improved sentiment heuristics, and a more involved relevance-weighting.

1 Introduction

Data was taken from the social anxiety subreddit where posts mentioned particular keywords relating to the outdoors. The task was to classify each text as either talking about the outdoors positively with regards to social anxiety, negatively, neutrally, or unrelated to the outdoors or anxiety (Xu et al.). For example this post: “Going for a walk in the rain can be really nice and refreshing so long as I'm wrapped up tight and dry in it.” should be classified as positive; “I felt this. One thing I'm always afraid of when going outside is to meet people that I know, especially those of my age.” should be classified as negative; “Yeah my eyes are very sensitive when the wind hits or the sun is out.” should be classified as neutral; and “..... Run like da wind” would be unrelated.

The approach we took was to classify sentences based on their sentiment (Section 2.2), weight each sentence based on its relevance (Section 2.3), and build a classification for the post based on the average score of each sentence (Section 2.4). We pursued such an approach because, if successful, it has the following advantages: firstly that it is computationally light, requiring minimal resources to be able to run; secondly it is intuitive, making it easy to explain the reasoning for any given classification it makes; and finally it is decomposable, so that each part can be iterated on and improved upon.

2 System description

Here we describe the steps we took to build our relevance-weighted sentiment analysis model.

2.1 Data cleaning and tokenization

Because the data was sourced from social media posts, preprocessing was necessary in order to make the data consistent and well-formed, so as to avoid inadvertent degradation of the sentiment and relevance models.

For example, a common feature in posts was the use of “ill” instead of “I'll”, which would artificially lower the sentiment of the sentence. Some posts would add unnecessary whitespace in the middle of words, such as “was n't” rather than “wasn't”, which prevented the sentiment analysis from properly identifying negations. And posts used different apostrophe characters, such as “’” and “'”, which were not all recognized by the models. We therefore implemented an initial data cleaning preprocessing step to correct these issues. We manually determined the possible cases and then ran detection using regex and string manipulation to transform the data into what we wanted. Some legitimate cases could be transformed erroneously, for example in the case of “ill”, but we took this as an acceptable error considering the prevalence of the usage where the poster meant “I'll”. Further work could be done on word grammatical analysis to reduce this possibility of error and to test if that would improve the results.

Another common feature of the posts was very long sentences, or sentences not delineated by the

Proceedings of The 9th Social Media Mining for Health Applications (#SMM4H) Workshop and Shared Tasks, pages 98–100
August 15, 2024

usual punctuation. Since our approach relied upon the tokenization of sentences, we therefore introduced a second preprocessing step, choosing a cut-off length of 40 words and ensuring sentences were no more than that number of words long. After scanning the data we determined that the vast majority of valid sentences were within 40 words or less, which is why we chose this value. When sentences were split in order to achieve this, we included an overlap of 5 words on each side, in order to preserve some of the sentiment-laden context around the 40-word split point. These values seemed to work better than parsing the posts without splitting the sentences, but further work could be done on either using natural language processing to try and determine sentence breaks, or testing different cut-off lengths and overlap windows to see what values achieve the best results.

2.2 Sentiment analysis

The sentiment of each sentence was calculated using the VADER sentiment analyzer (Hutto and Gilbert, 2014), which was pre-trained on the standard VADER lexicon which comes with the NLTK. This analyzer was pre-trained on Twitter (now called X) posts, and returns a sentiment analysis score between -1 (entirely negative) and 1 (entirely positive) for each sentence. The analyzer works by first assigning a default sentiment to each word, which it then modifies based on various heuristics — punctuation, capitalization, degree modification, contrastive conjunctions, and polarity flipping.

We used VADER because its training data is also social media posts, which use informal language and slang terms. However, we identified the following shortcomings when applying it to the current task. First, Twitter is a shorter-form social media platform than Reddit, the analyzer is not therefore designed to deal with the problem of long sentences (discussed in Section 2.1). Second, the analyzer is trained on general sentiment, and would have benefited from a finer-grained contextual training aimed specifically at social anxiety. For example, "staring at me" is considered neutral to the analyzer, but in the context of social anxiety this is something negative. Third, the heuristics introduced to handle negation are not rich enough to capture more nuanced forms of negation found in the longer-form text. For example, offers of advice can use negative words when describing a situation ("If you find walking in public *difficult*, try biking") despite offering a positive sentiment for the advised alternative (biking).

2.3 Relevance analysis

After testing various linear models, we determined that the Passive-Aggressive Regressor (Crammer et al., 2006) performed the best when computing the relevance of sentences. Support for these regressors is included in the SciKit-Learn Python library. We trained our regressor as follows. First, we applied a term frequency-inverse document frequency transformation on trigrams of the data (ignoring common English stop words). Second, we provided as labeled data the sentences of the training posts which contained one or more keywords, labeled as relevant (1) or irrelevant (0) based on whether the post as a whole was related (positive, neutral, or negative) or not (unrelated) as given by the task data. The resulting regressor returns a relevance score between 0 and 1 for any given text.

When classifying test data, the regressor is applied to the post as a whole to determine its relevance. If its relevance score is less than 0.25, then the post is classified as unrelated. Otherwise, it passes to the next stage, to which we now turn.

2.4 Relevance-weighted sentiment scoring

In order to calculate the overall sentiment of a post, we combined the sentiment analyzer s and the relevance regressor r described in the previous two sections. For each sentence t_i, we define the adjusted relevance R with the recurrence relation:

$$R(t_i) = \max\{r(t_i), 0.9 \cdot R(t_{i-1})\},$$

with $R(t_0) = rel(t_0)$ as the base case. This adjusted relevance allows us to model the fact that relevance can be inherited from previous sentences, with a topic being broached in one sentence and continued in later sentences without the same words necessarily appearing in them. The decay factor of 0.9 captures the intuition that the further from the original sentence we get, the less likely it is still what is being spoken about.

We use R to weight the sentiment of each sentence in the post, and then take the average of these weighted scores. As with r, R returns a relevance score between 0 and 1, so that this relevance-weighted sentiment score is calculated as:

$$\frac{\sum_{i=0}^{n} R(t_i) \cdot s(t_i)}{n}.$$

3 Evaluation

The resulting scores from our model were low. On the validation data, this approach scored as follows:

Label	Accuracy
Unrelated (0)	0.55
Positive (1):	0.33
Neutral (2)	0.18
Negative (3)	0.55
Total	0.45

And on the evaluation data, it scored as follows:

Score	Value
F1 score	0.358
Precision	0.365
Recall	0.411
Accuracy	0.396

As discussed in each section, the system and tools chosen did not fit well with the type of data provided. We consider the approach to be one that could work if the suggested improvements are made, such as improved data cleaning, a sentiment analyzer trained on a more suitable data set, improved sentiment heuristics, and a more involved relevance-weighting. As a comparison with other methods, here are the mean and median results of all the teams:

	Mean	Median
F1 score	0.5186	0.5795
Precision	0.5649	0.63
Recall	0.5379	0.5885
Accuracy	0.5746	0.627

4 Conclusion

We use a relevance-weight sentiment analysis approach for classifying the sentiment of Reddit posts with respect to social anxiety and the outdoors. Even though our particular implementation produced poor results, there are opportunities for improvement in almost every stage of the approach: improved data cleaning, a sentiment analyzer trained on a more suitable data set, improved sentiment heuristics, and a more involved relevance-weighting each improve the overall performance of such an approach.

References

Koby Crammer, Ofer Dekel, Joseph Keshet, Shai Shalev-Shwartz, Yoram Singer, and Manfred K Warmuth. 2006. Online passive-aggressive algorithms. *Journal of Machine Learning Research*, 7(3).

Clayton Hutto and Eric Gilbert. 2014. Vader: A parsimonious rule-based model for sentiment analysis of social media text. In *Proceedings of the international AAAI conference on web and social media*, volume 8, pages 216–225.

Dongfang Xu, Guillermo Lopez Garcia, Lisa Raithel, Rolland Roller, Philippe Thomas, Eiji Aramaki, Shuntaro Yada, Pierre Zweigenbaum, Sai Tharuni Samineni, Karen O'Connor, Yao Ge, Sudeshna Das, Abeed Sarker, Ari Klein, Lucia Schmidt, Vishakha Sharma, Raul Rodriguez-Esteban, Juan Banda, Ivan Flores Amaro, Davy Weissenbacher, and Graciela Gonzalez-Hernandez. Overview of the 9th social media mining for health applications (#SMM4H) shared tasks at ACL 2024. In *Proceedings of The 9th Social Media Mining for Health Research and Applications Workshop and Shared Tasks", month = Aug, year =*.

IITRoorkee@SMM4H 2024: Cross-Platform Age Detection in Twitter and Reddit Using Transformer-Based Model

Thadavarthi Vishnu Sri Sai Sankar, Dudekula Suraj, Mallamgari Nithin Reddy, Durga Toshniwal, and **Amit Agarwal**
Department of Computer Science and Technology
Indian Institute of Technology Roorkee
{t_vishnu,dudekula_s,mallamgari_nr,durga.toshniwal}@cs.iitr.ac.in

Abstract

This paper outlines the methodology for the automatic extraction of self-reported ages from social media posts as part of the Social Media Mining for Health (SMM4H) 2024 Workshop Shared Tasks. The focus was on Task 6: "Self-reported exact age classification with cross-platform evaluation in English." The goal was to accurately identify age-related information from user-generated content, which is crucial for applications in public health monitoring, targeted advertising, and demographic research. A number of transformer-based models were employed, including RoBERTa-Base, BERT-Base, BiLSTM, and Flan T5 Base, leveraging their advanced capabilities in natural language understanding. The training strategies included fine-tuning foundational pre-trained language models and evaluating model performance using standard metrics: F1-score, Precision, and Recall. The experimental results demonstrated that the RoBERTa-Base model significantly outperformed the other models in this classification task. The best results achieved with the RoBERTa-Base model were an F1-score of 0.878, a Precision of 0.899, and a Recall of 0.858.

1 Introduction

Social media data (e.g., Reddit and Twitter) plays a crucial role in health informatics, helping researchers understand public opinions on health-related issues. To engage researchers and students in analyzing social media data, the Social Media Mining for Health Applications (SMM4H) shared tasks workshop is organized by the University of Pennsylvania's Health Language Processing Lab.

The SMM4H-2024 workshop focuses on Large Language Models (LLMs) and the generalizability of social media NLP. This year's workshop includes seven shared tasks. Among these, we have worked on Task 6: "Self-reported exact age classification with cross-platform evaluation in English."

Our motivation for tackling this task stems from the observation that many patients express their health needs and medical concerns through social media. To enhance the research utility of social media data, it is essential to develop techniques for automatically identifying demographic information, such as user age, from these platforms. A detailed overview of the shared tasks in the 9th edition of the workshop can be found in (Xu et al., 2024)

The Task-6 presented in this workshop is a continuation of the work from SMM4H 2022 (Weissenbacher et al., 2022) workshop. Several papers have already been published addressing this task, highlighting its importance and the various approaches researchers have taken to solve it. For instance, (Claeser and Kent, 2022), (Kapur et al., 2022), (Tonja et al., 2022) and (Klein et al., 2022) explored different methodologies and achieved notable results.

The structure of this paper is as follows: Section 2 describes the motivation for the transformer based approaches, Section 3 describes the dataset and details of the classification task. Section 4 outlines the methodology and various experiments conducted. Section 5 discusses the results of these experiments. Finally, Section 6 concludes the paper, summarizing our findings and suggesting potential directions for future research.

2 Motivation for Using Transformer-Based Approaches

The task of extracting self-reported ages from social media posts, specifically tweets, poses significant challenges due to the similarity in content between posts that do and do not contain self-reported age information. This is illustrated by the word clouds generated from the labeled & unlabelled dataset for the top 50 bigrams, as shown in Figures 1a and 1b.

Proceedings of The 9th Social Media Mining for Health Applications (#SMM4H) Workshop and Shared Tasks, pages 101–105
August 15, 2024 ©2024 Association for Computational Linguistics

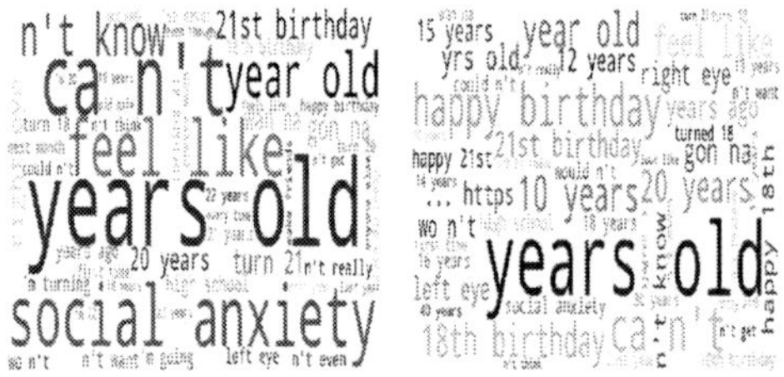

(a) Word Cloud for Labeled Dataset (Self-Reported Age)

(b) Word Cloud for Unlabeled Dataset (No Self-Reported Age)

Figure 1: Word Clouds for Labeled and Unlabeled Datasets

The word clouds from the labeled (self-reported age) and unlabeled (no self-reported age) datasets appear very similar, indicating that distinguishing between the two categories based solely on content can be difficult. Common phrases such as "years old," "happy birthday," and "social anxiety" are prevalent in both word clouds, which underscores the complexity of the task.

Given the subtle differences in the language used in these tweets, transformer-based models (Kalyan et al., 2021) are well-suited for this classification task. These models leverage deep learning techniques and advanced natural language understanding capabilities to capture nuanced patterns in text. Mathematically, the problem can be formulated as follows:

Given a set of tweets $T = \{t_1, t_2, \ldots, t_n\}$, the goal is to classify each tweet t_i into one of two classes:

$$y_i = \begin{cases} 1 & \text{if the age of the user can be determined} \\ 0 & \text{otherwise} \end{cases}$$

Transformer-based models, such as RoBERTa, BERT, BiLSTM, and T5, can be fine-tuned to learn the mapping function $f : T \rightarrow Y$ where $Y = \{y_1, y_2, \ldots, y_n\}$. These models use contextual embeddings and attention mechanisms to effectively distinguish between tweets that contain self-reported age information and those that do not.

3 Task Description and Dataset

The goal of this task is to develop an effective algorithm for the binary classification of tweets based on whether the exact age of the user can be determined from the tweet at the time it was posted. A tweet is labeled as "1" if the user's age can be determined from the text of the tweet at the time it was posted; otherwise, the tweet is labeled as "0." For this task, datasets are provided by the SMM4H-2024 workshop. The dataset includes posts from two social media platforms: Twitter and Reddit. The dataset contains approximately 13,200 tweets and 100,000 Reddit posts. This Dataset is divided into three parts:

The training data consists of 8,800 tweets from SMM4H22 and 100,000 unlabeled Reddit posts. For validation, we used 2,200 tweets from SMM4H22 and 1,000 Reddit posts related to dry eye disease, also from SMM4H22. The testing data includes 2,200 tweets from SMM4H22, 2,000 Reddit posts about dry eye disease from SMM4H22, and 12,482 Reddit posts about social anxiety. The Evaluation metrics used to evaluate in this task are standard metrics F1-Score, Precision, Recall.

4 Methods and Experiments

In this task, various models from the Huggingface toolkit (Wolf et al., 2020) were used for the automatic extraction of the exact age from tweets. The models used for the experiments are listed below:

RoBERTa-Base (Liu et al., 2019): This model was trained on a dataset containing 8,800 tweets. The tweets were tokenized using the model's subword tokenizer with a maximum token length of 128. The Adam optimizer (Kingma and Ba, 2017) was used for training with the following hyperparameters: learning rate = 2e-5, number of epochs = 10, and batch size = 64.

BERT-Base (Devlin et al., 2019): The BERT model from the Hugging Face library was used as a classifier. The create optimizer module from the hugging face toolkit library was used to fine-tune the BERT model. The learning rate was set to 2e-5, and the batch size was 16. The model is trained for upto 4 epochs

BiLSTM (Graves et al., 2013): BiLSTMs are effective for binary classification tasks as they contain two separate layers, one processing the input in the forward direction and the other in the backward direction. This bidirectional processing ensures the entire text sequence is effectively classified. The Sigmoid function was used as the activation function, and the Adam optimizer was used for training. The model was trained for up to 5 epochs.

Flan T5 Base (Chung et al., 2022): This model from the Hugging Face library was used for various NLP tasks, including summarizing, question answering, and text classification. For this task, the model was fine-tuned on the training dataset with a learning rate of 3e-4, a batch size of 8, and trained for 2 epochs.

5 Results and Discussion

Using the validation dataset provided by the organizers of the SMM4H-2024 workshop, the performance of the models was evaluated using the metrics Precision, Recall, and F1-Score. The results of the different models on the validation dataset are presented in Table 1. From Table 1, it is evident that the RoBERTa-Base model achieved the highest performance with an F1-Score of 0.88, Precision of 0.899, and Recall of 0.858, demonstrating the model's robustness in accurately identifying tweets where the user's exact age can be determined.

Subsequently, the predictions of the best-performing model, the RoBERTa-Base, were submitted on the test dataset. The results of these predictions are shown in Table 2. These results further validate the effectiveness of the RoBERTa-Base model in this classification task, reaffirming its suitability for practical applications in extracting demographic information from social media posts.

Additionally, Table 3 presents the predictions of our models on some sample tweets. This table lists predictions made by BERT, FlanTS, BiLSTM, and RoBERTa models, indicating whether they correctly identified the presence of self-reported age information. The correct labels for examples 1 and 3 are "0", indicating no self-reported age, while the correct labels for examples 2 and 4 are "1", indicating the presence of self-reported age.

In this context, RoBERTa demonstrates a high level of accuracy, correctly predicting the labels for three out of the four examples. Specifically, it successfully identified examples without self-reported age information (examples 1 and 3) and one with self-reported age (example 2). However, it incorrectly classified example 4, highlighting its occasional limitations in dealing with certain

Model	Precision	Recall	F1 Score
RoBERTa	0.89	0.86	0.88
BERT	0.90	0.87	0.87
BiLSTM	0.69	0.71	0.73
FlanT5	0.57	0.98	0.72

Table 1: Performance of our models on Task 6 Validation Dataset

Model	Precision	Recall	F1 Score
RoBERTa	0.899	0.858	0.878

Table 2: Performance of RoBERTa-Base model on Test Dataset

S.No	Example	Models
1	64 is for distance (use this for the glasses unless you are getting reading glasses).	BERT × FlanT5 × BiLSTM ✓ RoBERTa ✓
2	Well I'm 19 so later teens and earlier 20s sounds good to me but really I don't care, anyone can join	BERT ✓ FlanT5 ✓ BiLSTM × RoBERTa ✓
3	DMEK can give you 20/20 but not every time.	BERT ✓ FlanT5 ✓ BiLSTM × RoBERTa ✓
4	23, Indian male. Spend a lot of time in front of computer screens.	BERT × FlanTS ✓ BiLSTM ✓ RoBERTa ×

Table 3: Examples and Models' Predictions

nuances in the language.

BERT correctly predicted two examples but struggled with implicit age information. FlanTS also performed well on two examples but was less consistent than RoBERTa, especially with subtle context differences. BiLSTM, known for its sequential processing capability, accurately identified tweets without self-reported age information but had difficulty with context-dependent tweets.

Overall, Table 3 illustrates the comparative performance of these models, with RoBERTa generally showing superior accuracy but still facing challenges with certain tweet constructs. This comparative analysis underscores the importance of leveraging transformer-based models for their advanced contextual understanding capabilities.

6 Conclusion

This work presents our experiments on the binary classification of texts to determine whether they contain self-reported exact age information. We explored various transformer-based models, including RoBERTa-Base, BERT-Base, BiLSTM, and Flan T5 Base aiming to identify the most effective model for this binary classification task. Our results demonstrate that the RoBERTa-Base model outperforms the other models, achieving an F1-score of 0.878, a Precision of 0.899, and a Recall of 0.858.

The superior performance of the transformer-based models underscore their potential for practical applications in public health monitoring, targeted advertising, and demographic research. By leveraging the advanced natural language processing capabilities of transformer-based models, we were able to effectively capture nuanced patterns in text, thereby improving the accuracy of age classification tasks.

While the results are promising, there are still challenges to address, such as the occasional misclassification of tweets with subtle context differences and the token limit of transformer-based models. Our study highlights the importance of model robustness and the need for further research in this area. Future work could focus on developing effective chunking methods for input text to improve the classification accuracy and dealing with implicit age information and diverse linguistic constructs.

Overall, this study demonstrates the efficacy of transformer-based models in classifying social media posts based on the presence of self-reported age information, providing a foundation for more advanced analysis of user-generated content.

References

Hyung Won Chung, Le Hou, Shayne Longpre, Barret Zoph, Yi Tay, William Fedus, Eric Li, Xuezhi Wang, Mostafa Dehghani, Siddhartha Brahma, Albert Webson, Shixiang Shane Gu, Zhuyun Dai, Mirac Suzgun, Xinyun Chen, Aakanksha Chowdhery, Sharan Narang, Gaurav Mishra, Adams Yu, Vincent Zhao, Yanping Huang, Andrew Dai, Hongkun Yu, Slav Petrov, Ed H. Chi, Jeff Dean, Jacob Devlin, Adam Roberts, Denny Zhou, Quoc V. Le, and Jason Wei. 2022. Scaling instruction-finetuned language models. *arXiv preprint*.

Daniel Claeser and Samantha Kent. 2022. Fraunhofer FKIE @ SMM4H 2022: System description for shared tasks 2, 4 and 9. In *Proceedings of The Seventh Workshop on Social Media Mining for Health Applications, Workshop & Shared Task*, pages 103–107, Gyeongju, Republic of Korea. Association for Computational Linguistics.

Jacob Devlin, Ming-Wei Chang, Kenton Lee, and Kristina Toutanova. 2019. Bert: Pre-training of deep bidirectional transformers for language understanding. *Preprint*, arXiv:1810.04805.

Alex Graves, Abdel-rahman Mohamed, and Geoffrey Hinton. 2013. Speech recognition with deep recurrent neural networks.

Katikapalli Subramanyam Kalyan, Ajit Rajasekharan, and Sivanesan Sangeetha. 2021. Ammus : A survey of transformer-based pretrained models in natural language processing. *Preprint*, arXiv:2108.05542.

Keshav Kapur, Rajitha Harikrishnan, and Sanjay Singh. 2022. MaNLP@SMM4H'22: BERT for classification of Twitter posts. In *Proceedings of The Seventh Workshop on Social Media Mining for Health Applications, Workshop & Shared Task*, pages 42–43, Gyeongju, Republic of Korea. Association for Computational Linguistics.

Diederik P. Kingma and Jimmy Ba. 2017. Adam: A method for stochastic optimization. *Preprint*, arXiv:1412.6980.

Ari Klein, Arjun Magge, and Graciela Gonzalez. 2022. Reportage: Automatically extracting the exact age of twitter users based on self-reports in tweets. *PLOS ONE*, 17:e0262087.

Yinhan Liu, Myle Ott, Naman Goyal, Jingfei Du, Mandar Joshi, Danqi Chen, Omer Levy, Mike Lewis, Luke Zettlemoyer, and Veselin Stoyanov. 2019. Roberta: A robustly optimized bert pretraining approach. *Preprint*, arXiv:1907.11692.

Atnafu Lambebo Tonja, Olumide Ebenezer Ojo, Mohammed Arif Khan, Abdul Gafar Manuel Meque, Olga Kolesnikova, Grigori Sidorov, and Alexander Gelbukh. 2022. CIC NLP at SMM4H 2022: a BERT-based approach for classification of social media forum posts. In *Proceedings of The Seventh Workshop on Social Media Mining for Health Applications, Workshop & Shared Task*, pages 58–61, Gyeongju, Republic of Korea. Association for Computational Linguistics.

Davy Weissenbacher, Juan Banda, Vera Davydova, Darryl Estrada Zavala, Luis Gasco Sánchez, Yao Ge, Yuting Guo, Ari Klein, Martin Krallinger, Mathias Leddin, Arjun Magge, Raul Rodriguez-Esteban, Abeed Sarker, Lucia Schmidt, Elena Tutubalina, and Graciela Gonzalez-Hernandez. 2022. Overview of the seventh social media mining for health applications (#SMM4H) shared tasks at COLING 2022. In *Proceedings of The Seventh Workshop on Social Media Mining for Health Applications, Workshop & Shared Task*, pages 221–241, Gyeongju, Republic of Korea. Association for Computational Linguistics.

Thomas Wolf, Lysandre Debut, Victor Sanh, Julien Chaumond, Clement Delangue, Anthony Moi, Pierric Cistac, Tim Rault, Rémi Louf, Morgan Funtowicz, Joe Davison, Sam Shleifer, Patrick von Platen, Clara Ma, Yacine Jernite, Julien Plu, Canwen Xu, Teven Le Scao, Sylvain Gugger, Mariama Drame, Quentin Lhoest, and Alexander M. Rush. 2020. Transformers: State-of-the-art natural language processing. In *Proceedings of the 2020 Conference on Empirical Methods in Natural Language Processing: System Demonstrations*, pages 38–45, Online. Association for Computational Linguistics.

Dongfang Xu, Guillermo Lopez Garcia, Lisa Raithel, Rolland Roller, Philippe Thomas, Eiji Aramaki, Shuntaro Yada, Pierre Zweigenbaum, Sai Tharuni Samineni, Karen O'Connor, Yao Ge, Sudeshna Das, Abeed Sarker, Ari Klein, Lucia Schmidt, Vishakha Sharma, Raul Rodriguez-Esteban, Juan Banda, Ivan Flores Amaro, Davy Weissenbacher, and Graciela Gonzalez-Hernandez. 2024. Overview of the 9th social media mining for health applications (#SMM4H) shared tasks at ACL 2024. In *Proceedings of The 9th Social Media Mining for Health Research and Applications Workshop and Shared Tasks*, Bangkok, Thailand. Association for Computational Linguistics.

SMM4H'24 Task6 : Extracting Self-Reported Age with LLM and BERTweet: Fine-Grained Approaches for Social Media Text

Jaskaran Singh, Jatin Bedi,Maninder Kaur
Thapar Institute of Engineering and Technology
jsingh7_be21@thapar.edu, jatin.bedi@thapar.edu, manindersohal@thapar.edu

Abstract

The paper presents two distinct approaches to Task 6 of the SMM4H'24 workshop: extracting self-reported exact age information from social media posts across platforms. This research task focuses on developing methods for automatically extracting self-reported ages from posts on two prominent social media platforms: Twitter (now X) and Reddit. The work leverages two ways, one Mistral-7B-Instruct-v0.2 Large Language Model (LLM) and another pre-trained language model BERTweet, to achieve robust and generalizable age classification, surpassing limitations of existing methods that rely on predefined age groups. The proposed models aim to advance the automatic extraction of self-reported exact ages from social media posts, enabling more nuanced analyses and insights into user demographics across different platforms.

1 Introduction

The widespread use of social media platforms by individuals across demographics offers a unique opportunity to gain valuable insights into their health experiences and perspectives. Effectively leveraging social media data for research purposes necessitates the development of methods for automatically extracting demographic information, such as user age, with high accuracy. Existing methods for identifying user age on social media platforms often rely on categorizing users into predefined age groups. As computational analysis offers new possibilities for investigating complex subjects through social media data, models are being created to automatically identify demographic information, such as the age of users(Klein et al., 2022) (Sadeghi et al., 2024). Many studies have tackled age detection through automated methods, primarily using binary or multi-class classification of predetermined age categories(Chew et al., 2021). These approaches typically involve identifying users' ages from their posts, profile information, or external data sources and subsequently predicting their age groups based on various factors such as their social media activity profile details or a combination of both (Morgan-Lopez et al., 2017); however, the diversity and inconsistency in the number and scope of age groups used across studies indicate that these methods may not be universally applicable to all scenarios. Accurately pinpointing the exact age of social media users, instead of placing them into broad age categories, would enable the extensive use of social media data for applications needing precise age information. This would be particularly beneficial for identifying specific age-related risk factors in observational studies, which current binary or multi-class models fail to address. This work applies two strategies, one using Mistral-7B-Instruct-v0.2 Large Language Model (LLM) and another pre-trained language model BERTweet for automatically extracting self-reported exact age information from social media posts on two prominent platforms: Twitter and Reddit. The proposed models aim to address the issues of existing methods by directly identifying the user's exact age as expressed in the text, enabling the large-scale utilization of social media data for a wider range of research applications. `\documentclass` declaration and before `\begin{document}`) using `\usepackage{graphicx}`.

2 Data description

The training dataset consists of 8,800 labeled tweets and 100,000 unlabeled Reddit posts, primarily focused on age-related information. The labeled tweets indicate whether the user's exact age could be determined from the text, with "1" indicating explicit or inferred age and "0" otherwise. Validation set includes 2,200 labeled tweets and 1,000 Reddit posts about dry eye disease, while the testing set includes 2,200 labeled tweets, 2,000

Proceedings of The 9th Social Media Mining for Health Applications (#SMM4H) Workshop and Shared Tasks, pages 106–109
August 15, 2024 ©2024 Association for Computational Linguistics

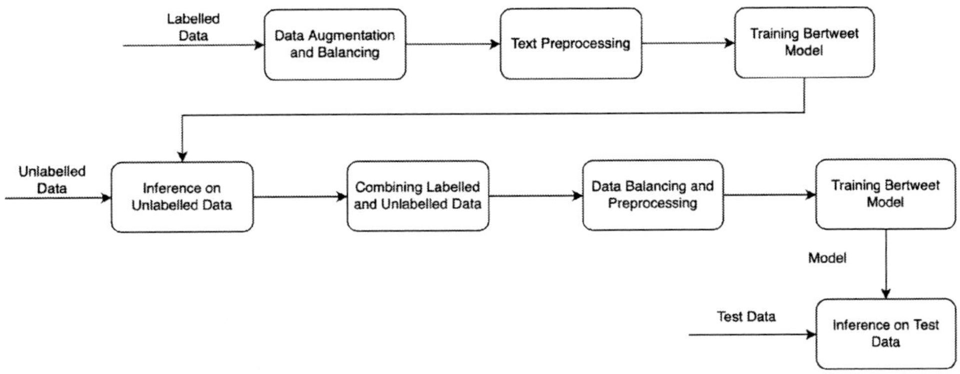

Figure 1: Flowchart of the methodology using BERTweet

Reddit dry eye disease posts, and 12,482 Reddit posts about social anxiety among individuals aged 13-25.

3 Methodology using LLM

This code utilizes a pre-trained LLM called "mistralai/Mistral-7B-Instruct-v0.2" (Jiang et al., 2023)from the Hugging Face library to classify social media posts based on whether the user's age can be inferred from the text content. The main steps are:

3.1 Preprocessing

Initially, preprocess the unlabeled_testing data by applying the *normalizeTweet()* function to each entry in the "text" column. This step ensured uniformity and prepared the text content for subsequent analysis.

3.2 Defining the Instructions for the LLM

The LLM instructions were defined through a string variable named "prompt." This string contains the instructions for the LLM. It explains the task (predicting age from social media posts) and provides examples of positive and negative cases for age reveal.

3.3 Building the Prediction Function

Define a function named y_pred.It creates a conversation-like structure with three parts: <User role: Provides the prompt explaining the task, Assistant role: Acknowledges understanding of the task. User role again: Provides the actual social media post enclosed in square brackets ([start] and [end]).> Apply the tokenizer to the conversation structure to convert it into a format suitable for the LLM (numerical representations). Generating a response from the LLM using the encoded conversation: max_new_tokens=1000 limits the generated response length. do_sample=True enables random sampling for potentially diverse outputs. Decodes the generated tokens back into human-readable text. Extracts the predicted label ("positive" or "negative") from the generated text, likely by splitting by "[/INST]." Use a try-except block to handle potential errors. Loop through each text entry (t) in the "text" column inside the try block. Append the predicted label from the LLM to a list y_pred.Inside the except block: If an error occurs, print the error message and "error occurred."

3.4 Creating the Output DataFrame

Create a DataFrame test with a " label " column containing the predicted labels from y_pred. Saves the DataFrame test to a CSV file named "output-test.csv."

4 Methodology using BERTweet

For the second attempt, a BERTweetbase (Nguyen et al., 2020) implemented using the Huggingface toolkit(Wolf et al., 2019) was exploited to extract the exact age from social media posts. The flow chart of the methodology using BERTweet is presented in Figure 1. The main steps of this approach are:

4.1 Balance the given labeled training dataset

The given labeled training data named “labeled_training.csv” comprising 8800 samples is imbalanced with 5965 for 0 labels and 2834 for 1 label. The following sequence of operations is applied to balance this data.

Model Used	Dataset	F1-score	Precision	Recall
Mistral-7B-Instruct-v0.2 (LLM)	validation data	0.725	0.773	0.835
	test data	0.793	0.716	0.889
BERTweet	Validation data	0.880	0.920	0.850
	test data	0.900	0.916	0.884

Table 1: Performance of our models on the validation and test sets for Task 6

4.1.1 Data augmentation

The *augmentation()* function is defined to perform data augmentation. In this case, it uses back-translation augmentation from English to German and back to English using the *naw.back_translation.BackTranslationAug* module from nlpaug. The code performs data augmentation using back translation to increase the diversity of the labeled training data, especially for the positive class (label = 1). this augmented text data is saved to a CSV file named augmented_text.csv

4.1.2 Concatenating Data

The two datasets, i.e., from CSV files named "labeled_training.csv" and augmented_text.csv, are concatenated together. This concatenates the original labeled training dataset and the augmented text dataset vertically (stacking them on top of each other) to create a single concatenated dataset.

4.1.3 Undersampling

Undersample the concatenated dataset using *RandomUnderSampler()* from the *learn* library to balance the class distribution by randomly removing instances from the majority class (label = 0) until both classes have an equal number of samples. The sampling strategy ensures that both classes have 5000 samples each. The balanced dataset is saved to a CSV file named balanced_data.csv.
Preprocess balanced labeled training dataset and labeled validation dataset

4.1.4 Tweet Normalization

The next step first preprocess the text data in both the training and validation sets (i.e., CSV files, 'balanced_data.csv' and 'labeled_validation.csv') using the *normalizeTweet()* function from the TweetNormalizer module. This function cleans up and standardizes the text data by handling tasks such as lowercasing, URL removal, punctuation removal, and other text normalization tasks specific to social media text.

4.1.5 Tokenization

After Tweet Normalization, the text data in both the training and validation sets is tokenized using the BERTweet tokenizer (AutoTokenizer.from_pretrained (checkpoint)). This tokenizer is specifically designed for tweet text and handles the tokenization of tweets, including special characters, hashtags, and mentions.

4.1.6 Preprocess unlabelled dataset

Preprocess unlabeled dataset using Tweet Normalization and tokenization as done previously.

4.1.7 Model Training

The model was trained using the balanced labeled training and validation datasets.
The trained model is applied to the tokenized unlabeled dataset for labeling it. The result is stored in the csv file 'unlabeled_labeled.csv'

4.1.8 Random undersampling

Random undersampling is performed on the CSV file 'unlabeled_labeled.csv' dataset to balance using RandomUnderSampler from imblearn.

4.1.9 Concatenating the data

The next step involves the concatenation of data of original "labeled_training.csv" and 'unlabeled_labeled.csv' and is saved as final_train.csv

4.1.10 Preprocessing

Preprocess this final training dataset('final_train.csv') and the validation dataset (labeled_validation.csv).using Tweet Normalization and tokenization as done previously

4.1.11 Train model and Model Evaluation

The BERTweet model is loaded, and the tokenizer is initialized. The model is trained on the training dataset and is used to make predictions on the validation dataset. The predictions are then evaluated using classification_report from sklearn.metrics.The trained model is evaluated on the validation dataset to assess its performance.

5 System description

For the first methodology, the pre-trained LLM 'Australia/ Mistral-7B-Instruct-v0.2' was configured to reduce memory usage (potentially using 4-bit weights) by setting load_in_4bit=True.". The Language Model (LLM) generates a response from the encoded conversation, with the maximum length of the response limited to 1000 tokens by setting max_new_tokens=1000. Additionally, enabling do_sample=True allows for random sampling, potentially yielding diverse outputs.
For the second strategy, the textual data was pre-processed using the *normalizeTweet()* function from the TweetNormalizer module. Following Tweet Normalization, text data in both the training and validation sets undergo tokenization using the BERTweet tokenizer (AutoTokenizer.from_pretrained(checkpoint)) with a maximum sequence length of 512 tokens. We optimized the model using the BERTweet model's training parameters as *Experiment name:* between-test, *learning rate*: 2e-5, *number of training epochs*: 5, *Checkpoint saving strategy*: No checkpoints saved, and *batch size per device as* 64. The experiment used Nvidia A100 GPU with Python as the programming language.

6 Results

The performance of the final trained models was assessed using the Task 6 validation set prior to evaluating and submitting the prediction file on the test set. The performance of the models was evaluated using the F1-score for the positive class (i.e., posts annotated as "1"). The results of the validation set for Mistral-7B-Instruct-v0.2 (LLM) and BERTweet–base are reported in Table 1.As indicated in Table 1, the top-performing model on the Task 6 validation set is the BERTweet-base model, which achieved an F1-score of 0.880. BERTweet-base model showed better results in predicting positive class(1) than Mistral-7B-Instruct-v0.2 (LLM). When evaluating the performance of the models on the test set, the BERTweet–base model achieved an F1-score of 0.90 on the test set, with precision and recall values of 0.916 and 0.884, as seen in Table 1. These findings highlight the superior performance of the BERTweet-base model in accurately predicting the positive class (label "1") compared to Mistral-7B-Instruct-v0.2 (LLM) on both the validation and test datasets for Task 6.

7 Conclusion

This work uses Mistral-7B-Instruct and BERTweet models for precise age extraction from social media posts. The limitations of traditional approaches were overcome by directly inferring user age from text. Evaluation of validation and test sets reveals significant performance disparities between the two models. BERTweetbase outperforms Mistral-7B-Instruct with an F1-score of 0.90 for the positive class on the test set, showcasing its superior efficacy in age prediction. These results underscore the efficacy of advanced language models in enhancing demographic analysis and research applications. In future work, refining prompt engineering techniques and utilizing advanced models such as Llama 3 can be exploited to enhance performance. Additionally, exploring ensembling methods with models like Bernie can offer even greater accuracy and robustness in demographic predictions.

References

Robert Chew, Caroline Kery, Laura Baum, Thomas Bukowski, Annice Kim, Mario Navarro, et al. 2021. Predicting age groups of reddit users based on posting behavior and metadata: classification model development and validation. *JMIR Public Health and Surveillance*, 7(3):e25807.

Albert Q Jiang, Alexandre Sablayrolles, Arthur Mensch, Chris Bamford, Devendra Singh Chaplot, Diego de las Casas, Florian Bressand, Gianna Lengyel, Guillaume Lample, Lucile Saulnier, et al. 2023. Mistral 7b. *arXiv preprint arXiv:2310.06825*.

Ari Z Klein, Arjun Magge, and Graciela Gonzalez-Hernandez. 2022. Reportage: Automatically extracting the exact age of twitter users based on self-reports in tweets. *PloS one*, 17(1):e0262087.

Antonio A Morgan-Lopez, Annice E Kim, Robert F Chew, and Paul Ruddle. 2017. Predicting age groups of twitter users based on language and metadata features. *PloS one*, 12(8):e0183537.

Dat Quoc Nguyen, Thanh Vu, and Anh Tuan Nguyen. 2020. Bertweet: A pre-trained language model for english tweets. *arXiv preprint arXiv:2005.10200*.

Reyhaneh Sadeghi, Ahmad Akbari, and Mohammad Mehdi Jaziriyan. 2024. Exaauac: Arabic twitter user age prediction corpus based on language and metadata features.

Thomas Wolf, Lysandre Debut, Victor Sanh, Julien Chaumond, Clement Delangue, Anthony Moi, Pierric Cistac, Tim Rault, Rémi Louf, Morgan Funtowicz, et al. 2019. Huggingface's transformers: State-of-the-art natural language processing. *arXiv preprint arXiv:1910.03771*.

AAST-NLP@#SMM4H'24: Finetuning Language Models for Exact Age Classification and Effect of Outdoor Spaces on Social Anxiety

Ahmed El-Sayed and **Omar Nasr** and **Noha S. Tawfik**
Arab Academy for Science, Technology & Maritime Transport
{ahmedelsayedhabashy,omarnasr52}@gmail.com, noha.abdelsalam@aast.edu

Abstract

This paper evaluates the performance of "AAST-NLP" in the Social Media Mining for Health (SMM4H) Shared Tasks 3 and 6, where more than 20 teams participated in each. We leveraged state-of-the-art transformer-based models, including Mistral, to achieve our results. Our models consistently outperformed both the mean and median scores across the tasks. Specifically, an F1-score of **0.636** was achieved in classifying the impact of outdoor spaces on social anxiety symptoms, while an F1-score of **0.946** was recorded for the classification of self-reported exact ages

1 Introduction

The widespread use of social media platforms has encouraged the employment of Natural Language Processing (NLP) in all domains, specifically in health-related applications (Correia et al., 2020). These platforms are considered unfiltered, real-time, valuable sources for analysis As millions of users openly share their personal health narratives and experiences on a daily basis. The Social Media Mining for Health Applications (SMM4H-2024) workshop provides a chance to develop natural language processing (NLP) models that automatically extract meaningful information from social media data through seven shared tasks in various contexts, including pharmaceutical, social, health, and clinical. In this paper, we describe our team's participation in both Task 3, Self-reported Exact Age Classification, and Task 6, Multi-class Classification of the Effects of Outdoor Spaces on Social Anxiety Symptoms in Reddit. The former task explores the extraction of patient demographics, enabling large-scale observational studies. Similarly, the latter investigates outdoor spaces mentioned in social media and aims to qualitatively assess their effects and relation to Social Anxiety Disorder (SAD).

2 Tasks & Datasets Description

2.1 Task 3

Social Media Mining for Health (SMM4H-24) (Xu et al., 2024) organized a total of 7 tasks, each with its own theme targeted at social media mining tasks. Task 3 involves categorizing posts mentioning specific outdoor keywords into four groups based on their effects on social anxiety symptoms: positive effect, neutral or no effect, negative effect, or unrelated mentions. The dataset includes posts from r/socialanxiety subreddit, filtered for users aged 12-25 and mentions 80 green/blue space keywords. The dataset consists of 1800 training examples, 600 validation examples and 600 testing examples. The training dataset was severely imbalanced with 1131 being labeled as unrelated, 395 as neutral, 160 as positive and 114 negative.

2.2 Task 6

Task 6 involves classifying social media posts according to self exact-age reporting.This task expands the scope of social media data usage by accurately extracting the precise ages reported by users rather than categorizing them into ranges of age groups as is typically done. This task builds upon the tasks established in SMM4H 2022 (Weissenbacher et al., 2022), which addressed a similar thematic task. The key distinction lies in adapting this year's task towards cross-platform evaluation, thereby extending the scope and applicability of the original concept. The dataset includes posts from X (previously Twitter) and Reddit. In addition to the annotated datasets used in training, validation and testing phases, an additional 100000 unlabeled Reddit posts from r/AskDocs where provided. The provided dataset is imbalanced, with 5966 posts provided having no exact age mention and 2834 posts with exact age mentions. The validation set follows a similar distribution with 2435 and 1765 posts for the respective classes.

Proceedings of The 9th Social Media Mining for Health Applications (#SMM4H) Workshop and Shared Tasks, pages 110–113
August 15, 2024

3 Proposed System

3.1 Data Preprocessing

3.1.1 Task 3

The social media data in task 3 consisted of multi-sentence posts, many of which were not relevant to our objective. This required preprocessing the data to filter meaningful sentences. Consequently, we investigated three distinct methodologies for sentence filtering. These methodologies encompassed isolating solely the sentence containing the specified keyword, retrieving the sentence featuring the specified keyword along with two additional sentences (one preceding and one succeeding it), and extracting two preceding and two succeeding sentences in conjunction with the sentence containing the keyword. Our objective was to extract the sentence containing the necessary context to classify it into the correct category. Adding extra sentences, besides those containing the specified keyword, aimed to provide additional context that could enhance our model's capability. Our findings show that utilizing solely the sentence containing the keyword yielded superior performance in both F1-Score and training time. Moreover, we eliminated emojis, hyperlinks, punctuation, extra spaces, and line breaks.

3.1.2 Task 6

Similar to task 3, the posts included numerous sentences lacking relevance to our designated task. Opposed to experimenting with different sentence combinations, we opted to extract only the sentences containing digits as these were most likely to include age information. The common abbreviations used on social media, such as "yo" for years old and "bday" for birthday," were addressed passively. Additionally, certain social media patterns were challenging for the model to understand and required further resolution for training. These included:

- Samples starting with words matching the pattern `(\d+)([FMfm])` indicated a male or female mentioning their age.
- Some samples contained a reversed version of this pattern, so `([FMfm])(\d+)` was used instead.
- Samples with the pattern `(\d+)y` implied a person stating their age in years.
- Samples contained `(\d{2})s` implied an age group rather than an exact age mention yet it was often misclassified so they were replaced by their written counterparts foe examples "20s" would be replaced by twenties.

Finally, All of the posts had their emojis, hyperlinks and punctuation marks removed. Task 6 organizers provided an extra substantial volume of unlabeled Reddit posts with sentences that included 2-digit numbers. We employed a rule-based approach to label the provided in order to increase the size of our training dataset. The rules used were directly extracted from the task annotation guidelines and domain knowledge. For instance, any sentence that had a digit that matched the regex pattern `(\d+)([FMfm])` would be labeled as containing an exact age mention. The Reddit platform, in particular, has a number of abbreviations that correspond to exact-age mentions such as "21F" (which would man a 21 years old female). Those abbreviations were resolved through the regex patterns. The abbreviations were resolved using regex through the following patterns `(\d+)([FMfm])`, `([FMfm])(\d+)`, `(\d+)([FMfm])` and `(\d+)(?:\.,|;)`. Additionally, emojis, punctuation, and hyperlinks were removed.

3.2 Language Models

For both tasks, multiple Language Models renowned for their State of the Art (SOTA) performance on Natural Language Processing (NLP) tasks were experimented with, including RoBERTa (Liu et al., 2019), BERTweet (Nguyen et al., 2020), and XLM-RoBERTa (Conneau et al., 2020). Dice Loss (Li et al., 2020) was utilized in fine-tuning our language models due to their proven performance in tackling NLP tasks in recent studies. It is particularly effective when handling imbalanced datasets because it emphasizes the correct classification of underrepresented classes and maximizes the overlap between predicted probabilities and actual outputs. This leads to improved model performance, where minority classes are crucial compared to other loss functions. For task 6, We also experimented with Mistral-7b (Jiang et al., 2023) through HuggingFace's quantized version.[1]. To achieve better performance with Mistral; we employed prompt engineering, which involves, in other words, crafting strategic prompts to guide AI systems, ensuring effective outputs (Chen et al., 2023). It includes

[1]https://huggingface.co/unsloth/mistral-7b-bnb-4bit

selecting language, context, and constraints to generate relevant responses. Several schemes were tested, but the one used for the testing set was based on Chain-of-Thought (Wei et al., 2023) by feeding the language model samples from the training and validation sets following an iterative approach to avoid misclassified examples.

3.3 Hyperparamters and Training

The hyperparameters reported in table 1 represent those employed in training the model used to submit the final results for both tasks

Hyperparameter	Task 3	Task 6
Epochs	30	30
Learning Rate	2e-5	1e-5
Batch Size	8	16
Max length	200	80
Optimizer	Adam	Adam
Early Stopping Patience	5	5
Reduce On Plateau	2	2
Loss Function	Dice Loss	Dice Loss

Table 1: Training Hyperparameters. Parameters shown for RoBERTa and BERTweet-Large/RoBERTa-Large for tasks 3 and 6, respectively.

The training procedure was conducted using Kaggle's [2] free-to-use platform, which provides 29 GB of RAM, a 16 GB NVIDIA P100 GPU, and Python. The autofit functionality from ktrain (Maiya, 2022) was utilized, incorporating a triangular learning rate policy (Smith, 2017).

4 Results

4.1 Task 3

Our experiments concluded that Roberta-Large[3] had the best performance of the three models when evaluated during the validation stage and was then used in the testing phase.

Table 2 illustrates our model's performance on the test set. Our F1 and recall results exceed the mean and median of the other submissions by a large margin, showcasing our model's robustness and efficiency. Our model's precision also surpasses the mean and median, though by a narrower margin.

	Precision	Recall	F1-Score
Validation	0.68	0.65	0.66
Mean	0.5649	0.5379	0.5186
Median	0.63	0.5885	0.5795
RoBERTa	**0.631**	**0.644**	**0.635**

Table 2: Task 3 Results.

4.2 Task 6

For task 6, the results revealed that BERTweet exhibited superior performance compared to the other two models, An Ensemble of BERTweet and RoBERTa were used in the testing phase. Moreover, high results were achieved through prompt engineering in our zero-shot experiments with Mistral despite not utilizing any pre-training.

Ensembling via majority voting was used for the final submission on the test set, the three models used were all given equal weights. Table 3 illustrates the results obtained on the test set.

	Precision	Recall	F1-Score
Validation	0.93	0.96	0.94
Mean	0.924	0.926	0.924
Median	**0.934**	0.949	0.936
Ensemble	0.932	**0.959**	**0.946**

Table 3: Task 6 Results.

5 Error Analysis

In our error analysis of the validation set for Task 3, we observed that one of the main reasons for performance degradation was the misclassification of unrelated and neutral instances. Upon checking samples of these instances manually, we identified that there exist some examples to be mislabelled or at least confusing even on the human level. Eliminating such ambiguous data in following versions of the task will definitely increase the training data quality and reflect such quality on the models' performance. Additionally, in Task 6, mislabeled examples in both the training and validation sets pose a risk to the effectiveness of the training and evaluation processes. One clear example of that is the post with ID: 8812 in the validation set which was wrongfully classified as having no exact age mention.

[2]https://www.kaggle.com/

[3]https://huggingface.co/FacebookAI/roberta-large

6 Conclusion

In this work, approaches were presented to address the challenges of self-reported exact age classification and classification of effects of outdoor spaces on social anxiety symptoms. For the first task, an ensemble of RoBERTa, BERTweet, and Mistral was employed to handle the problem. Despite the achieved performance, it is believed that there is room for improvement in the approach in the future. Areas of potential improvement include refining the preprocessing procedure and fine-tuning Mistral language model.

References

Banghao Chen, Zhaofeng Zhang, Nicolas Langrené, and Shengxin Zhu. 2023. Unleashing the potential of prompt engineering in large language models: a comprehensive review. *Preprint*, arXiv:2310.14735.

Alexis Conneau, Kartikay Khandelwal, Naman Goyal, Vishrav Chaudhary, Guillaume Wenzek, Francisco Guzmán, Edouard Grave, Myle Ott, Luke Zettlemoyer, and Veselin Stoyanov. 2020. Unsupervised cross-lingual representation learning at scale. *Preprint*, arXiv:1911.02116.

Rion Brattig Correia, Ian B. Wood, Johan Bollen, and Luís M. Rocha. 2020. Mining social media data for biomedical signals and Health-Related behavior. *Annual review of biomedical data science*, 3(1):433–458.

Albert Q. Jiang, Alexandre Sablayrolles, Arthur Mensch, Chris Bamford, Devendra Singh Chaplot, Diego de las Casas, Florian Bressand, Gianna Lengyel, Guillaume Lample, Lucile Saulnier, Lélio Renard Lavaud, Marie-Anne Lachaux, Pierre Stock, Teven Le Scao, Thibaut Lavril, Thomas Wang, Timothée Lacroix, and William El Sayed. 2023. Mistral 7b. *Preprint*, arXiv:2310.06825.

Xiaoya Li, Xiaofei Sun, Yuxian Meng, Junjun Liang, Fei Wu, and Jiwei Li. 2020. Dice loss for data-imbalanced nlp tasks. *Preprint*, arXiv:1911.02855.

Yinhan Liu, Myle Ott, Naman Goyal, Jingfei Du, Mandar Joshi, Danqi Chen, Omer Levy, Mike Lewis, Luke Zettlemoyer, and Veselin Stoyanov. 2019. Roberta: A robustly optimized bert pretraining approach. *Preprint*, arXiv:1907.11692.

Arun S. Maiya. 2022. ktrain: A low-code library for augmented machine learning. *Preprint*, arXiv:2004.10703.

Dat Quoc Nguyen, Thanh Vu, and Anh Tuan Nguyen. 2020. Bertweet: A pre-trained language model for english tweets. *Preprint*, arXiv:2005.10200.

Leslie N. Smith. 2017. Cyclical learning rates for training neural networks. *Preprint*, arXiv:1506.01186.

Jason Wei, Xuezhi Wang, Dale Schuurmans, Maarten Bosma, Brian Ichter, Fei Xia, Ed Chi, Quoc Le, and Denny Zhou. 2023. Chain-of-thought prompting elicits reasoning in large language models. *Preprint*, arXiv:2201.11903.

Davy Weissenbacher, Juan Banda, Vera Davydova, Darryl Estrada Zavala, Luis Gasco Sánchez, Yao Ge, Yuting Guo, Ari Klein, Martin Krallinger, Mathias Leddin, Arjun Magge, Raul Rodriguez-Esteban, Abeed Sarker, Lucia Schmidt, Elena Tutubalina, and Graciela Gonzalez-Hernandez. 2022. Overview of the seventh social media mining for health applications (#SMM4H) shared tasks at COLING 2022. In *Proceedings of The Seventh Workshop on Social Media Mining for Health Applications, Workshop & Shared Task*, pages 221–241, Gyeongju, Republic of Korea. Association for Computational Linguistics.

Dongfang Xu, Guillermo Lopez Garcia, Lisa Raithel, Rolland Roller, Philippe Thomas, Eiji Aramaki, Shuntaro Yada, Pierre Zweigenbaum, Sai Tharuni Samineni, Karen O'Connor, Yao Ge, Sudeshna Das, Abeed Sarker, Ari Klein, Lucia Schmidt, Vishakha Sharma, Raul Rodriguez-Esteban, Juan Banda, Ivan Flores Amaro, Davy Weissenbacher, and Graciela Gonzalez-Hernandez. 2024. Overview of the 9th social media mining for health (#SMM4H) research and applications workshop and shared tasks at ACL 2024. In *Proceedings of The 9th Social Media Mining for Health Research and Applications Workshop and Shared Tasks*, Bangkok, Thailand. Association for Computational Linguistics.

CogAI@SMM4H 2024: Leveraging BERT-based Ensemble Models for Classifying Tweets on Developmental Disorders

Liza Dahiya and **Rachit Bagga**
Computer Science & Engineering
Indian Institute of Technology, Bombay
{lizadahiya, rachitbagga}@cse.iitb.ac.in

Abstract

This paper presents our work for the Task 5 of the Social Media Mining for Health Applications 2024 Shared Task - Binary classification of English tweets reporting children's medical disorders. In this paper, we present and compare multiple approaches for automatically classifying tweets from parents based on whether they mention having a child with attention-deficit/hyperactivity disorder (ADHD), autism spectrum disorders (ASD), delayed speech, or asthma. We use ensemble of various BERT-based models trained on provided dataset that yields an F1 score of **0.901** on the test data.

1 Introduction

The prevalence of child developmental disorders, including attention-deficit/hyperactivity disorder (ADHD), autism spectrum disorders (ASD), delayed speech, and asthma, presents significant challenges for families and healthcare systems worldwide. Understanding the experiences and needs of parents raising children with these conditions is crucial for informing support interventions and public health policies. In response to this challenge, social media data can be leveraged (Kim et al., 2020) after classifying tweets from parents mentioning if their child has one of these developmental disorders. This task holds significant promise for advancing our understanding of the prevalence and correlates of these conditions, particularly in relation to pregnancy exposures. Some studies have been done previously to assess the relationship between prenatal exposure to various substances during pregnancy to the risk of developing these disorders but many had low statistical power (Castro et al., 2016; Linnet et al., 2003). In this paper, we present a comprehensive analysis of multiple approaches for classifying tweets related to parental experiences with ADHD, ASD, delayed speech, and asthma. Our study focuses on a binary classification task, aiming to discern whether tweets mention these developmental disorders or not. Through this work, we aim to contribute to the advancement of epidemiologic research (Cortese and Coghill, 2018; Rowland et al., 2002) and provide valuable insights into parental experiences with developmental disorders.

In this paper, we provide a detailed explanation of our system architecture, describe our experiments and share results obtained during this task.

2 Methods

2.1 Dataset

The dataset provided was divided into three partitions: training, validation, and test sets consisting of 7398, 389, and 1947 tweets respectively. Tweets that describe a child with ADHD, ASD, delayed speech, or asthma are annotated **1**, and those that simply mention a disorder are annotated **0**.

Text	Label
Finally a dr has diagnosed my 3.5yr old with asthma. Now he will be on chronic medicine and we can hopefully keep him healthy and thriving.	1
Can u give any tips to "live with it" please. I think my son has ADD. Trying to help him	0
Flying tomorrow...during a pandemic with a nonverbal 3 year old. We could use some prayers, please.ðŸ˜ðŸ¥´	1

Table 1: Examples of Tweets with Annotations

2.2 Pre-processing

In the preprocessing of the data, two steps were undertaken to ensure the quality & balance of the dataset.

2.2.1 Data Cleaning

Since the twitter data contains large amounts of noise, we applied several data cleaning procedures. This included the removal of URLs, mentions (ex: @USER), and hashtags (ex: #funny). Additionally, handling of emojis was crucial. Emojis were replaced with their corresponding textual representations to remove potentially distracting elements and simplify the text for further analysis.

Proceedings of The 9th Social Media Mining for Health Applications (#SMM4H) Workshop and Shared Tasks, pages 114–116
August 15, 2024 ©2024 Association for Computational Linguistics

Model	F1			Recall			Precision		
	M1	M2	M3	M1	M2	M3	M1	M2	M3
MentalBERT	0.885	0.897	**0.917**	0.854	0.868	0.896	0.896	0.905	0.923
PsychBERT	0.843	0.855	**0.869**	0.821	0.835	0.854	0.866	0.874	0.879
TwitterBERT	0.885	**0.902**	0.897	0.854	0.868	0.896	0.896	0.905	0.923
DistilBERT	0.877	0.885	**0.887**	0.848	0.853	0.864	0.863	0.875	0.882

Table 2: Scores on the validation set of tweets, where *M1* is trained without preprocessing, *M2* is trained with augmented data, and *M3* is trained with cleaned and augmented data.

2.2.2 Data Augmentation

The technique used for data augmentation was "gender-based" text augmentation. Within the texts labelled as **1**, all instances of the term "son" (and corresponding pronouns "he," "his," and "him") were systematically substituted with "daughter" (and corresponding pronouns "she" and "her").The distribution of true labels increased from 30.82% before augmentation to **42.37%.** after augmentation improving the class imbalance situation.

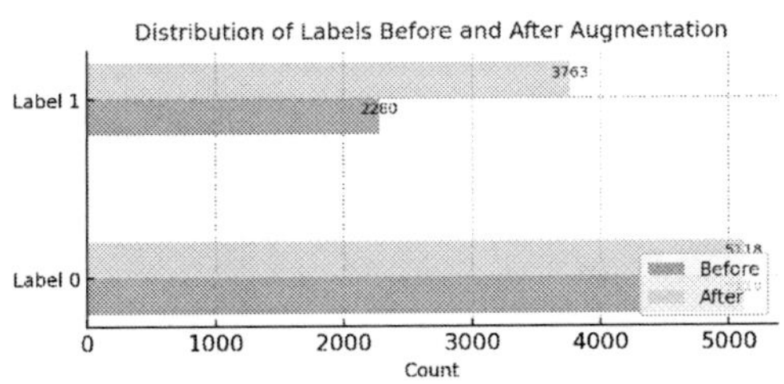

Figure 1: Data Distribution before and after Augmentation

This augmentation technique was straightforward to implement and has shown to improve performance by including additional context.

3 Experiments and Results

During our model exploration, we fine-tuned multiple pre-trained models including MentalBERT (Ji et al., 2022), PsychBERT (Vajre et al., 2021), TwitterBERT (Zhang et al., 2022), and DistilBERT (Sanh et al., 2019). We trained each of them for 200 epochs with a learning rate of 1e-5 and a batch size of 32. Each of these models offers unique advantages that could be beneficial for our purpose.

MentalBERT and PsychBERT are tailored for mental health and biomedical text datasets which could potentially capture features relevant to our task of identifying tweets related to "developmental disorders". TwitterBERT's training on Twitter data makes it adept at handling the informal language of tweets which can improve classification accuracy while DistilBERT's computational efficiency can make it suitable for scalable deployment, an advantage in handling large volumes of Twitter data.

We systematically evaluated the performance of each of these models by adopting an incremental approach. Initially, we trained the models using the raw data without any pre-processing (**M1**). Subsequently, we performed training after augmenting the data (**M2**). Finally, we conducted training using cleaned data with data augmentation (**M3**). This comprehensive approach allowed us to assess the impact of classification performance of each model, providing valuable insights into their robustness and effectiveness for our specific task.

Our final model then was an ensemble of all these methods assigned weights according to their "F1-scores" on validation dataset. We hardcoded weights of 0.6 score to MentalBERT, 0.2 score to TwitterBERT, 0.1 score to each PsychBERT and DistilBERT. This was decided based on their individual performance as can be seen in **Table 2**.

Statistic	F_1-score	Precision	Recall
Our	0.901	0.885	0.917
Mean	0.822	0.818	0.838
Median	0.901	0.885	0.917

Table 3: Performance on Test Data for Task 5

4 Conclusion

In this task, we experimented with various BERT models and presented the results in Table 2. Table 3 presents our results along with the mean and median results obtained during the challenge, depicting a final F1-score of **0.901**. In future, we would like to experiment with more ensemble modelling methods, use better data augmentation techniques and For greater scope of research, a potential direction could involve incorporating multimodal data sources, such as images or audio recordings, to enrich the understanding of parental experiences and enhance the accuracy of classification models.

References

VM Castro, SW Kong, CC Clements, R Brady, AJ Kaimal, AE Doyle, EB Robinson, SE Churchill, IS Kohane, and RH Perlis. 2016. Absence of evidence for increase in risk for autism or attention-deficit hyperactivity disorder following antidepressant exposure during pregnancy: a replication study. *Translational psychiatry*, 6(1):e708–e708.

Samuele Cortese and David Coghill. 2018. Twenty years of research on attention-deficit/hyperactivity disorder (adhd): looking back, looking forward. *BMJ Ment Health*, 21(4):173–176.

Shaoxiong Ji, Tianlin Zhang, Luna Ansari, Jie Fu, Prayag Tiwari, and Erik Cambria. 2022. MentalBERT: Publicly Available Pretrained Language Models for Mental Healthcare. In *Proceedings of LREC*.

Eric S Kim, Peter James, Emily S Zevon, Claudia Trudel-Fitzgerald, Laura D Kubzansky, and Francine Grodstein. 2020. Social media as an emerging data resource for epidemiologic research: characteristics of regular and nonregular social media users in nurses' health study ii. *American Journal of Epidemiology*, 189(2):156–161.

Karen Markussen Linnet, Søren Dalsgaard, Carsten Obel, Kirsten Wisborg, Tine Brink Henriksen, Alina Rodriguez, Arto Kotimaa, Irma Moilanen, Per Hove Thomsen, Jørn Olsen, et al. 2003. Maternal lifestyle factors in pregnancy risk of attention deficit hyperactivity disorder and associated behaviors: review of the current evidence. *American Journal of Psychiatry*, 160(6):1028–1040.

Andrew S Rowland, Catherine A Lesesne, and Ann J Abramowitz. 2002. The epidemiology of attention-deficit/hyperactivity disorder (adhd): a public health view. *Mental retardation and developmental disabilities research reviews*, 8(3):162–170.

Victor Sanh, Lysandre Debut, Julien Chaumond, and Thomas Wolf. 2019. Distilbert, a distilled version of bert: smaller, faster, cheaper and lighter. *ArXiv*, abs/1910.01108.

Vedant Vajre, Mitch Naylor, Uday Kamath, and Amarda Shehu. 2021. Psychbert: a mental health language model for social media mental health behavioral analysis. In *2021 IEEE International Conference on Bioinformatics and Biomedicine (BIBM)*, pages 1077–1082. IEEE.

Xinyang Zhang, Yury Malkov, Omar Florez, Serim Park, Brian McWilliams, Jiawei Han, and Ahmed El-Kishky. 2022. Twhin-bert: A socially-enriched pre-trained language model for multilingual tweet representations. *arXiv preprint arXiv:2209.07562*.

ADE Oracle at #SMM4H 2024: A Two-Stage NLP System for Extracting and Normalizing Adverse Drug Events from Tweets

Andrew S. Davis Billy Dickson Sandra Kübler
Department of Linguistics
Indiana University
{ad7,dicksonb,skuebler}@iu.edu

Abstract

This study describes the approach of Team ADE Oracle for Task 1 of the Social Media Mining for Health Applications (#SMM4H) 2024 shared task. Task 1 challenges participants to detect adverse drug events (ADEs) within English tweets and normalize these mentions against the Medical Dictionary for Regulatory Activities standards. Our approach utilized a two-stage NLP pipeline consisting of a named entity recognition model, retrained to recognize ADEs, followed by vector similarity assessment with a RoBERTa-based model. Despite achieving a relatively high recall of 37.4% in the extraction of ADEs, indicative of effective identification of potential ADEs, our model encountered challenges with precision. We found marked discrepancies between recall and precision between the test set and our validation set, which underscores the need for further efforts to prevent overfitting and enhance the model's generalization capabilities for practical applications.

1 Introduction

This paper outlines Team ADE Oracle's participation in the 9th Social Media Mining for Health Research and Applications (#SMM4H) 2024 workshop's Task 1 (Xu et al., 2024), which involved extracting and normalizing adverse drug events (ADEs) from tweets into Medical Dictionary for Regulatory Activities (MedDRA) terms[1]. The task complexity increased in 2024 by combining ADE detection with normalization, a challenge heightened by the informal and diverse language used on social media (Xu et al., 2024). Addressing ADEs through social media enhances pharmacovigilance, providing critical data for public health interventions (Huynh et al., 2016; Alimova and Tutubalina, 2019; Vydiswaran et al., 2019; Magge et al., 2021; Liu et al., 2022; Lee et al., 2023). Our approach employed a spaCy-based NLP pipeline, retraining a Named Entity Recognition (NER) module to extract ADEs, and a RoBERTa model for aligning text with MedDRA standards (Weissenbacher et al., 2022), navigating the trade-offs between recall and precision. While our system effectively identified many ADEs, the prevalence of false positives points to a need for further refinement to enhance the accuracy and utility of our methods for public health surveillance.

2 Dataset

This study employed the #SMM4H 2024 Task 1 dataset, comprising 30,949 tweets distributed across 18,185 training, 965 validation, and 11,799 test tweets (Klein et al., 2024; Xu et al., 2024).

3 System Description

Our methodology for the #SMM4H 2024 Task 1 involves a two-stage process: We use an NER package to extract ADEs, followed by the normalization of these entities against the MedDRA using vector similarity techniques (Yazdani et al., 2023a,b).

3.1 Preprocessing

For preprocessing the dataset, we implemented a two-step approach to optimize data for training. Initially, all labeled entities representing ADEs were converted to lowercase to ensure consistency and address case discrepancies between labels and their occurrences in the tweet text. Subsequently, we employed the tokenizer[2] from the blank, spaCy "en" model[3] to tokenize the text (Dai et al., 2017).

3.2 NER for ADE Extraction

We chose to use the blank spaCy model "en" for training a customized NER model tailored

[1] `https://www.meddra.org`

[2] `https://spacy.io/api/tokenizer`

[3] `https://spacy.io/usage/models`

Proceedings of The 9th Social Media Mining for Health Applications (#SMM4H) Workshop and Shared Tasks, pages 117–120
August 15, 2024

to the extraction of ADE entities due to its robust handling of English syntax and adaptability to the specialized domain of pharmacovigilance (Dai et al., 2017; Jiang et al., 2022). Specifically, we trained the model's span categorizer[4] component to identify and label ADE spans within tweets effectively. The span categorizer comprises two main components: a suggester function and a labeler model. The suggester function, employing the spacy.ngram_suggester.v1, was selected to propose candidate spans with designated lengths—specifically one to five tokens. These candidates, which may overlap, are presented in a ragged array format comprising two columns that denote the start and end positions of each span. Subsequently, the labeler model evaluates each candidate span, assigning the ADE label as appropriate based on the predictive outcomes.

This model was trained on the 18,185 labeled tweets of the official training set. Optimization was achieved over 49 epochs with a batch size of 8 and a dropout rate of 0.5, selecting the best-performing iteration for our analyses.

3.3 Vectorization and Normalization

For the normalization stage, we employed the base model RoBERTa to vectorize the ADE entities and MedDRA entries (Liu et al., 2019; Gencoglu, 2020; Weissenbacher et al., 2022). We did not further fine-tune the base RoBERTa model, as our focus was solely on utilizing its semantic representation capabilities. We extracted and vectorized ADE entities from the validation set using our span categorizer, and vectorized the textual descriptions of MedDRA adverse event terms. The MedDRA vectors were stored in a vector database from Facebook's Faiss library, which is designed for efficient similarity searching of dense vectors at scale (Johnson et al., 2019; Douze et al., 2024). We then iterated through our extracted entities and used Euclidean distance (L2 distance) to identify the closest match between each ADE entity vector and the MedDRA term vectors in the database.

3.4 Evaluation Metrics

The performance of our NER model in identifying ADEs, along with the pipeline's effectiveness in matching ADEs to MedDRA terms, was evaluated using the official metrics of #SMM4H 2024 Task 1, specifically F1, precision, and recall.

[4] `https://spacy.io/api/spancategorizer`

Task & Metric	F1	P	R
ADE Extraction	16.6	15.1	18.4
ADE Normalization	8.4	7.5	9.4

Table 1: Validation Set Scores for ADE Tasks

Task & Metric	F1	P	R
ADE Extr. official	13.2	8.0	37.4
ADE Norm. official	8.2	5.0	23.7
ADE Norm. unseen IDs	1.4	0.7	10.0

Table 2: Comprehensive Test Set Scores for ADE Tasks

4 Results

Our system consisting of the NER model for ADE extraction and RoBERTa for the normalization task is evaluated in Tables 1 and 2 on the validation set and the official test set, respectively.

4.1 ADE Extraction

Table 2 shows that the ADE extraction model achieved an F1 of 13.2 on the test set, with precision and recall scores of 8.0% and 37.4%, respectively. These results highlight the model's higher success in recall, indicating its effectiveness in identifying ADE mentions. However, the model's low precision of 8.0% highlights a significant challenge in specificity. The model's unexpectedly high recall on the test set compared to the validation set, where recall and precision were more balanced, indicates differences in the distribution and complexity of the validation and test set.

4.2 ADE Normalization

For ADE normalization, the official scores in Table 2 show a precision of 5.0%, a recall of 23.7%, and an F1 of 8.2. Additionally, when evaluating the model's performance on previously unseen MedDRA IDs, it returned significantly lower metrics (precision: 0.7%, recall: 10.0%, F1: 1.4). This considerable drop suggests challenges in generalizing to new, unseen ADE terms, reflecting potential limitations in the model's generalizing capability. The results on the validation set were somewhat consistent, with an F1-score of 8.4 and slightly lower precision and recall of 7.5% and 9.4%, respectively.

4.3 Discussion

Our results point to the challenges inherent in biomedical NLP tasks, especially in balancing precision and recall and generalizing to new data. The low precision observed shows issues with generalizability beyond training data in NER. The results may also reflect the complexities of social media language, complicating ADE detection and normalization.

Moreover, the differences in results between validation and official test data underline the importance of robust cross-validation strategies to mimic real-world performance and prevent overfitting. Further efforts need to focus on integrating domain-specific knowledge bases to heighten normalization accuracy and better manage new ADE identifiers.

5 Conclusion

Our contribution to #SMM4H 2024 Task 1 consists of an NER model retrained to identify ADEs and a similarity-based RoBERTa model to normalize them. The findings from our system underline the challenges and opportunities presented by the use of NLP in detecting and normalizing ADEs from social media. Despite achieving high recall, our model's low precision highlights a significant challenge in accurately identifying relevant ADEs amid the informal language prevalent on platforms like Twitter. Furthermore, the task has demonstrated that while our current methodology is capable of initial identification, it falls short in scenarios involving generalizing to data different from the training data, which is crucial for practical applications.

For future work, we will investigate enhancing model precision through advanced linguistic analysis, employing models pre-tuned on ADE datasets, fine-tuning RoBERTa for vectorization of ADE entities and MedDRA entries, and incorporating additional ADE data.

References

Ilseyar Alimova and Elena Tutubalina. 2019. Detecting adverse drug reactions from biomedical texts with neural networks. In *Proceedings of the 57th Annual Meeting of the Association for Computational Linguistics: Student Research Workshop*, pages 415–421, Florence, Italy. Association for Computational Linguistics.

Xiang Dai, Sarvnaz Karimi, and Cecile Paris. 2017. Medication and adverse event extraction from noisy text. In *Proceedings of the Australasian Language Technology Association Workshop 2017*, pages 79–87, Brisbane, Australia.

Matthijs Douze, Alexandr Guzhva, Chengqi Deng, Jeff Johnson, Gergely Szilvasy, Pierre-Emmanuel Mazaré, Maria Lomeli, Lucas Hosseini, and Hervé Jégou. 2024. The Faiss library. https://arxiv.org/abs/2401.08281.

Oguzhan Gencoglu. 2020. Sentence transformers and Bayesian optimization for adverse drug effect detection from Twitter. In *Proceedings of the Fifth Social Media Mining for Health Applications Workshop & Shared Task*, pages 161–164, Barcelona, Spain (Online). Association for Computational Linguistics.

Trung Huynh, Yulan He, Alistair Willis, and Stefan Rueger. 2016. Adverse drug reaction classification with deep neural networks. In *Proceedings of COLING 2016, the 26th International Conference on Computational Linguistics: Technical Papers*, pages 877–887, Osaka, Japan. The COLING 2016 Organizing Committee.

Hang Jiang, Yining Hua, Doug Beeferman, and Deb Roy. 2022. Annotating the Tweebank corpus on named entity recognition and building NLP models for social media analysis. In *Proceedings of the Thirteenth Language Resources and Evaluation Conference*, pages 7199–7208, Marseille, France. European Language Resources Association.

Jeff Johnson, Matthijs Douze, and Hervé Jégou. 2019. Billion-scale similarity search with GPUs. *IEEE Transactions on Big Data*, 7(3):535–547.

Ari Z Klein, Juan M Banda, Yuting Guo, Ana Lucia Schmidt, Dongfang Xu, Ivan Flores Amaro, Raul Rodriguez-Esteban, Abeed Sarker, and Graciela Gonzalez-Hernandez. 2024. Overview of the 8th social media mining for health applications (# smm4h) shared tasks at the amia 2023 annual symposium. *Journal of the American Medical Informatics Association*, 31(4):991–996.

Seunghee Lee, Hyekyung Woo, Chung Chun Lee, Gyeongmin Kim, Jong-Yeup Kim, and Suehyun Lee. 2023. Drug_snsminer: standard pharmacovigilance pipeline for detection of adverse drug reaction using sns data. *Scientific Reports*, 13(1):3779.

Xi Liu, Han Zhou, and Chang Su. 2022. PingAnTech at SMM4H task1: Multiple pre-trained model approaches for adverse drug reactions. In *Proceedings of The Seventh Workshop on Social Media Mining for Health Applications, Workshop & Shared Task*, pages 4–6, Gyeongju, Republic of Korea. Association for Computational Linguistics.

Yinhan Liu, Myle Ott, Naman Goyal, Jingfei Du, Mandar Joshi, Danqi Chen, Omer Levy, Mike Lewis, Luke Zettlemoyer, and Veselin Stoyanov. 2019. Roberta: A robustly optimized bert pretraining approach. *arXiv preprint arXiv:1907.11692*.

Arjun Magge, Elena Tutubalina, Zulfat Miftahutdinov, Ilseyar Alimova, Anne Dirkson, Suzan Verberne, Davy Weissenbacher, and Graciela Gonzalez-Hernandez. 2021. Deepademiner: a deep learning pharmacovigilance pipeline for extraction and normalization of adverse drug event mentions on twitter. *Journal of the American Medical Informatics Association*, 28(10):2184–2192.

V.G.Vinod Vydiswaran, Grace Ganzel, Bryan Romas, Deahan Yu, Amy Austin, Neha Bhomia, Socheatha Chan, Stephanie Hall, Van Le, Aaron Miller, Olawunmi Oduyebo, Aulia Song, Radhika Sondhi, Danny Teng, Hao Tseng, Kim Vuong, and Stephanie Zimmerman. 2019. Towards text processing pipelines to identify adverse drug events-related tweets: University of Michigan @ SMM4H 2019 task 1. In *Proceedings of the Fourth Social Media Mining for Health Applications (#SMM4H) Workshop & Shared Task*, pages 107–109, Florence, Italy. Association for Computational Linguistics.

Davy Weissenbacher, Juan Banda, Vera Davydova, Darryl Estrada Zavala, Luis Gasco Sánchez, Yao Ge, Yuting Guo, Ari Klein, Martin Krallinger, Mathias Leddin, Arjun Magge, Raul Rodriguez-Esteban, Abeed Sarker, Lucia Schmidt, Elena Tutubalina, and Graciela Gonzalez-Hernandez. 2022. Overview of the seventh social media mining for health applications (#SMM4H) shared tasks at COLING 2022. In *Proceedings of The Seventh Workshop on Social Media Mining for Health Applications, Workshop & Shared Task*, pages 221–241, Gyeongju, Republic of Korea. Association for Computational Linguistics.

Dongfang Xu, Guillermo Lopez Garcia, Lisa Raithel, Rolland Roller, Philippe Thomas, Eiji Aramaki, Shuntaro Yada, Pierre Zweigenbaum, Sai Tharuni Samineni, Karen O'Connor, Yao Ge, Sudeshna Das, Abeed Sarker, Ari Klein, Lucia Schmidt, Vishakha Sharma, Raul Rodriguez-Esteban, Juan Banda, Ivan Flores Amaro, Davy Weissenbacher, and Graciela Gonzalez-Hernandez. 2024. Overview of the 9th social media mining for health applications (#SMM4H) shared tasks at ACL 2024. In *Proceedings of The 9th Social Media Mining for Health Research and Applications Workshop and Shared Tasks*, Bangkok, Thailand.

Anthony Yazdani, Hossein Rouhizadeh, David Vicente Alvarez, and Douglas Teodoro. 2023a. Ds4dh at# smm4h 2023: zero-shot adverse drug events normalization using sentence transformers and reciprocal-rank fusion. *arXiv preprint arXiv:2308.12877*.

Anthony Yazdani, Hossein Rouhizadeh, Alban Bornet, and Douglas Teodoro. 2023b. Conorm: Context-aware entity normalization for adverse drug event detection. *medRxiv*, pages 2023–09.

BrainStorm @ iREL at #SMM4H 2024: Leveraging Translation and Topical Embeddings for Annotation Detection in Tweets

Manav Chaudhary[1], **Harshit Gupta**[1], **Vasudeva Varma**[1]
[1]IIIT Hyderabad

Abstract

The proliferation of LLMs in various NLP tasks has sparked debates regarding their reliability, particularly in annotation tasks where biases and hallucinations may arise. In this shared task, we address the challenge of distinguishing annotations made by LLMs from those made by human domain experts in the context of COVID-19 symptom detection from tweets in Latin American Spanish. This paper presents BrainStorm @ iREL's approach to the SMM4H 2024 Shared Task, leveraging the inherent topical information in tweets, we propose a novel approach to identify and classify annotations, aiming to enhance the trustworthiness of annotated data.

1 Introduction

Data annotation, essential for improving machine learning models, involves labeling raw data with relevant information. However, this process is often costly and time-consuming. In recent times, the field of Natural Language Processing (NLP) has seen a transformative shift with the widespread adoption of Large Language Models (LLMs) like GPT-4 (OpenAI (2024)), Gemini (Gemini Team (2023)) and BLOOM (Le Scao et al. (2023)). These advanced models have shown remarkable capabilities in automating data annotation (Tan et al. (2024)), aiding in a crucial yet labor-intensive step in machine learning workflows. However, despite their impressive performance, the integration of LLMs in annotation tasks has sparked a debate within the research community. Proponents highlight their efficiency and consistency, while skeptics point to potential issues such as underlying biases and hallucinations.

While many recent efforts have focused on distinguishing between human and machine-generated text (Hans et al. (2024), Gambetti and Han (2023), Abburi et al. (2023)), detecting whether annotations are done by LLMs offers a novel perspective on AI detection. The advent of powerful LLMs, while driving innovation, poses risks of increased spread of untruthful news, fake reviews, and biased opinions, highlighting the need for a variety of detection technologies.

This paper addresses Task 7 of the SMM4H-2024 (Xu et al. (2024)): The 9th Social Media Mining for Health Research and Applications Workshop and Shared Tasks, focusing on the identification of data annotations made by LLMs versus those made by human domain experts. Our objective is to develop methods for distinguishing between annotations made by LLMs and those by human experts in the context of COVID-19 tweets in Latin American Spanish. This task is crucial for evaluating the generalizability and reliability of LLMs in real-world applications, particularly in health-related NLP tasks.

2 Methodology

Our approach to identifying whether a tweet was labeled as containing COVID-19 symptoms by an LLM or a human domain expert involves several key steps. We begin by preparing the dataset and leveraging both original and translated tweet texts to evaluate the performance of different models. Additionally, we incorporate topical embeddings to enhance the distinction between human and LLM annotations.

2.1 Dataset Preparation

The dataset consists of three columns: indexN, TweetText, and label. The TweetText column contains tweets written in Latin American Spanish, and the label column indicates whether the tweet was annotated by a human (human) or by GPT-4 (machine). The task is to determine if a tweet labeled as containing COVID-19 symptoms was annotated by an LLM or a human.

Given the bilingual nature of our approach, we

Proceedings of The 9th Social Media Mining for Health Applications (#SMM4H) Workshop and Shared Tasks, pages 121–123
August 15, 2024

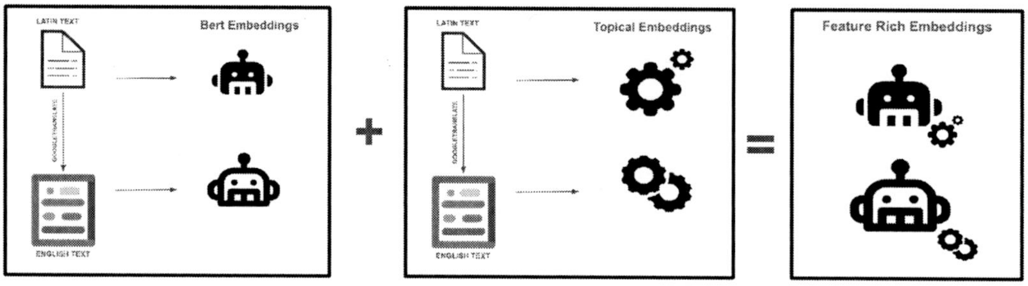

Figure 1: Diagram illustrating our method. The process starts with data translation from Latin American Spanish to English. These two datasets are used to generate BERT embeddings, followed by topical embeddings using BERTopic. These two embeddings are combined to give a new feature-rich embedding to be used for training our models.

first translate the Latin American Spanish tweets into English using Google Translate. This step enables us to apply and compare models trained on different languages and specific to tweet data.

2.2 Models Used

We compare the performance of two models:

- **dccuchile/bert-base-spanish-wwm-cased (Cañete et al. (2020))**: A BERT model pre-trained on Spanish text, which we use to process the original Spanish tweets. Note that we also use this model as our baseline, achieving a score of 0.50 on the test set.
- **vinai/bertweet-covid19-base-cased (Nguyen et al. (2020))**: A BERT model pre-trained specifically on English COVID-19 related tweets, which we use to process the translated English tweets. Note that an ablation study using just the translated tweets and the BERTweet model have been left for future exploration.

By comparing these models, we aim to leverage the strengths of language-specific and domain-specific pre-training.

2.3 Topical Embeddings with BERTopic

To improve the annotations further, we incorporate topical embeddings. The text data in both languages undergoes topic modeling using the BERTopic (Grootendorst (2022)) library. BERTopic extracts latent topics from the text using BERT embeddings. This step assigns a topic label to each tweet in both Spanish and English versions. During tokenization, the embeddings of these topic labels are appended to the tokenized representations of the tweets. Using a custom architecture, the topic embeddings are concatenated with the pooled output of the models, and the resulting combined representation is passed through a classification layer to predict the tweet's label.

The rationale behind this is that tweets written by humans have an intrinsic topical coherence that can be captured and distinguished from machine annotations. Our hypothesis is that human-annotated tweets are more contextually consistent and thematically structured compared to those annotated by an LLM.

In our approach, we treat human annotations as the gold standard—the absolute truth. This means we assume that any tweet labeled by a human is correctly annotated. Conversely, we recognize that tweets labeled by the LLM may include both correct and incorrect annotations.

Table 1: Classification Results on the Test Set

Model	Score
Baseline Spanish	0.50
Topical Spanish	0.50
Topical English	0.51

The motivation for using Topic Modeling is based on the nature of the tweets themselves. Since the tweets are written by humans, there is an inherent topical structure that a model can learn. By utilizing topical embeddings, we enhance the model's ability to capture this structure, thus improving its performance in identifying whether the annotations were made by a human or an LLM.

3 Results

We evaluated the models on the test set using the accuracy score provided by the organizers on CodaLab. We observe that Topical Spanish (with BERTopic) achieved a score of 0.50, indicating that the incorporation of topical embeddings did not improve the performance over the baseline in the original Spanish tweets.

Topical English (translated tweets with BERTopic) achieved a score of 0.51, showing a marginal improvement over the baseline, suggesting some potential in the use of translated tweets and topical embeddings.

While the results indicate only slight improvements, they underscore the challenges inherent in distinguishing between human and LLM annotations in this specific context.

4 Conclusion

This study explored the feasibility of distinguishing between human and LLM annotations in COVID-19 symptom detection from tweets in Latin American Spanish. By leveraging both language-specific and domain-specific models, along with topical embeddings, we aimed to enhance the accuracy of annotation classification. Our findings reveal that while topical embeddings and the use of translated tweets offer some promise, the improvements are marginal. The results suggest that more sophisticated techniques or additional features might be necessary to achieve significant enhancements in performance.

The slight improvement observed with translated English tweets suggests that the method has potential when combined with domain-specific models like BERTweet, pointing to the importance of further exploring multilingual and domain-adaptive approaches. There is a need to conduct detailed ablation studies to isolate the impact of various components, such as the translation process, topical embeddings, and different pre-trained models. An investigation into more advanced topic modeling techniques or the integration of other context-aware embeddings will also help.

By addressing these areas, we can further enhance the reliability of distinguishing between human and machine annotations, ultimately contributing to more trustworthy NLP systems in critical domains like healthcare.

References

Harika Abburi, Kalyani Roy, Michael Suesserman, Nirmala Pudota, Balaji Veeramani, Edward Bowen, and Sanmitra Bhattacharya. 2023. A simple yet efficient ensemble approach for ai-generated text detection. *arXiv preprint arXiv:2311.03084*.

José Cañete, Gabriel Chaperon, Rodrigo Fuentes, Jou-Hui Ho, Hojin Kang, and Jorge Pérez. 2020. Spanish pre-trained bert model and evaluation data. In *PML4DC at ICLR 2020*.

Alessandro Gambetti and Qiwei Han. 2023. Combat ai with ai: Counteract machine-generated fake restaurant reviews on social media. *arXiv preprint arXiv:2302.07731*.

Google Gemini Team. 2023. Gemini: a family of highly capable multimodal models. *arXiv preprint arXiv:2312.11805*.

Maarten Grootendorst. 2022. Bertopic: Neural topic modeling with a class-based tf-idf procedure. *arXiv preprint arXiv:2203.05794*.

Abhimanyu Hans, Avi Schwarzschild, Valeriia Cherepanova, Hamid Kazemi, Aniruddha Saha, Micah Goldblum, Jonas Geiping, and Tom Goldstein. 2024. Spotting llms with binoculars: Zero-shot detection of machine-generated text. *arXiv preprint arXiv:2401.12070*.

Teven Le Scao et al. 2023. Bloom: A 176b-parameter open-access multilingual language model.

Dat Quoc Nguyen, Thanh Vu, and Anh Tuan Nguyen. 2020. BERTweet: A pre-trained language model for English Tweets. In *Proceedings of the 2020 Conference on Empirical Methods in Natural Language Processing: System Demonstrations*, pages 9–14.

OpenAI. 2024. Gpt-4 technical report. *Preprint*, arXiv:2303.08774.

Zhen Tan, Alimohammad Beigi, Song Wang, Ruocheng Guo, Amrita Bhattacharjee, Bohan Jiang, Mansooreh Karami, Jundong Li, Lu Cheng, and Huan Liu. 2024. Large language models for data annotation: A survey. *Preprint*, arXiv:2402.13446.

Dongfang Xu, Guillermo Lopez Garcia, Lisa Raithel, Rolland Roller, Philippe Thomas, Eiji Aramaki, Shuntaro Yada, Pierre Zweigenbaum, Sai Tharuni Samineni, Karen O'Connor, Yao Ge, Sudeshna Das, Abeed Sarker, Ari Klein, Lucia Schmidt, Vishakha Sharma, Raul Rodriguez-Esteban, Juan Banda, Ivan Flores Amaro, Davy Weissenbacher, and Graciela Gonzalez-Hernandez. 2024. Overview of the 9th social media mining for health (#SMM4H) research and applications workshop and shared tasks at ACL 2024. In *Proceedings of The 9th Social Media Mining for Health Research and Applications Workshop and Shared Tasks*, Bangkok, Thailand. Association for Computational Linguistics.

UKYNLP@SMM4H2024: Language Model Methods for Health Entity Tagging and Classification on Social Media (Tasks 4 & 5)

Motasem S Obeidat
University of Kentucky
obeidat.s.motasem@uky.edu

Vinu H Ekanayake
University of Kentucky
vinu.ekanayake@uky.edu

Md Sultan Al Nahian
University of Kentucky
mna245@uky.edu

Ramakanth Kavuluru
University of Kentucky
ramakanth.kavuluru@uky.edu

Abstract

We describe the methods and results of our submission to the 9th Social Media Mining for Health Research and Applications (SMM4H) 2024 shared tasks 4 and 5. Task 4 involved extracting the clinical and social impacts of non-medical substance use and task 5 focused on the binary classification of tweets reporting children's medical disorders. We employed encoder language models and their ensembles, achieving the top score on task 4 and a high score for task 5.

1 Introduction

In today's digital landscape, social media platforms have evolved beyond mere communication channels, transforming into vital sources of real-time, user-generated data that reflect a wide array of public experiences and concerns. This transformation is particularly pertinent in the realm of public health, where social media discussions provide invaluable insights into both prevalent health issues and the personal impacts of various conditions. The critical analysis of these online dialogues is essential for understanding and addressing two significant public health challenges: nonmedical substance use and children's health disorders.

Task 4 and Task 5, conducted as part of the 9th Social Media Mining for Health Research and Applications (SMM4H) workshop, illustrate the potential of leveraging social media for health research. Task 4 focuses on extracting the clinical and social impacts of nonmedical substance use from Reddit discussions. Understanding these impacts is vital for developing nuanced interventions and educational programs aimed at mitigating the adverse outcomes of substance misuse. The analysis of user discussions can reveal the multifaceted consequences of such behavior, informing targeted interventions and effective medication strategies.

Task 5 addresses the binary classification of tweets related to children's health disorders, distinguishing between tweets from users reporting a child with disorders such as ADHD, ASD, delayed speech, or asthma from those that merely mention these conditions without indicating a personal effect. This task underlines the necessity for innovative data collection methods that can complement traditional epidemiological approaches, which often face logistical and financial barriers. By analyzing Twitter data, researchers can access a broader dataset, uncovering patterns and insights with speed and scale infeasible in conventional studies. This capability is crucial for developing targeted public health interventions and building support services that meet the needs of families dealing with children's health disorders.

2 Datasets

2.1 Task 4: Extracting clinical and social impacts of nonmedical substance use

The Task 4 dataset consists of 26,126 Reddit posts (60% for training, 20% for validation, and 20% for testing/evaluation). Only 318 posts (approximately 1.22%) were annotated for clinical impacts (health, physical condition, and mental well-being) or social impacts (effects on social relationships, community dynamics, and broader societal issues) (Ge et al., 2024). Thus this dataset has high imbalance in the sense that most posts would not contain any task specific entities.

2.2 Task 5: Binary classification of tweets reporting children's medical disorders

Task 5 uses a dataset of tweets, specifically targeting discussions where users reported their pregnancy and mentioned health conditions affecting their children — ADHD, ASD, delayed speech, or asthma. For binary classification, tweets were labeled as "1" if they reported a user having a child with a disorder and "0" if they merely mentioned

Proceedings of The 9th Social Media Mining for Health Applications (#SMM4H) Workshop and Shared Tasks, pages 124–129
August 15, 2024 ©2024 Association for Computational Linguistics

Method (# parameters)	Strict Precision	Strict Recall	Strict F1
ALBERT CW0 (BiLSTM) (223M)	15.4	25.8	19.3
BERT CW100 (No BiLSTM) (110M)	11.5	23.9	15.5
BERT CW0 (No BiLSTM) (110M)	14.3	26.4	18.6
BERT CW0 (BiLSTM) (110M)	17.4	28.3	21.5

Table 1: Task 4 strict validation results for clinical and social impact recognition

the disorder. A total of 10,734 tweets were collected and divided into three sets: 7,398 tweets for training, 389 for validation, and 1,947 for testing. (Klein et al., 2024)

3 Methodology

For Task 4, the main NLP task is named entity recognition (NER), for which we use a span-based encoder-only model that enumerates contiguous spans of tokens up to a certain length (we used 8) and classifies each of them as any of the allowed entity types. We used the Princeton University Relation Extraction (PURE) (Zhong and Chen, 2021) pipeline's entity model component, which is originally inspired by other prior efforts (Wadden et al., 2019). Span-level representations for each token are first derived from pre-trained language models, such as the BERT base uncased model (Devlin et al., 2019) and the ALBERT xx-large model (Lan et al., 2020), both English models. The span representation is the concatenation of these encoder-only model outputs for the first and last token embeddings along with an embedding for span length. A feed forward network processes these span representations to compute the probability distribution of entity types. As a variation, we also incorporated a bidirectional LSTM (BiLSTM) layer, with 150 units in each direction (totaling 300 units), between the span embeddings and the classification layers in the BERT and ALBERT models; this was shown to improve results in prior experiments and was also reported by Li et al. (2021).

The dataset for Task 4 was used with varying batch sizes and context windows. Here, a context window denotes the extent of textual context surrounding the target sentence that is being considered during the entity extraction process. We did some basic preprocessing of the messages such as lower casing and converting it into the format expected by our span based approach. For hyperparameters used in task 4, please refer to Table 5 in the Appendix.

For task 5, we fine-tuned pretrained encoder-only language models for tweet classification. Specifically, we used a DeBERTa v2 xlarge model (He et al., 2021), a DeBERTa v3 Large model, already fine-tuned on the Multi-NLI (MNLI) dataset (Manakul et al., 2023), and a RoBERTa (Liu et al., 2019) base model, previously fine-tuned on a general tweet dataset (Loureiro et al., 2022). All models were further fine-tuned on the task 5 dataset, which focuses on children's disorders such as ADHD, ASD, speech delays, and asthma. The best-performing models were obtained by systematically adjusting key hyperparameters such as the number of epochs, batch size, and learning rate. The specific hyperparameters used for the models are detailed in Table 8 in the Appendix. To combine the potential complementary predictive capabilities of each model, we integrated them into 3-model majority vote ensemble.

4 Results

4.1 Task 4 findings

To evaluate the impact of the additional BiLSTM layer modification, the models were assessed using the F1 score, both with and without it. Experiments were conducted across several batch sizes and a batch size of four was determined to be optimal. Table 1 presents the strict F1 scores on the validation dataset, and Table 2 presents the relaxed, strict, and token-level F1 scores obtained on the test dataset, utilizing two different context window sizes: 0 and 100 (indicated as CW0 and CW100). The first three entries in each table represent the results that were officially submitted. The final row is a post-evaluation entry. Relaxed and token-level F1 scores on the validation dataset are provided in Appendix Tables 6 and 7, respectively. For token-level F1 scores, the micro-average F1 is reported.

For the validation results in Table 1, we used strict F1 scores to evaluate the models. The ALBERT model, configured with a context window of

Method (# parameters)	Relaxed F1	Strict F1	Token-level F1
ALBERT CW0 (BiLSTM) (223M)	40.4	14.4	53.2
BERT CW100 (No BiLSTM) (110M)	46.2	17.1	53.1
BERT CW0 (No BiLSTM) (110M)	45.9	16.1	48.8
BERT CW0 (BiLSTM) (110M)	47.8	14.5	52.4

Table 2: Test results for Task 4 (clinical/social impact spotting) considering relaxed, strict, and token-level F1 scores

Method (# parameters)	Precision	Recall	F1 Score
DeBERTa v3 large (MNLI) (304M)	94.2	97.0	95.6
RoBERTa base (Twitter) (100M)	93.9	91.1	92.5
DeBERTa v2 XL (900M)	93.5	95.6	94.5
Ensemble method (1.3B)	93.5	95.6	94.5

Table 3: Task 5 validation results for classification of tweets with parental disclosure of childhood disorders

zero and augmented by a BiLSTM layer, reached an F1 score of 19.3. In contrast, extending the context window to 100 for the BERT model without the inclusion of a BiLSTM layer decreased performance, with the F1 dropping to 15.5. The BERT model, without a BiLSTM layer and with a context window of zero, recorded an F1 score of 18.6. Finally, the BERT model with a BiLSTM layer and a context window of zero achieved the highest F1 score of 21.5 among the methods tested. These findings suggest that adding a BiLSTM layer does improve performance in the examined scenarios.

For the test results presented in Table 2, relaxed F1 scores were used for comparison as specified by the SMM4H testing guidelines. The results reveal a marginal benefit from integrating BiLSTM layers. Specifically, the ALBERT model augmented with a BiLSTM layer achieved a score of 40.4, which is substantially lower than the scores achieved by both configurations of the BERT model without the BiLSTM layer, which scored 46.2 and 45.9 for context windows of 100 and zero, respectively. Furthermore, the BERT model configured with a BiLSTM layer and a zero context window registered a score of 47.8, surpassing the performance of its counterpart without the BiLSTM. These results suggest that although BiLSTM layers have the potential to enhance feature representation and temporal dependencies, their effectiveness is likely dependent on the specific model architectures and the contextual requirements of the task. Although the ALBERT architecture and training regimen were introduced to be more efficient and performant compared to BERT models, that did not turnout to be the case for this task. Our best test score (row 2 of Table 2) is also the top score in the shared task. It is important to note nontrivial variations in the ranking of best models (as per strict F1 scores) based on validation and test scores — potentially due to smaller datasets, validation results may not be strong indicators of what works best in the end.

4.2 Task 5 findings

Extensive testing was conducted to determine the optimal values for parameters such as learning rate, batch size, and number of epochs for each model used. Tables 3 (validation) and 4 (test) present the results obtained for the two DeBERTa models, the RoBERTa model, and the ensemble model that combines all three. In the validation phase, the DeBERTa v3 Large model, fine-tuned on the MNLI dataset, exhibited superior performance, achieving an F1-score of 95.6. The DeBERTa v2 XL and the ensemble models also performed notably well, each achieving an F1-score of 94.5. The RoBERTa Base model, fine-tuned on Twitter data, achieved an F1 score of 92.5.

Method	F1 Score
DeBERTa v3 large (MNLI)	92.4
RoBERTa base (Twitter)	89.5
DeBERTa v2 XL	94.8
Ensemble method	94.2

Table 4: Task 5 test results

In the testing phase, these models were applied to the test data for both internal evaluation and competition submissions. For the competition, we submitted results from the DeBERTa v3 Large model, which secured an F1-score of 92.4 with the test dataset. In the post-evaluation phase, results from the other models were analyzed and submitted. The DeBERTa v2 XL model achieved the highest performance with an F1-score of 94.8, followed by the ensemble model with an F1-score of 94.2 and the RoBERTa Base model with an F1-score of 89.5. The DeBERTa v3 and RoBERTa models dipped over 3% from validation to test F1-score. But the DeBERTa v2 XL and ensemble models more or less stayed the same potentially due to their larger model capacities. This aspect needs further investigation to see whether there is a generalizable explanation for this or if this is purely an artefact of the task 5 dataset.

5 Concluding Remarks

In this report, we reviewed our approaches to SMM4H 2024 tasks 4 and 5, which mostly focused on applying encoder-only language models to NER and text classification. We were officially informed of our first place in task 4 results and it appears our post-evaluation scores for task 5 are near the top. We conducted preliminary error analyses that mostly pointed to informal/casual language (that does not adhere to grammatical norms) as a prominent trait characterizing both false positives and false negatives. While the precision and recall are around the same range for task 5, for the NER task 4 we notice recall is at least ten points higher than precision; the difference was close to 12 points in row 2 (Table 1). Reducing this gap without compromising too much on precision is a promising general strategy we intend to pursue in the future. This could be done by lowering the output probability threshold as a starting point. More sophisticated changes to the loss function and potential post-processing strategies may be needed to obtain further improvements.

We did not attempt to use decoder-only large language models (LLMs) (e.g., Mistral and Llama) because in our lab's prior explorations (Gupta et al., 2023), with ample training data, encoder-only models always fared better for information extraction (IE) tasks. This was also observed by other researchers who focused on IE tasks that need language understanding capabilities more than generative skills. However, it is worth reassessing this with bigger LLMs (7B and more parameters) to see if they excel at supervised NER and classification tasks. Our working hypothesis is that LLMs may show benefits when dealing with shorter entity NER tasks (single or two token entities) but encoder-only models might still be the best option for handling longer entities that were more common in task 4. Additionally, we also plan to exploit even bigger LLMs (e.g, GPT-4) to generate synthetic examples that augment training data for both tasks to see if that benefits the overall performance. However, it is not clear how augmentation can be carried out for NER to retain longer named entities intact while changing the overall sentence with decent coherence and preserved meaning. Effective augmentation for classification may be relatively easier to achieve. These are some future directions we hope to pursue soon.

Acknowledgement

This work is supported by the NIH National Institute on Drug Abuse through grant R01DA057686. The content is solely the responsibility of the authors and does not necessarily represent the official views of the NIH.

References

Jacob Devlin, Ming-Wei Chang, Kenton Lee, and Kristina Toutanova. 2019. BERT: Pre-training of deep bidirectional transformers for language understanding. In *Proceedings of the 2019 Conference of the North American Chapter of the Association for Computational Linguistics: Human Language Technologies, Volume 1 (Long and Short Papers)*, pages 4171–4186, Minneapolis, Minnesota. Association for Computational Linguistics.

Yao Ge, Sudeshna Das, Karen O'Connor, Mohammed Ali Al-Garadi, Graciela Gonzalez-Hernandez, and Abeed Sarker. 2024. Reddit-impacts: A named entity recognition dataset for analyzing clinical and social effects of substance use derived from social media. *arXiv preprint arXiv:2405.06145*.

Shashank Gupta, Xuguang Ai, and Ramakanth Kavuluru. 2023. Comparison of pipeline, sequence-to-sequence, and gpt models for end-to-end relation extraction: experiments with the rare disease use-case. *arXiv preprint arXiv:2311.13729*.

Pengcheng He, Xiaodong Liu, Jianfeng Gao, and Weizhu Chen. 2021. Deberta: Decoding-enhanced bert with disentangled attention. In *International Conference on Learning Representations*.

AZ Klein, JA Gutiérrez Gómez, LD Levine, and G Gonzalez-Hernandez. 2024. Using longitudinal twitter data for digital epidemiology of childhood health outcomes: An annotated data set and deep neural network classifiers. *J Med Internet Res*, 26:e50652.

Zhenzhong Lan, Mingda Chen, Sebastian Goodman, Kevin Gimpel, Piyush Sharma, and Radu Soricut. 2020. ALBERT: A lite BERT for self-supervised learning of language representations. In *8th International Conference on Learning Representations, ICLR 2020, Addis Ababa, Ethiopia, April 26-30, 2020*. OpenReview.net.

Fei Li, ZhiChao Lin, Meishan Zhang, and Donghong Ji. 2021. A span-based model for joint overlapped and discontinuous named entity recognition. In *Proceedings of the 59th Annual Meeting of the Association for Computational Linguistics and the 11th International Joint Conference on Natural Language Processing (Volume 1: Long Papers)*, pages 4814–4828.

Yinhan Liu, Myle Ott, Naman Goyal, Jingfei Du, Mandar Joshi, Danqi Chen, Omer Levy, Mike Lewis, Luke Zettlemoyer, and Veselin Stoyanov. 2019. Roberta: A robustly optimized bert pretraining approach. *Preprint*, arXiv:1907.11692.

Daniel Loureiro, Francesco Barbieri, Leonardo Neves, Luis Espinosa Anke, and Jose Camacho-collados. 2022. TimeLMs: Diachronic language models from Twitter. In *Proceedings of the 60th Annual Meeting of the Association for Computational Linguistics: System Demonstrations*, pages 251–260, Dublin, Ireland. Association for Computational Linguistics.

Potsawee Manakul, Adian Liusie, and Mark John Francis Gales. 2023. Selfcheckgpt: Zero-resource black-box hallucination detection for generative large language models. *ArXiv*, abs/2303.08896.

David Wadden, Ulme Wennberg, Yi Luan, and Hannaneh Hajishirzi. 2019. Entity, relation, and event extraction with contextualized span representations. In *Proceedings of the 2019 Conference on Empirical Methods in Natural Language Processing and the 9th International Joint Conference on Natural Language Processing (EMNLP-IJCNLP)*, pages 5784–5789.

Zexuan Zhong and Danqi Chen. 2021. A frustratingly easy approach for entity and relation extraction. In *Proceedings of the 2021 Conference of the North American Chapter of the Association for Computational Linguistics: Human Language Technologies*, pages 50–61, Online. Association for Computational Linguistics.

Appendix

Hyperparameter	**Value**
head_hidden_dim	150
width_embedding_dim	150
max_span_length	8
lstm_hidden_dim	150 (300 units with BiLSTM)
train_batch_size	4
learning_rate	1e-5
task_learning_rate	5e-4
context_window	0 and 100
warmup_proportion	0.1

Table 5: Task 4 hyperparameters and configurations

Method (# parameters)	Relaxed Precision	Relaxed Recall	Relaxed F1 Score
ALBERT CW0 (BiLSTM) (223M)	31.6	52.6	39.5
BERT CW100 (No BiLSTM) (110M)	24.0	50.0	32.5
BERT CW0 (No BiLSTM) (110M)	28.1	51.7	36.4
BERT CW0 (BiLSTM) (110M)	32.1	52.4	39.9

Table 6: Task 4 relaxed validation results for clinical and social impact recognition

Method (# parameters)	Token-level Precision	Token-level Recall	Token-level F1
ALBERT CW0 (BiLSTM) (223M)	45.6	33.0	38.3
BERT CW100 (No BiLSTM) (110M)	46.4	38.7	42.2
BERT CW0 (No BiLSTM) (110M)	49.4	34.2	40.5
BERT CW0 (BiLSTM) (110M)	48.6	35.3	40.9

Table 7: Task 4 token-level validation results for clinical and social impact recognition

Method (# parameters)	# Epochs	Learning Rate	Batch Size
DeBERTa v3 Large (MNLI) (304M)	3	1e-5	8
RoBERTa base (Twitter) (100M)	3	2e-5	8
DeBERTa v2 XL (900M)	3	1e-5	16

Table 8: Hyperparameters for the models used in Task 5

LHS712_ADENotGood at #SMM4H 2024 Task 1: Deep-LLMADEminer: A deep learning and LLM pharmacovigilance pipeline for extraction and normalization of adverse drug event mentions on Twitter

Yifan Zheng,[1] Jun Gong,[1] Shushun Ren,[2] Dalton Simancek,[3] V.G.Vinod Vydiswaran[3]
[1]College of Pharmacy, [2]Department of Biostatistics, [3]Department of Learning Health Sciences
University of Michigan, Ann Arbor
{yifzheng, jungong, shushunr, daltonsi, vgvinodv}@umich.edu

Abstract

Adverse drug events (ADEs) pose major public health risks, with traditional reporting systems often failing to capture them. Our proposed pipeline, called Deep-LLMADEminer, used natural language processing approaches to tackle this issue for #SMM4H 2024 shared task 1. Using annotated tweets, we built a three part pipeline: RoBERTa for classification, GPT-4-turbo for span extraction, and BioBERT for normalization. Our models achieved F1-scores of 0.838, 0.306, and 0.354, respectively, offering a novel system for Task 1 and similar pharmacovigilance tasks.

1 Introduction

Adverse drug events (ADEs) are significant public health challenges, contributing to substantial morbidity and mortality (Watson et al., 2019). Effective pharmacovigilance, essential for ensuring the safe use of medications, struggles with the under-reporting of adverse drug reactions (ADRs), with estimates indicating that over 90% of ADRs go unreported (Hazell and Shakir, 2006). Social media offers a novel avenue for real-time, patient-centered insights into ADRs, supplementing traditional data sources. We developed the **Deep-LLMADEminer** pipeline for the SMM4H-2024 Task 1 (Xu et al., 2024) to extract and normalize ADEs from English-language tweets. This study aims to assess the performance of the three-part pipeline in extracting and normalizing ADEs from tweets.

2 The Deep-LLMADEminer pipeline

In step 1, we train a classifier to detect the presence of ADEs in tweets. In step 2, we train a large language model (LLM) to extract ADE entities and their spans from tweet text. Finally, in step 3, we train a classifier to map the extracted ADE entities to formal IDs in the MedDRA ontology[1], a standardized hierarchical medical terminology (Fig. 1). The #SMM4H 2024 Task 1 dataset (Xu et al., 2024) comprised of 17,974 training tweets annotated with 1,711 ADE mentions and 959 validation tweets annotated with 87 ADE mentions. Additionally, 11,799 test tweets were provided for model evaluation.

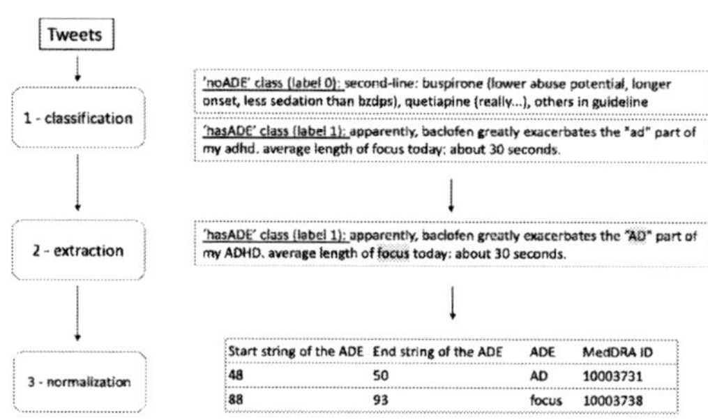

Figure 1: Deep-LLMADEminer pipeline

2.1 Step 1: ADE Classification

Using the RoBERTa-base model (Liu et al., 2019), we developed a binary classification system to identify tweets containing adverse drug event (ADE) mentions. The preprocessing involved removing HTML tags, URLs, user mentions, hashtags (converted to plain text), special non-ASCII characters, punctuation, and excess whitespace, and converting all text to lowercase. We fine-tuned RoBERTa on a labeled dataset, where each tweet was tokenized and encoded into input IDs, attention masks, and token-type IDs. Key training parameters were: epochs=8, maximum sequence length=256 tokens, and batch size=16, learning rate=1e-5. The output was processed through a linear layer to classify tweets as containing or not containing ADR mentions. Our workflow for step 1 included data loading, preprocessing, training, validation with

[1]MedDRA® the Medical Dictionary for Regulatory Activities terminology is the international medical terminology developed under the auspices of the International Council for Harmonisation of Technical Requirements for Pharmaceuticals for Human Use (ICH).

Proceedings of The 9th Social Media Mining for Health Applications (#SMM4H) Workshop and Shared Tasks, pages 130–132
August 15, 2024 ©2024 Association for Computational Linguistics

cross-entropy loss, and evaluation of accuracy.

2.2 Step 2: ADE Span Extraction

In step 2 of ADE span extraction, we employed the GPT-4 model via the OpenAI API (Achiam et al., 2023) to develop a text span detection method for extracting adverse drug events (ADEs) from tweets. Data preprocessing involved merging datasets that contained tweets with ADE mentions. Various prompts were experimented with to enhance the model's detection capabilities. We provided the model with 50 examples of tweets from training examples using around 10 different prompts. We explored the impacts of linguistic variations in our prompts to optimize the detection of adverse drug events from tweets. Specifically, we conducted experiments with 10 distinct prompts that varied primarily in verb usage, terminology, and the one-shot example. These variations included the use of different verbs such as "identify," "extract," and combinations of both. Additionally, we experimented with terminological changes, alternating between "adverse drug reactions" and "adverse drug effects" to assess any differences in model performance. For our experimental setup, each prompt was tested with a one-shot example tailored to illustrate the specific wording of the prompt. This approach allowed us to evaluate the model's responsiveness to linguistic nuances in a controlled manner. We did not record the results for each prompt but the most effective following prompt format is derived from testing these prompts. The final specific system message was used to guide the model: *"Identify and extract the text of adverse drug effects from the tweets in square brackets."* An example-based few-shot learning approach provided the model with specific examples to cover a wide range of ADE instances. For instance, a prompt used was: *"[I feel like a pile of crap #sick #cold #stomach reacting to some antibiotics. I will never again take #ciprofloxacin #withdrawal gives you chills],"* with the model extracting and formatting the response accordingly. The parameter `temperature` is set to 0 to minimize randomness, fostering deterministic responses from the model. We employ a `top_p` value of 0.95, which allows the model to consider a broader set of possible responses, enhancing the diversity of the output while still focusing on the most probable ones. Both `frequency_penalty` and `presence_penalty` are set to 0, indicating no additional constraints on the frequency or presence of terms in the generated text, thus not artificially influencing the model's natural language processing capabilities. These settings collectively ensure that our model interactions are precisely tailored to maximize accuracy and consistency in identifying and extracting relevant text spans for pharmacovigilance analysis. We also implemented a function to find and return the start and end indices of detected ADE texts.

2.3 Step 3: ADE Normalization

For ADE normalization, we fine-tuned BioBERT (Lee et al., 2020) to map the extracted ADE mentions from tweets to their respective MedDRA Preferred Terms (PTs), making it a multiclass classification task. The preprocessing involved converting all ADE mentions to PT ID levels, utilizing a comprehensive dictionary containing approximately 80,000 entries. This facilitated the accurate alignment of lower-level term (LLT) IDs with PT IDs. The training configurations were: epochs=8, batch size=16, and learning rate=5e-5. Our methodological pipeline comprised data loading, preprocessing to ensure consistent ID levels, and model training on processed data. Subsequent evaluation on validation data assessed the model's performance, ensuring effective normalization of tweets to corresponding PT IDs for enhanced pharmacovigilance. Other attempts to further enhance performance included fine-tuning GPT3.5 Turbo with 1,711 normalization examples. However, time constraints prevented a full evaluation.

3 Results

Models	Accuracy	F1	Precision	Recall
		Validation Set (n=932)		
(1) RoBERTa	0.955	0.838	0.817	0.862
		Evaluation Test Set		
(2) GPT-4	-	0.306	0.378	0.338
(3) BioBERT	-	0.354	0.395	0.321

Table 1: Performance on the unseen validation set for step 1 and the evaluation test set for steps 2 and 3.

Table 1 includes the performance metrics of RoBERTa on the unseen validation set (n=932) for step 1. We ultimately selected RoBERTa for ADE Classification with an F1-score of 0.838. It also shows the performance metrics on the evaluation dataset for steps 2 and 3. We employed models from GPT-4 for ADE span extraction, achieving an F1-score of 0.306, and utilized BioBERT for ADE normalization, which achieved an F1-score of 0.354. The effectiveness of ADE normalization in step 3 is influenced by the quality of the extracted ADE spans in the preceding task. Therefore, the relatively modest F1 score in step 2 directly impacted the overall performance in step 3.

To facilitate further development and reproducability, we have shared the implementation code for our system participation on GitHub.[2]

4 Conclusion

In our submission to the #SMM4H 2024 Task 1, we evaluated models for ADE classification, span extraction, and normalization steps using RoBERTa, GPT-4, and BioBERT, respectively. Despite some model errors in identification and resource limitations, our methods remained efficient and cost-effective.. Future work will focus on refining these models and addressing data imbalance to improve ADE detection and reporting. Specifically, for step 1, we plan to implement weighting strategies to correct dataset imbalance for more balanced and improved outcomes. For step 2, increasing the number of training examples will potentially boost model accuracy. For step 3, we aim to fine-tune hyperparameters based on validation datasets to enhance model performance.

Limitations

Our approach was not without limitations. In step 1, our methods did not address the potential issue of database imbalance; in step 2, we limited ourselves to one-shot learning to minimize costs; and in step 3, we avoided fine-tuning due to the high costs and lengthy training times associated with such processes. Additionally, we utilized the Unified Medical Language System (UMLS) to obtain synonyms for enhancing our normalization efforts in step 3. However, due to the large dataset size and limited training time, we couldn't fully leverage this approach to improve our performance. Addressing these issues and exploring these additional strategies can potentially lead to improved overall performance. Finally, the notable performance drop in Task 2 bottlenecks performance in Task 3, which challenges the utility of the overall pipeline. Further optimization should prioritize Task 2 performance for the more practical utility of the full end-to-end system.

[2]The software code, written in Python, is available for research use at: https://github.com/NLP4HealthUMich/Deep-LLMADEminer

Ethics Statement

The authors recognize that the tweets provided as part of SMM4H 2024 tasks reference health symptoms and medication use and consider these as private data that is publicly accessible under the ongoing consent provided by Twitter/X's User Agreement Terms and Conditions. The tweets for this study were sourced from the SMM4H 2024 shared task coordinators and were accessed strictly for research purposes. The data were only utilized to participate in the shared task and for no other uses.

References

Josh Achiam, Steven Adler, Sandhini Agarwal, Lama Ahmad, Ilge Akkaya, Florencia Leoni Aleman, Diogo Almeida, Janko Altenschmidt, Sam Altman, Shyamal Anadkat, et al. 2023. Gpt-4 technical report. *arXiv preprint arXiv:2303.08774*.

Lorna Hazell and Saad AW Shakir. 2006. Under-reporting of adverse drug reactions. *Drug safety*, 29(5):385–396.

Jinhyuk Lee, Wonjin Yoon, Sungdong Kim, Donghyeon Kim, Sunkyu Kim, Chan Ho So, and Jaewoo Kang. 2020. Biobert: a pre-trained biomedical language representation model for biomedical text mining. *Bioinformatics*, 36(4):1234–1240.

Yinhan Liu, Myle Ott, Naman Goyal, Jingfei Du, Mandar Joshi, Danqi Chen, Omer Levy, Mike Lewis, Luke Zettlemoyer, and Veselin Stoyanov. 2019. Roberta: A robustly optimized bert pretraining approach. *arXiv preprint arXiv:1907.11692*.

Sarah Watson, Ola Caster, Paula A Rochon, and Hester den Ruijter. 2019. Reported adverse drug reactions in women and men: aggregated evidence from globally collected individual case reports during half a century. *EClinicalMedicine*, 17:100188. DOI: 10.1016/j.eclinm.2019.10.001.

Dongfang Xu, Guillermo Lopez Garcia, Lisa Raithel, Rolland Roller, Philippe Thomas, Eiji Aramaki, Shuntaro Yada, Pierre Zweigenbaum, Sai Tharuni Samineni, Karen O'Connor, Yao Ge, Sudeshna Das, Abeed Sarker, Ari Klein, Lucia Schmidt, Vishakha Sharma, Raul Rodriguez-Esteban, Juan Banda, Ivan Flores Amaro, Davy Weissenbacher, and Graciela Gonzalez-Hernandez. 2024. Overview of the 9th social media mining for health applications (#SMM4H) shared tasks at ACL 2024. In *Proceedings of The 9th Social Media Mining for Health Research and Applications Workshop and Shared Tasks*, Bangkok, Thailand. Association for Computational Linguistics.

HaleLab_NITK@SMM4H'24: Binary Classification of English Tweets reporting Children's Medical Disorders

Ritik Mahajan and **Sowmya Kamath S.**
Healthcare Analytics and Language Engineering (HALE) Lab,
Department of Information Technology,
National Institute of Technology Karnataka, Surathkal, Mangalore 575025 INDIA
ritik.232it026@nitk.edu.in sowmyakamath@nitk.edu.in

Abstract

This paper describes the work undertaken as part of the SMM4H-2024 shared task, specifically Task 5, which involves the binary classification of English tweets reporting children's medical disorders. The primary objective is to develop a system capable of automatically identifying tweets from users who report their pregnancy and mention children with specific medical conditions, such as attention-deficit/hyperactivity disorder (ADHD), autism spectrum disorders (ASD), delayed speech, or asthma, while distinguishing them from tweets that merely reference a disorder without much context. Our approach leverages advanced natural language processing techniques and machine learning algorithms to accurately classify the tweets. The system achieved an overall F1-score of 0.87, highlighting its robustness and effectiveness in addressing the classification challenge posed by this task.

1 Introduction

The proliferation of social media platforms such as Twitter (now known as X), Reddit, and Facebook has led to an unprecedented surge in user-generated content. Millions of individuals publicly share their thoughts, experiences, and health-related information online, which presents a unique opportunity to analyze and investigate public health trends and issues. Among these platforms, Twitter stands out as a particularly valuable source of rich information for both the general public and researchers. By analyzing tweets, researchers can gain insights into various health-related phenomena, track the spread of diseases, monitor public sentiment toward health policies, and identify emerging health concerns. The real-time, streaming nature of Twitter data makes it an indispensable tool for public health surveillance and research, facilitating a deeper understanding of health behaviors and outcomes on a global scale (Bachina et al., 2021).

The reporting of children's medical disorders stands out as a crucial area of study, given the importance of early detection, diagnosis, and treatment in pediatric healthcare. Many children are diagnosed with disorders that can profoundly impact their daily lives and may persist throughout their lifetimes. Conditions such as attention-deficit/hyperactivity disorder (ADHD), autism spectrum disorders (ASD), delayed speech, and asthma are frequently mentioned by parents and caregivers on social media. Studying these discussions on Twitter provides valuable insights into the prevalence and public perception of these disorders (Saini and Yadav, 2022). It also helps in understanding the challenges faced by families and the effectiveness of various interventions, through analysis of patterns and trends in symptom reporting, treatment experiences, and support mechanisms, which are essential for improving healthcare strategies and policies (Unnikrishnan et al., 2023). Furthermore, by monitoring these conversations, researchers can identify gaps in awareness and education, potentially guiding more targeted and effective public health campaigns.

2 SMM4H'24 Task 5 - Description

Task 5 is a binary classification task that involves automatically distinguishing tweets posted by users who have reported their pregnancy on Twitter and specifically mention children with ADHD, ASD, delayed speech, or asthma (annotated as "1"), from tweets that merely mention a disorder (annotated as "0"). This task enables the large-scale utilization of Twitter, not only for epidemiologic studies but also to explore parents' experiences and directly target support interventions.

The dataset (Klein et al., 2023) consists of 7,398 English language tweets for training, 389 tweets for validation, and 10,000 tweets for testing purposes. Tweets in which parents explicitly mention that

Proceedings of The 9th Social Media Mining for Health Applications (#SMM4H) Workshop and Shared Tasks, pages 133–135
August 15, 2024 ©2024 Association for Computational Linguistics

their child is suffering from ADHD, ASD, delayed speech, or asthma are annotated as '1'. In contrast, other tweets are annotated as '0', which may or may not have mention of a disorder. By applying natural language processing (NLP) techniques and machine learning algorithms to this dataset, we aim to develop a robust model capable of accurately identifying and categorizing tweets related to children's medical disorders.

3 Methodology

This binary classification task involves automatically distinguishing tweets posted by users who had reported their pregnancy on Twitter and mentioned that their children had ADHD, ASD, delayed speech, or asthma in other tweets. Various preprocessing techniques were employed on the tweets during the data processing phase to ensure they were standardized and prepared for analysis The names of disorders and digits were standardized. Additionally, terms referring to a child, such as, *son*, *child*, *daughter*, etc., were unified to the common term "*child*" since the focus is on identifying tweets about child disorders posted by their parents. Furthermore, URLs, usernames, hashtags, emojis, and smileys were eliminated using the tweet-preprocessor library in Python (Van Rossum and Drake Jr, 1995). Common abbreviations and contractions found in tweets, such as "*lol*", "*thx*", "*btw*", and "*we're*" were expanded to their full forms, while non-alphanumeric characters and extra white spaces were pruned. The text was converted to lowercase, elongated words were corrected by keeping the occurrence of repeated characters to two, and lemmatization was performed using the Spacy lemmatizer (Honnibal and Montani, 2017). These steps standardized the text format and improved its suitability for subsequent analysis and modeling.

Experiments were carried out using various Transformer-based models to model the tweets. RoBERTa-base (Liu et al., 2019) was chosen for its ability to capture contextual information effectively and was implemented using the Huggingface toolkit (Wolf et al., 2019) for classifying tweets mentioning children's medical disorders. After preprocessing the textual data as elaborated earlier, the text sequences were subjected to tokenization using the RoBERTa tokenizer, with the maximum text length set to 128. Model optimization was achieved using the Adam optimizer with a batch size of 8 and a learning rate of 1e-5. The training was carried out for up to 10 epochs, with early stopping triggered by validation set performance and a patience value set at 4 epochs. Moreover, a dropout rate of 0.3, determined iteratively, was applied to regularize the model. The model architecture is built on RoBERTa-base, incorporating an additional hidden dense layer with a Rectified Linear Unit (ReLU) activation function. A sigmoid activation function is used in the output layer. The experiments were conducted on Google Colab using a T4 GPU as the hardware accelerator.

4 Results and Discussion

The system's performance was evaluated on both the validation and test sets. Initially, the performance on the validation set was compared to assess the model's effectiveness before evaluating it on the test set for submission. Efforts towards hyperparameter tuning helped achieve optimal performance on the validation set. Key hyperparameters, such as dropout rates and the number of hidden layers, were varied systematically to enhance the model's performance. This method enabled systematic exploration of hyperparameter configurations to determine the most effective settings based on validation set performance. The evaluation of the model's performance was based on the F1-score metric. The F1-score is a critical metric for this binary classification task because it balances precision and recall, offering a unified measure that considers both false positives and false negatives. This is particularly vital for distinguishing tweets about children with specific medical disorders from general mentions of disorders, ensuring that the model accurately identifies relevant tweets and minimizes the misclassification of non-relevant ones. Precision and recall scores are reported alongside the F1-score to comprehensively evaluate the model's performance in terms of the accuracy of positive predictions and the model's ability to capture all relevant instances. The results obtained from evaluating the system on the validation set are reported in Table 1.

Table 1: System Performance on *test* and *val* datasets

Dataset	F1-score	Precision	Recall
validation	0.88	0.89	0.88
test	0.87	0.86	0.88

The system achieved an F1-score of 0.87 on the test set. There is a substantial difference in performance between our RoBERTa classifier (0.868) and the RoBERTa baseline classifier (0.927) in (Klein et al., 2024). The baseline classifier is built on the RoBERTa-large pre-trained model and has been tested on a set of 1,947 tweets, while the proposed classifier leverages the RoBERTa-base pre-trained model and is tested on a set of 10,000 tweets. Achieving an F1-score of 0.87 on the test set demonstrates that the model generalizes well to unseen data, performing consistently with high accuracy. This suggests that the system effectively learns and captures the underlying patterns and features in the tweets related to children's medical disorders.

5 Concluding Remarks

In this article, an approach to accurately distinguishing between tweets that mention specific disorders in the context of parenting and those that merely reference a disorder, using advanced NLP techniques and Transformed-based models is presented. Evaluation on both the validation and test sets demonstrates the system's reliability, with consistent F1-scores indicating its effectiveness in generalizing to unseen data. Moving forward, we aim to explore further refinements to the model architecture, incorporating additional features, and expanding the training dataset to enhance the system's performance.

References

Sony Bachina, Spandana Balumuri, and Sowmya Kamath. 2021. Ensemble albert and roberta for span prediction in question answering. In *Proceedings of the 1st workshop on document-grounded dialogue and conversational question answering (DialDoc 2021)*, pages 63–68.

Matthew Honnibal and Ines Montani. 2017. spaCy 2: Natural language understanding with Bloom embeddings, convolutional neural networks and incremental parsing. To appear.

Ari Z Klein, José Agustín Gutiérrez Gómez, Lisa D Levine, and Graciela Gonzalez-Hernandez. 2024. Using longitudinal twitter data for digital epidemiology of childhood health outcomes: An annotated data set and deep neural network classifiers. *J Med Internet Res*, 26:e50652.

Ari Z Klein, Shriya Kunatharaju, Karen O'Connor, and Graciela Gonzalez-Hernandez. 2023. Pregex: rule-based detection and extraction of twitter data in pregnancy. *Journal of Medical Internet Research*, 25:e40569.

Yinhan Liu, Myle Ott, Naman Goyal, Jingfei Du, Mandar Joshi, Danqi Chen, Omer Levy, Mike Lewis, Luke Zettlemoyer, and Veselin Stoyanov. 2019. Roberta: A robustly optimized bert pretraining approach. *arXiv preprint arXiv:1907.11692*.

Gurdeep Saini and Naveen Yadav. 2022. Ensemble neural models for depressive tendency prediction based on social media activity of twitter users. In *Security, Privacy and Data Analytics: Select Proceedings of ISPDA 2021*, pages 211–226. Springer.

Reshma Unnikrishnan et al. 2023. Efficient parameter tuning of neural foundation models for drug perspective prediction from unstructured socio-medical data. *Engineering Applications of Artificial Intelligence*, 123:106214.

Guido Van Rossum and Fred L Drake Jr. 1995. Python tutorial.

Thomas Wolf, Lysandre Debut, Victor Sanh, Julien Chaumond, Clement Delangue, Anthony Moi, Pierric Cistac, Tim Rault, Rémi Louf, Morgan Funtowicz, et al. 2019. Huggingface's transformers: State-of-the-art natural language processing. *arXiv preprint arXiv:1910.03771*.

Team Yseop at #SMM4H 2024: Multilingual Pharmacovigilance Named Entity Recognition and Relation Extraction

Anubhav Gupta
Yseop
agupta@yseop.com

Abstract

This paper describes three RoBERTa based systems. The first one recognizes adverse drug events (ADE) in English tweets and links them with MedDRA concepts. It scored F1-norm of **40** for the Task 1. The next one extracts pharmacovigilance related named entities in French and scored a F1 of **0.4132** for the Task 2a. The third system extracts pharmacovigilance related named entities and their relations in Japanese. It obtained a F1 of **0.5827** for the Task 2a and **0.0301** for the Task 2b. The French and Japanese systems are the best performing system for the Task 2[1].

1 Introduction

As of 2008, the European Commission proposed certain measures[2] to protect the public from the harm caused by Adverse Drug Reaction (ADR). Nonetheless, Koyama et al. (2023) observed a global increase in Adverse Drug Event (ADE) related deaths. Thus, pharmacovigilance, i.e. activities involving detection, comprehension, and prevention of adverse effects due to medication is an important subject. With the arrival of internet, the general public started using it to seek and share health related information[3] (Gerber and Eiser, 2001). With the help of natural language processing systems, this publicly available data can be analysed to extract information related to side effects. As a result, it can play a key role in strengthening pharmacovigilance reporting systems.

Note: Even though, the "ICH E2A Clinical safety data management: definitions and standards for expedited reporting"[4] distinguishes between ADR and ADE, in this paper we will use the terms interchangeably to refer to the unintended consequences of taking a prescribed medication.

SMM4H-2024: The 9th Social Media Mining for Health Research and Applications Workshop and Shared Tasks — Large Language Models (LLMs) and Generalizability for Social Media NLP (Xu et al., 2024) proposed 7 shared tasks. We participated in two of them:

- Task 1: Extraction and normalization of adverse drug events (ADEs) in English tweets
- Task 2: Cross-Lingual Few-Shot Relation Extraction for Pharmacovigilance in French, German, and Japanese

2 Task 1 - English

In this task, we have to identify ADEs in a short text, called "tweets", written in English. If ADEs are present, then they have to be mapped to Preferred Terms Id (ptId) from the Medical Dictionary for Regulatory Activities (MedDRA)[5].

Due to an error on our part, we did this task with the SMM4H -2023 Task 5 dataset, shared on the task's Google Groups on 06/02/2024.

2.1 Dataset

The training set had 17385 tweets, out of which only 1239 mentioned any ADE. The tweets and the annotations were provided in two separate files. An example of a tweet:

```
SMM4H2022uCZV2SRsCe4vzjFm
@USER__________ have to go to a doc
now to see why i'm still gaining.
stupid paxil made me gain like 50
pounds ?? and now i have to lose it
```

[1]All the models will shared on https://huggingface.co/yseop and the code will be available on https://github.com/yseop/YseopLab

[2]https://ec.europa.eu/commission/presscorner/detail/cs/MEMO_08_782

[3]https://web.archive.org/web/20150924101434/https://www.pushdoctor.co.uk/Resources/PushDoctor-Health-report.pdf

[4]https://database.ich.org/sites/default/files/E2A_Guideline.pdf

[5]https://www.meddra.org/

Proceedings of The 9th Social Media Mining for Health Applications (#SMM4H) Workshop and Shared Tasks, pages 136–141
August 15, 2024

The annotation file had the spans (start and end position in the tweet) and lowest level term[6] id (`llt`) of each ADE mention:

```
SMM4H2022uCZV2SRsCe4vzjFm
ADE 61 68 gaining 10047896
```

A MedDRA dictionary (**llt.asc**) containing the mapping between ADE, `llt`, and preferred term id (`ptid`) was also provided.

For the tweets in quotes we found that the spans were off by 1, we corrected that for the training.

2.2 Model

We augmented data with English texts from the SM-ADE sub-task (Wakamiya et al., 2023) to train a binary classifier that can distinguish between the tweets having ADE from those without ADE. The resulting classifier could get a F-1 score 0.73 on the validation set. This was less than the F1 achieved by Gupta and Rayar (2023)'s multilingual Bert model with similar dataset. Therefore, we did not use the classifier.

We fine-tuned RoBERTa[7] (Liu et al., 2019) for token classification[8] using Huggingface (Wolf et al., 2020). We preprocessed the training data by aligning the annotations with tokens. The first token of an ADE entity was labelled `B-MISC` and remaining tokens were labelled `I-MISC`. The non-ADE tokens in the input text were labelled `O`. While training, the model was evaluated on the validation set using seqeval's[9] `f1_score`.

We kept aside 20% of the training data for validation using scikit-learn's (Pedregosa et al., 2011) stratified `train_test_split` and fine-tuned the model on a) 80% of the train set, and b) on augmented training data consisting of the 80% of the provided training corpus and the non-ADE English data from the SM-ADE sub-task. There was not much of the difference in the performance (see Figure 1), so we did not submit the model trained on the augmented data.

The ADEs detected by the first model (see Table 1 for the parameters used) were searched in the MedDRA dictionary with the help of Sentence-Transformers'[10] (Reimers and Gurevych, 2019) `multi-qa-mpnet-base-dot-v1` model. The results (sub1.zip) were submitted to the leader board.

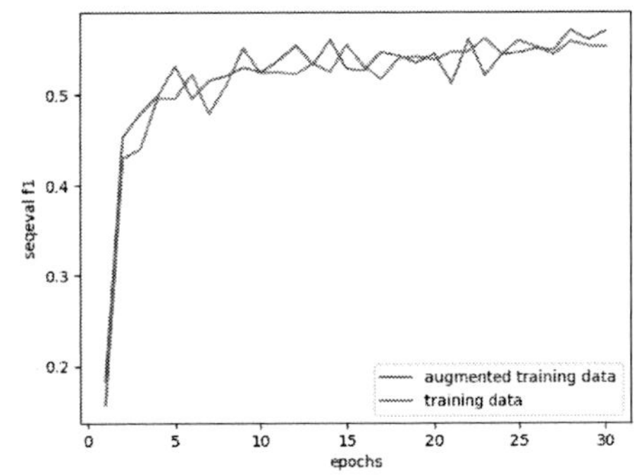

Figure 1: Comparison of the raw training data with the augmented one.

2.3 Test Results

The model obtained a `F1-norm` score of **40.0** compared to **43.9** by the baseline provided by DeepADEMiner (Magge et al., 2021). `-Norm` metrics are calculated by comparing the normalized ADE, i.e. `llt`. Table 2 compares the performance of the model with the baseline and the median of all the submissions.

3 Task 2 - French and Japanese

The goals of this task were Named Entity Recognition (NER) and Relation Extraction (RE). The corpus consisted of, mainly, German and Japanese text taken from online forums and Twitter/X. It also had four French documents translated from German texts.

3.1 Dataset

The French dataset had 4 documents and the Japanese had 392. The documents were annotated in the brat standoff format[11] (Stenetorp et al., 2011). There are 3 entities and 2 relations:

- Entities:
  - DISORDER, a symptom not necessarily related to a drug
  - DRUG
  - FUNCTION, bodily functions
- Relations:

[6] https://www.meddra.org/how-to-use/basics/hierarchy

[7] https://huggingface.co/FacebookAI/roberta-base

[8] https://huggingface.co/docs/transformers/en/tasks/token_classification

[9] https://github.com/chakki-works/seqeval

[10] https://www.sbert.net/index.html

[11] https://brat.nlplab.org/standoff.html

Training Parameters	Task 1	Task 2a	Task 2b
tokenizer max length		512	
learning rate		1e-05	
weight decay	0.001	0.0	0.0
epochs	30	50	15
batch size		16	
machine		ml.g5.xlarge	

Table 1: Non default hyperparameters used for fine-tuning.

System	F1-Norm	P-Norm	F1-NER	F1-Norm-Unseen
`sub1.zip`	40	39.6	47.2	29.5
Median	29.3	33.9	37.6	14.1
Baseline	43.9	39.3	48.1	32.3

Table 2: Performance on the Task 1 test set.

- CAUSED, the first entity causes the second entity
- TREATMENT_FOR, the first entity remedies the second one

An example of French data with annotation:

```
Salut <user>, pour moi, ça a commencé
à l'âge de <pi>. J'ai suivi une
thérapie et une cure pendant deux ans,
...
Prends soin de toi. <user>

T4   DRUG 123 130    pilules
...
R3   CAUSED Arg1:T10 Arg2:T15
R13  TREATMENT_FOR Arg1:T24 Arg2:T26
```

An example of Japanese data with annotation:

```
881844583344201728:
昼間のレクサプロが副作用ひどくて未
だに気持ち悪い

T1    DRUG 22 27        レクサプロ
T2    DISORDER 28 31    副作用
T3    DISORDER 38 43    気持ち悪い
R1    CAUSED Arg1:T1 Arg2:T2
```

There were some inconsistencies in the data, for example the Japanese term 錠 was not annotated as `DRUG` in most of the documents. Whereas the French equivalent **pilules** (see the example above) and other common nouns such as 薬 and 製品 were.

In the Japanese training, the span of certain entities was updated as shown in the Table 3.

File	Entity	Span
ja_twjp_020-040_0	T1	36 42
ja_twjp_200-220_1	T15	47 55
ja_twjp_240-260_8	T8	140 142
ja_twjp_320-340_4	T1	109 114
ja_twjp_340-360_14	T13	144 151
ja_twjp_440-460_19	T8	73 77
ja_twjp_460-480_18	T6	20 26

Table 3: Annotation files that were corrected.

3.2 Task 2a - French Model

Since there was not enough training data, we decided to use `Mistral-7B-Instruct-v0.1`[12] (Jiang et al., 2023) and DrBERT-CASM2[13] via the `medkit`[14] library. DrBERT (Labrak et al., 2023) was fine-tuned on CASM2 corpus for NER task to produce DrBERT-CASM2. The CASM2 is a private corpus that contains documents from CAS (Grabar et al., 2018).

The Mistral LLM is prompted with the prompt described in Appendix A and parameters `do_sample=False` and `max_new_tokens=256`. If the entities returned by the LLM are in the text they are added to the list of candidates. Then, all the entities extracted by DrBERT-CASM2 are added to the candidates. Lastly, the entities in the candidate list are used to find other substrings in the text.

[12] `https://huggingface.co/mistralai/Mistral-7B-Instruct-v0.1`

[13] `https://huggingface.co/camila-ud/DrBERT-CASM2`

[14] `https://medkit.readthedocs.io/en/stable/index.html`

3.3 Task 2a - Japanese Model

Baseline: If we label all occurrences of 副作用 in the training set as `DISORDER`; 薬 and 製品 as `DRUG`, we get a Macro F1 of 0.1658. If we also annotate all consecutive sequence of katakana characters that are not present in the `JMdict_e`[15] (Japanese–Multilingual Dictionary) as `DRUG`, the Macro F1 becomes 0.2842.

We kept aside the 20% of the provided train dataset as validation set using scikit-learn's stratified `train_test_split`. Then on the remaining dataset we fine-tuned `daisaku-s/medtxt_ner_roberta`[16] as token classification task. This model was previously trained on MedTxt-CR dataset (Yada et al., 2022). The seqeval F1 score was better than the baseline and hence it was submitted to the leader-board.

3.4 Task 2b - Japanese Model

The training data contains:

- 390 examples of DRUG causing DISORDER
- 100 examples of DRUG TREATMENT_FOR DISORDER
- 98 examples of DISORDER causing DISORDER
- 20 examples of DRUG causing FUNCTION
- 8 examples of DISORDER causing FUNCTION
- 3 examples of DRUG TREATMENT_FOR FUNCTION

out of 8497 possible relations.

From the train set, we created a new corpus for relation classification. Similar to Zhong and Chen (2021), we extracted, for each pair of entities, the text between them (entities included). If there is no relation between the pair, it is annotated as 'O', otherwise the label in the train set was used. For the example in Figure 2, we take the text span 抗うつ剤に関しては抵抗がありましたが、安酒を煽るより**100**倍は建設的な精神状態 and annotate it as CAUSED.

We kept aside the 20% of the new corpus as validation set. Then on the remaining dataset

Figure 2: Example of CAUSED relation.

Task	Precision	Recall	F1
Task 2a - Fr	0.6068	0.3133	0.4132
Task 2a - Ja (dev)	0.5873	0.3581	0.4449
Task 2a - Ja	0.5752	0.5903	0.5827
Task 2b - Ja (dev)	0.1852	0.0564	0.0865
Task 2b - Ja	0.0226	0.0449	0.0301

Table 4: Performance on the Task 2 test set.

we fine-tuned `daisaku-s/medtxt_ner_roberta` as sequence classification task[17].

3.5 Test Results

The results are presented in Table 4. One of the reasons for the bad performance of the Japanese model is tokenization error. For example, in the text `ja_twjp_060-080_19` one of the entities is **DISORDER 47 50** 副作用, however at the given span, the `daisaku-s/medtxt_ner_roberta` tokenizer returns の副作用 as the single token.

4 Conclusion

In Task 1, despite training on the wrong dataset we managed to be in the top 50 percentile. The difficult part was normalization of the ADE using MedDRA dictionary, as a result F1-Norm was lower than F1-NER. For the Task 2, using a model adapted to the clinical domain helped to get the best results. The Task 2b (relation extraction) was challenging, given that the winning team obtained an overall F1 score of 0.0189. For future, We would explore approaches such as GLiNER (Zaratiana et al., 2023) and XMC (D'Oosterlinck et al., 2024) to improve NER in Task 1 and Task 2a.

References

Karel D'Oosterlinck, Omar Khattab, François Remy, Thomas Demeester, Chris Develder, and Christopher Potts. 2024. In-context learning for extreme multi-label classification. *arXiv preprint arXiv:2401.12178*.

Ben S Gerber and Arnold R Eiser. 2001. The patient-physician relationship in the internet age: future prospects and the research agenda. *Journal of medical Internet research*, 3(2):e842.

[15] `http://www.edrdg.org/wiki/index.php/JMdict-EDICT_Dictionary_Project`

[16] `https://huggingface.co/daisaku-s/medtxt_ner_roberta`

[17] `https://huggingface.co/docs/transformers/en/tasks/sequence_classification`

Natalia Grabar, Vincent Claveau, and Clément Dalloux. 2018. CAS: French corpus with clinical cases. In *Proceedings of the Ninth International Workshop on Health Text Mining and Information Analysis*, pages 122–128, Brussels, Belgium. Association for Computational Linguistics.

Anubhav Gupta and Frédéric Rayar. 2023. Frag at the ntcir-17 mednlp-sc task. In *Proceedings of the 17th NTCIR Conference on Evaluation of Information Access Technologies*, Tokyo, Japan. NII Institutional Repository.

Albert Q. Jiang, Alexandre Sablayrolles, Arthur Mensch, Chris Bamford, Devendra Singh Chaplot, Diego de las Casas, Florian Bressand, Gianna Lengyel, Guillaume Lample, Lucile Saulnier, Lélio Renard Lavaud, Marie-Anne Lachaux, Pierre Stock, Teven Le Scao, Thibaut Lavril, Thomas Wang, Timothée Lacroix, and William El Sayed. 2023. Mistral 7b. *Preprint*, arXiv:2310.06825.

Toshihiro Koyama, Shunya Iinuma, Michio Yamamoto, Takahiro Niimura, Yuka Osaki, Sayoko Nishimura, Ko Harada, Yoshito Zamami, and Hideharu Hagiya. 2023. International trends in adverse drug event-related mortality from 2001 to 2019: An analysis of the world health organization mortality database from 54 countries. *Drug Saf*, 47(3):237–249.

Yanis Labrak, Adrien Bazoge, Richard Dufour, Mickael Rouvier, Emmanuel Morin, Béatrice Daille, and Pierre-Antoine Gourraud. 2023. DrBERT: A Robust Pre-trained Model in French for Biomedical and Clinical domains. In *Proceedings of the 61th Annual Meeting of the Association for Computational Linguistics (ACL'23), Long Paper*, Toronto, Canada. Association for Computational Linguistics.

Yinhan Liu, Myle Ott, Naman Goyal, Jingfei Du, Mandar Joshi, Danqi Chen, Omer Levy, Mike Lewis, Luke Zettlemoyer, and Veselin Stoyanov. 2019. Roberta: A robustly optimized BERT pretraining approach. *CoRR*, abs/1907.11692.

Arjun Magge, Elena Tutubalina, Zulfat Miftahutdinov, Ilseyar Alimova, Anne Dirkson, Suzan Verberne, Davy Weissenbacher, and Graciela Gonzalez-Hernandez. 2021. DeepADEMiner: a deep learning pharmacovigilance pipeline for extraction and normalization of adverse drug event mentions on Twitter. *Journal of the American Medical Informatics Association*, 28(10):2184–2192.

Fabian Pedregosa, Gaël Varoquaux, Alexandre Gramfort, Vincent Michel, Bertrand Thirion, Olivier Grisel, Mathieu Blondel, Peter Prettenhofer, Ron Weiss, Vincent Dubourg, Jake Vanderplas, Alexandre Passos, David Cournapeau, Matthieu Brucher, Matthieu Perrot, and Édouard Duchesnay. 2011. Scikit-learn: Machine learning in Python. *Journal of Machine Learning Research*, 12:2825–2830.

Nils Reimers and Iryna Gurevych. 2019. Sentence-bert: Sentence embeddings using siamese bert-networks. In *Proceedings of the 2019 Conference on Empirical Methods in Natural Language Processing*. Association for Computational Linguistics.

Pontus Stenetorp, Goran Topić, Sampo Pyysalo, Tomoko Ohta, Jin-Dong Kim, and Jun'ichi Tsujii. 2011. Bionlp shared task 2011: Supporting resources. In *Proceedings of BioNLP Shared Task 2011 Workshop*, pages 112–120, Portland, Oregon, USA. Association for Computational Linguistics.

Shoko Wakamiya, Lis Kanashiro Pereira, Lisa Raithel, Hui-Syuan Yeh, Peitao Han, Seiji Shimizu, Tomohiro Nishiyama, Gabriel Herman Bernardim Andrade, Noriki Nishida, Hiroki Teranishi, Narumi Tokunaga, Philippe Thomas, Roland Roller, Pierre Zweigenbaum, Yuji Matsumoto, Akiko Aizawa, Sebastian Möller, Cyril Grouin, Thomas Lavergne, Aurélie Névéol, Patrick Paroubek, Shuntaro Yada, and Eiji Aramaki. 2023. Ntcir-17 MedNLP-SC social media adverse drug event detection: Subtask overview. In *Proceedings of the 17th NTCIR Conference on Evaluation of Information Access Technologies*, Tokyo, Japan. NII Institutional Repository.

Thomas Wolf, Lysandre Debut, Victor Sanh, Julien Chaumond, Clement Delangue, Anthony Moi, Pierric Cistac, Tim Rault, Rémi Louf, Morgan Funtowicz, Joe Davison, Sam Shleifer, Patrick von Platen, Clara Ma, Yacine Jernite, Julien Plu, Canwen Xu, Teven Le Scao, Sylvain Gugger, Mariama Drame, Quentin Lhoest, and Alexander M. Rush. 2020. Transformers: State-of-the-art natural language processing. In *Proceedings of the 2020 Conference on Empirical Methods in Natural Language Processing: System Demonstrations*, pages 38–45, Online. Association for Computational Linguistics.

Dongfang Xu, Guillermo Lopez Garcia, Lisa Raithel, Rolland Roller, Philippe Thomas, Eiji Aramaki, Shuntaro Yada, Pierre Zweigenbaum, Sai Tharuni Samineni, Karen O'Connor, Yao Ge, Sudeshna Das, Abeed Sarker, Ari Klein, Lucia Schmidt, Vishakha Sharma, Raul Rodriguez-Esteban, Juan Banda, Ivan Flores Amaro, Davy Weissenbacher, and Graciela Gonzalez-Hernandez. 2024. Overview of the 9th social media mining for health applications (#SMM4H) shared tasks at ACL 2024. In *Proceedings of The 9th Social Media Mining for Health Research and Applications Workshop and Shared Tasks*, Bangkok, Thailand. Association for Computational Linguistics.

Shuntaro Yada, Yuta Nakamura, Shoko Wakamiya, and Eiji Aramaki. 2022. Real-mednlp: Overview of real document-based medical natural language processing task subtasks. In *Proceedings of the 16th NTCIR Conference on Evaluation of Information Access Technologies*, Tokyo, Japan. NII Institutional Repository.

Urchade Zaratiana, Nadi Tomeh, Pierre Holat, and Thierry Charnois. 2023. Gliner: Generalist model for named entity recognition using bidirectional transformer. *Preprint*, arXiv:2311.08526.

Zexuan Zhong and Danqi Chen. 2021. A frustratingly easy approach for entity and relation extraction. In *North American Association for Computational Linguistics (NAACL)*.

A Mistral Prompt Template

The LLM input used for NER is a few-shot prompt containing two examples. It uses the text from fr_1025_lifeline_v1_FR_1971_1_1647857960 and fr_1069_lifeline_v1_FR_6168_1_1648459053 as examples. The entities and the relations for each example is used as shown below:

<s>[INST] From the medical report in French below, extract all the mentions of entities DRUG, DISORDER and the relationships CAUSED and TREATMENT_FOR in brat format.
TEXT 1[/INST] DRUG pilules

DISORDER angoisses
FUNCTION règles
...
DISORDER humeur
FUNCTION hormones
CAUSED Arg1:pilules Arg2:angoisses
...
TREATMENT_FOR Arg1:Insidon Arg2:humeur

</s>[INST] From the medical report in French below, extract all the mentions of entities DRUG, DISORDER and the relationships CAUSED and TREATMENT_FOR in brat format.
TEXT 2 [/INST] DRUG mirtazapine

DISORDER problèmes de sommeil
FUNCTION dors
...
CAUSED Arg1:antiémétiques Arg2:somnolence
...
TREATMENT_FOR Arg1:zolpidem Arg2:troubles du sommeil

</s>[INST] From the medical report in French below, extract all the mentions of entities DRUG, DISORDER and the relationships CAUSED and TREATMENT_FOR in brat format.
TEXT FROM THE TEST SET [/INST]

KUL@SMM4H2024: Optimizing Text Classification with Quality-Assured Augmentation Strategies

Sumam Francis
KU Leuven
sumam.francis@kuleuven.be

Marie-Francine Moens
KU Leuven
sien.moens@kuleuven.be

Abstract

This paper presents our models for the Social Media Mining for Health 2024 shared task, specifically Task 5, which involves classifying tweets reporting a child with childhood disorders (annotated as "1") versus those merely mentioning a disorder (annotated as "0"). We utilized a classification model enhanced with diverse textual and language model-based augmentations. To ensure quality, we used semantic similarity, perplexity, and lexical diversity as evaluation metrics. Combining supervised contrastive learning and cross-entropy-based learning, our best model, incorporating R-drop and various LM generation-based augmentations, achieved an impressive F1 score of 0.9230 on the test set, surpassing the task mean and median scores.

1 Introduction

The Social Media Mining for Health (SMM4H-24) (Xu et al., 2024) shared task 5 aims to explore data sources for assessing the link between pregnancy exposures and childhood disorders in real-time and at scale. Many children face lifelong disorders, with 17% in the U.S. diagnosed with developmental disabilities and 8% with asthma. This task involves a binary classification to automatically distinguish tweets from users who reported their pregnancy and have a child with childhood developmental disorders (annotated as "1"), from tweets that merely mention a disorder without evidence of a diagnosis (annotated as "0").

SMM4H Task 5 includes datasets of English tweets with annotations on the presence or absence of childhood disorders. Recent studies (Wu et al., 2021; Francis and Moens, 2023; Liu et al., 2022) have shown that data augmentation enhances training data diversity and model robustness. It's crucial to ensure the quality of these augmentations to maintain original meanings and diversify training data. Augmented data using model-based augmentations together with regularised dropout (R-drop) (Wu et al., 2021) serve as regularization methods, mitigating overfitting in machine learning models. Model-based augmentations involve generating new data samples using pre-trained language models (LMs) that have rich semantic and syntactic knowledge stored in their parameters. This can create variations of the original data while preserving its semantic content.

2 Methodology

This section outlines our methodology consisting of 3 steps: 1. enhancing textual data with LM-based augmentation techniques, 2. assessing augmentation quality, and, 3. finetuning a transformer-based (Vaswani et al., 2017) model Bertweet (Nguyen et al., 2020) integrating supervised contrastive loss, cross-entropy loss, and regularised dropout (R-Drop) (Wu et al., 2021) in the training process.

2.1 LM-based Data Augmentation

The textual augmentations leveraging LMs used to enhance training data are **masked language modeling (MLM)**: where we mask certain tokens or spans in the input text and predict them using context from surrounding tokens (Devlin et al., 2019). We used the BERT-large model for MLM, which predicts the masked tokens based on the surrounding context. Tokens to be masked are selected based on their importance and identified using POS tags[1]. VERB and ADJECTIVE tags are masked, while NOUN phrases are left intact to preserve vital classification information. The next method is **text replacement using LMs:** This technique replaces specific words or phrases in the input text with semantically similar alternatives predicted by an LM, generating diverse and meaningful variations of the input text. For text replacement, the GPT-

[1] https://www.nltk.org/

Proceedings of The 9th Social Media Mining for Health Applications (#SMM4H) Workshop and Shared Tasks, pages 142–145
August 15, 2024 ©2024 Association for Computational Linguistics

2 (Radford et al., 2019) model was employed to predict semantically similar alternatives, differing from MLM as it directly replaces words rather than predicting masked tokens. The third approach is **back translation:** where the English-French Marian MT[2] translation model is used to translate input sentences to French and then back to English, creating nuanced paraphrasing (Sennrich et al., 2016).

2.2 Evaluation Metrics for Augmentation Quality

Ensuring the quality of generated augmentations is essential to ensure they retain the original meaning and enhance the training data. We use several methods to evaluate augmentation quality. **Semantic similarity:** measures the degree of similarity between original and augmented text based on meaning, using Sentence BERT (Reimers and Gurevych, 2019) to calculate embedding-based cosine similarity. **Perplexity:** measures how well an LM predicts text samples, with lower perplexity indicating better performance and more meaningful text. We use a pre-trained GPT-2 (Radford et al., 2019) model to calculate the perplexity. **Lexical diversity**: measures the variety of vocabulary used in the text, with higher lexical diversity indicating a richer expression and ensuring useful augmentations. We use token overlap as a measure to calculate the diversity (McCarthy and Jarvis, 2010). Further, these data are used to train the classification model incorporating R-drop.

Table 1: Precision (P), Recall (R) and F1 scores (F1) on the validation set of the SMM4H2024 Task 5 with BERTweet model.

Augmentation	F1	P	R
-	0.9230	0.9078	0.9301
+ LM-aug	0.9309	0.9143	0.9471
+ LM-aug+aug-ql	0.9358	0.9225	**0.9497**
+ LM-aug+aug-ql+R-drop	**0.9398**	**0.9403**	0.9333

Table 2: Precision (P), Recall (R) and F1 scores (F1) on the test set of the SMM4H2024 Task 5 with BERTweet model.

Augmentation	F1	P	R
+ LM-aug+aug-ql+R-drop	0.923	0.906	0.940
Posteval			
+ LM-aug(pos) +aug-ql+R-drop	0.938	0.927	0.949
Task mean results	0.822	0.818	0.838
Task median results	0.901	0.885	0.917

3 Experiments and Results

The dataset comprises a training set ($7,398$ tweets), a validation set (389 tweets), and a test set ($10,000$ tweets). For pre-processing, we removed URLs, retweets, mentions, extra spaces, non-ASCII words, and characters. We also lower-cased the text, trimmed white spaces, and inserted spaces between punctuation marks.

For classification, each model was fine-tuned over 10 epochs with a learning rate of $5e-5$ using the Adam optimizer. The batch size is set to 32, and the maximum sequence length to 128. We use PyTorch and HuggingFace [3] library for training the BERTweet large model (Nguyen et al., 2020), applying both cross-entropy loss and supervised contrastive loss (Khosla et al., 2020). Model checkpoints are saved every 200 step based on the validation set's F1-score. The BERTweet loss function was adapted to include KL divergence for R-drop regularization. Training data was enriched with LM augmentations from section 3 (LM-aug). Quality checks are integrated into the augmentation pipeline to filter out poor quality augmentations (aug-ql).

Threshold values for the augmentation quality checks were empirically determined based on the performance of the validation set. This involved experimenting with different thresholds and observing their impact on the model's performance metrics (Precision, Recall, F1 scores) on the validation data. We set the following threshold values for augmentation quality check: semantic similarity > 0.7, perplexity < 100, and $0.4 <$ lexical diversity < 0.75. The semantic similarity threshold (> 0.7) ensures that the augmented text retains a high degree of similarity in meaning to the original text. The perplexity threshold (< 100) ensures that the augmented text is coherent and grammatically correct. We calculated the perplexity using a pre-trained GPT-2 model. Lower perplexity indicates better language model prediction quality. When experimenting with higher thresholds the generated text was less meaningful. A threshold of 100 was chosen because it balanced coherence with augmentation diversity. The lexical diversity range ($0.4 - 0.75$) ensures a balance between the diversity and relevance of the vocabulary used in the augmented text. A range of 0.4 to 0.75 was set to avoid too much similarity (which would defeat the purpose of augmentation) and too much dif-

[2]Helsinki-NLP/opus-mt-en-fr

[3]https://huggingface.co/models

ference (which might change the context or meaning). These thresholds were optimized iteratively, with each adjustment followed by re-evaluating the model's performance on the validation set to ensure the thresholds contributed positively to the model's overall effectiveness.

Incorporating LM augmentations and R-drop into BERTweet yielded improved performance for detecting childhood disorder diagnoses in tweets (see Tables 1 and 2) compared to baseline model setup. Augmentations diversified the data, enhancing the model's robustness and generalization. The metrics for evaluating augmentation quality further improved model performance by ensuring high-quality training data. The diversity from augmented data reduced the risk of overfitting and the model's reliance on specific patterns. Contextually generated LM-augmented examples facilitated better language understanding. The integration of R-drop with supervised contrastive loss and cross-entropy loss further promoted the learning of more generalized features by capturing various aspects of the data, enhancing the precision of the model. In the post-evaluation, the LM augmentations were performed only on the positive class (pos) which improved results. The results surpassed the shared task's mean and median scores, demonstrating the effectiveness of this approach.

4 Conclusion

In this work, we developed a classification model enhanced with R-drop and LM-based augmentations to mitigate label imbalance and avoid overfitting, thereby improving performance. We incorporated evaluation metrics like semantic similarity, perplexity, and lexical diversity to ensure the quality of the augmentations, adding only those that meet set thresholds. Our approach of integrating data augmentation and rigorous filtering strategies showed superior performance compared to the baselines we set up. This can be attributed to the enriched dataset, which reduces overfitting and enhances generalization.

Similar performance levels were achieved by the BERTweet model in (Klein et al., 2024), highlighting that the choice of model architecture and hyperparameter tuning is crucial. However, our LM-based augmentations provide a significant edge in data diversity and model robustness. Furthermore, by augmenting only the positive class, our model demonstrated even more significant improvements in the overall F1 scores. The use of LM-based augmentations introduced meaningful variations in the training data, helping the model learn to generalize better from a more diverse set of examples. The rigorous filtering strategies ensured that only high-quality augmented data were used, preserving the original meaning and enhancing the model's ability to handle diverse inputs. Our approach significantly enhances model generalizability, achieving an impressive F1 score of 0.923 on the test set.

References

Jacob Devlin, Ming-Wei Chang, Kenton Lee, and Kristina Toutanova. 2019. BERT: pre-training of deep bidirectional transformers for language understanding. In *Proceedings of the 2019 Conference of the North American Chapter of the Association for Computational Linguistics: Human Language Technologies, NAACL-HLT*. Association for Computational Linguistics.

Sumam Francis and Marie-Francine Moens. 2023. Text augmentations with r-drop for classification of tweets self reporting covid-19. *CoRR*, abs/2311.03420.

Prannay Khosla, Piotr Teterwak, Chen Wang, Aaron Sarna, Yonglong Tian, Phillip Isola, Aaron Maschinot, Ce Liu, and Dilip Krishnan. 2020. Supervised contrastive learning. *CoRR*, abs/2004.11362.

Ari Z Klein, José Agustín Gutiérrez Gómez, Lisa D Levine, and Graciela Gonzalez-Hernandez. 2024. Using longitudinal twitter data for digital epidemiology of childhood health outcomes: An annotated data set and deep neural network classifiers. *J Med Internet Res*, 26.

Zhiwei Liu, Yongjun Chen, Jia Li, Man Luo, Philip S. Yu, and Caiming Xiong. 2022. Improving contrastive learning with model augmentation. *CoRR*, abs/2203.15508.

Philip M. McCarthy and Scott Jarvis. 2010. Mtld, vocd-d, and hd-d: A validation study of sophisticated approaches to lexical diversity assessment. *Behavior Research Methods*, 42:381–392.

Dat Quoc Nguyen, Thanh Vu, and Anh Tuan Nguyen. 2020. Bertweet: A pre-trained language model for english tweets. In *Proceedings of the 2020 Conference on Empirical Methods in NLP*. ACL.

Alec Radford, Jeff Wu, Rewon Child, David Luan, Dario Amodei, and Ilya Sutskever. 2019. Language models are unsupervised multitask learners.

Nils Reimers and Iryna Gurevych. 2019. Sentence-bert: Sentence embeddings using siamese bert-networks. In *Proceedings of the 2019 Conference on Empirical Methods in Natural Language Processing and the 9th International Joint Conference on Natural*

Language Processing, EMNLP-IJCNLP 2019, Hong Kong, China, November 3-7, 2019, pages 3980–3990. Association for Computational Linguistics.

Rico Sennrich, Barry Haddow, and Alexandra Birch. 2016. Improving neural machine translation models with monolingual data. In *Proceedings of the 54th Annual Meeting of the Association for Computational Linguistics, ACL 2016, August 7-12, 2016, Berlin, Germany, Volume 1: Long Papers*. The Association for Computer Linguistics.

Ashish Vaswani, Noam Shazeer, Niki Parmar, Jakob Uszkoreit, Llion Jones, Aidan N. Gomez, Lukasz Kaiser, and Illia Polosukhin. 2017. Attention is all you need. In *Advances in Neural Information Processing Systems 30: Annual Conference on Neural Information Processing Systems*.

Lijun Wu, Juntao Li, Yue Wang, Qi Meng, Tao Qin, Wei Chen, Min Zhang, Tie-Yan Liu, et al. 2021. R-drop: Regularized dropout for neural networks. *Advances in Neural Information Processing Systems*, 34.

Dongfang Xu, Guillermo Lopez Garcia, Lisa Raithel, Rolland Roller, Philippe Thomas, Eiji Aramaki, Shuntaro Yada, Pierre Zweigenbaum, Karen O'Connor, Yao Ge, Sudeshna Das, Abeed Sarker, Ari Klein, Lucia Schmidt, Vishakha Sharma, Raul Rodriguez-Esteban, Juan Banda, Ivan Flores Amaro, Sai Tharuni Samineni, and Graciela Gonzalez-Hernandez. 2024. Overview of the 9th social media mining for health applications (#SMM4H) shared tasks at ACL 2024. In *Proceedings of the 9th Social Media Mining for Health Applications Workshop and Shared Task*.

LHS712NV at #SMM4H 2024 Task 4: Using BERT to classify Reddit posts on non-medical substance use

Valeria Fraga,[1] Neha Nair,[1] Dalton Simancek,[2] V.G.Vinod Vydiswaran[2,1]
[1]School of Information, [2]Department of Learning Health Sciences
University of Michigan, Ann Arbor
{vfraga, nehakkn, daltonsi, vgvinodv}@umich.edu

Abstract

This paper summarizes our participation in the Shared Task 4 of #SMM4H 2024. Task 4 was a named entity recognition (NER) task identifying clinical and social impacts of non-medical substance use in English Reddit posts. We employed the Bidirectional Encoder Representations from Transformers (BERT) model to complete this task. Our team achieved an F1-score of 0.892 on a validation set and a relaxed F1-score of 0.191 on the test set.

1 Introduction

Substance use, whether prescription or illicit, poses a significant public health challenge, contributing to addiction, overdose, and various associated health concerns (Lo et al., 2020 Apr). Understanding the clinical and social ramifications of non-medical substance use is crucial for enhancing the treatment of substance use disorder, informing intervention strategies, and developing preventive measures (Xu et al., 2024). This paper addresses a named entity recognition (NER) task focused on identifying two key entity types: clinical and social impacts. Clinical impacts refer to the effects of substance use on individuals' health and well-being, while social impacts encompass broader societal consequences (Xu et al., 2024).

The Social Media Mining for Health (#SMM4H) shared tasks aim to leverage natural language processing (NLP) techniques to extract valuable health insights from social media data. In #SMM4H 2024, we participated in Task 4 on extraction of the clinical and social impacts of non-medical substance use from Reddit posts. We were particularly motivated to participate in this task because of the pressing public health concern surrounding non-medical substance use. Understanding the clinical and social impacts of such usage is crucial for developing effective interventions and prevention strategies. By leveraging advanced, deep-learning based NLP models, we aimed to contribute to this understanding and advance the field of health informatics through innovative data analysis techniques.

2 System Description

2.1 Data Preprocessing

The #SMM4H 2024 Task 4 is a named entity recognition task with a goal to identify two entities – *Clinical Impacts* and *Social Impacts* – in Reddit posts. Individual sequences of tokens are tagged with one of these two labels or *No Label (-)* to denote neither of the two desired classes. (Ge et al., 2024) To perform this task, we use the Bidirectional Encoder Representations from Transformers (BERT) model (Devlin et al., 2019), which is highly effective at capturing contextual information from both the previous and subsequent sequences of text tokens. The process begins with data preprocessing, where the dataset is first split into training and test sets. Then, the text is tokenized, with each token being assigned one of the following labels: *Clinical Impacts*, *Social Impacts*, or *No Label (-)*.

2.2 BERT Fine-tuning and Evaluation

The model is then fine-tuned using the HuggingFace Transformers library, with training parameters such as batch size, learning rate, and the number of epochs specified. The selection of BERT for this task is underscored by its ability to comprehend the contextual nuances of words, enabling accurate identification of named entities. Fine-tuning pre-trained BERT models often results in enhanced performance on downstream tasks such as NER, making it a popular choice in natural language processing applications.

During training, four performance metrics, viz. precision, recall, F1 score, and accuracy, are computed to assess the model's performance. Subsequently, the model is evaluated on a separate validation dataset to gauge its effectiveness. Predictions are generated, and the four metrics are calculated and displayed, along with a visualization

Proceedings of The 9th Social Media Mining for Health Applications (#SMM4H) Workshop and Shared Tasks, pages 146–148
August 15, 2024

Epoch	Precision	Recall	F1	Accuracy
1	0.929	0.988	0.958	0.929
2	0.938	0.983	0.960	0.937
3	0.940	0.980	0.960	0.938

Table 1: Model evaluation of training data

Run	Precision	Recall	F1	Accuracy
1	0.892	0.892	0.892	0.892

Table 2: Model evaluation of validation data

of the confusion matrix. Finally, the trained model is tested on a separate dataset, and the predictions are saved for further analysis or task submission.

3 Results

3.1 Model Performance on training and validation

The model's performance was evaluated using training and validation datasets. The training phase comprised three epochs, each reporting training loss, validation loss, precision, recall, F1 score, and accuracy metrics (Table 1). Additionally, results on the validation data (performance metrics in Table 2 and confusion matrix in Figure 1) provided further insights into the model's performance. These results showed that the model did not identify any entities in the Social Impacts class and tended to label entities to the majority class (No label).

3.2 Test Data Analysis

Although the model seemed to perform well overall with the training and validation data, it did not perform as well with the test data. With the test data, the model had a token-level F1 score of 17.15% and a relaxed F1 score of 19.12%.

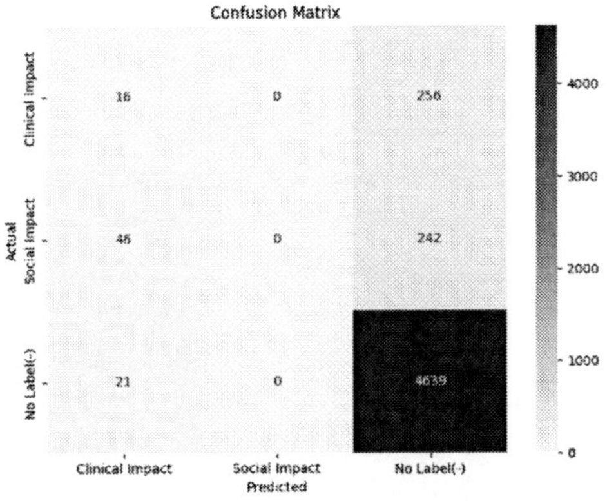

Figure 1: Confusion matrix of validation data

Labels	Accuracy (%)
No label	99.5
Clinical Impacts	5.9
Social Impacts	0.0

Table 3: The percent accuracy of the model's predicted labels

These findings collectively underscore the model's efficacy in classifying instances with *No label (-)* accurately while indicating room for improvement, particularly in distinguishing *Social Impact* instances (Table 3). Further optimization efforts may be warranted to enhance the model's performance across all classes.

4 Conclusion

This paper discusses our submission for #SMM4H 2024 Task 4 on named entity recognition (NER) for identifying clinical and social impacts of substance use in Reddit posts using Large Language Models (LLMs), particularly the BERT model. While the BERT model demonstrated high overall performance, there were notable disparities in its efficacy across different metrics and datasets. Challenges remain in accurately identifying *Social Impact* instances, highlighting the need for further refinement. Despite these challenges, the BERT model shows promise in automating the detection of clinical impacts in Reddit posts. To facilitate continued development, we have shared the implementation code for our system participation on GitHub.[1]

Limitations

In spite of the promising results obtained, it is crucial to acknowledge several limitations inherent in our study. The utilization of Reddit data presents inherent unpredictability due to the dynamic and unregulated nature of user-generated content. This variability may introduce noise and biases into our model, potentially impacting its generalizability to broader contexts. The size of the training data poses a potential constraint. With only 800 posts available for training, the dataset may not capture the full spectrum of patterns and nuances present in the data, thereby limiting the model's ability to discern complex relationships and generalize effectively. Inconsistencies in model performance raise concerns about reliability and robustness. Despite employing advanced deep-learning

[1] The software code, written in Python, is available at: https://github.com/NLP4HealthUMich/SMM4H2024-Task4

based techniques such as the BERT model, we observed fluctuating F1 scores, ranging from 0.4 to 0.9. These inconsistencies suggest potential issues in model stability or sensitivity to varying conditions, warranting further investigation and refinement. Finally, we did not use the "entity or not" column in the training of our model or investigate the results of other models besides the BERT model, which could have added much-needed context to the model and resulted in better performance. In light of these limitations, a cautious interpretation of the results is advised. Future research should address these constraints to enhance the validity and applicability of the findings and explore the performance of other models on the data.

Ethics Statement

Firstly, we acknowledge that the dataset provided to us was already anonymized. We prioritize the privacy and confidentiality of Reddit users, and we are committed to maintaining the anonymity of individuals whose posts are included in our dataset. We are dedicated to ensuring that our research contributes positively to understanding and mitigating substance use disorders. By identifying the clinical and social impacts of substance use, we aim to provide valuable insights that can inform intervention strategies and support efforts to address this pressing public health issue.

The Reddit posts for this study were sourced through the #SMM4H 2024 shared task coordinators and were accessed under the guidance of an academic advisor, strictly for research purposes. The data was exclusively utilized for participation in #SMM4H 2024 Task 4 and no other uses. We acknowledge our study's limitations, including the potential biases and uncertainties inherent in working with anonymized data. We remain transparent about these limitations and encourage future research to address them through robust methodologies and data validation techniques.

Overall, our research is conducted with the utmost integrity and respect for ethical considerations. We are dedicated to advancing our understanding of natural language processing techniques for health insights while upholding ethical standards and promoting the well-being of individuals and communities affected by substance use.

References

Jacob Devlin, Ming-Wei Chang, Kenton Lee, and Kristina Toutanova. 2019. BERT: Pre-training of deep bidirectional transformers for language understanding. In *Proceedings of the 2019 Conference of the North American Chapter of the Association for Computational Linguistics: Human Language Technologies, Volume 1 (Long and Short Papers)*, pages 4171–4186, Minneapolis, Minnesota. Association for Computational Linguistics.

Yao Ge, Sudeshna Das, Karen O'Connor, Mohammed Ali Al-Garadi, Graciela Gonzalez-Hernandez, and Abeed Sarker. 2024. Reddit-impacts: A named entity recognition dataset for analyzing clinical and social effects of substance use derived from social media. DOI:10.48550/arXiv.2405.06145.

T. Wing Lo, Jerf W. K. Yeung, and Cherry H. L. Tam. 2020 Apr. Substance abuse and public health: A multilevel perspective and muliple responses. *Int J Environ Res Public Health*, 17(7):2610. DOI: 10.3390/ijerph17072610.

Dongfang Xu, Guillermo Lopez Garcia, Lisa Raithel, Rolland Roller, Philippe Thomas, Eiji Aramaki, Shuntaro Yada, Pierre Zweigenbaum, Sai Tharuni Samineni, Karen O'Connor, Yao Ge, Sudeshna Das, Abeed Sarker, Ari Klein, Lucia Schmidt, Vishakha Sharma, Raul Rodriguez-Esteban, Juan Banda, Ivan Flores Amaro, Davy Weissenbacher, and Graciela Gonzalez-Hernandez. 2024. Overview of the 9th social media mining for health applications (#SMM4H) shared tasks at ACL 2024. In *Proceedings of The 9th Social Media Mining for Health Research and Applications Workshop and Shared Tasks*, Bangkok, Thailand. Association for Computational Linguistics.

712forTask7 at #SMM4H 2024 Task 7: Classifying Spanish Tweets Annotated by Humans versus Machines with BETO Models

Hafizh R. Yusuf,[1] David Belmonte,[2] Dalton Simancek,[3] V.G.Vinod Vydiswaran[3,1]
[1]School of Information, [2]Department of Psychiatry, [3]Department of Learning Health Sciences
University of Michigan, Ann Arbor
{hafizhry, dbelmont, daltonsi, vgvinodv}@umich.edu

Abstract

The goal of Social Media Mining for Health (#SMM4H) 2024 Task 7 was to train a machine learning model that is able to distinguish between annotations made by humans and those made by a Large Language Model (LLM). The dataset consisted of tweets originating from #SMM4H 2023 Task 3, wherein the objective was to extract COVID-19 symptoms in Latin-American Spanish tweets. Due to the lack of additional annotated tweets for classification, we reframed the task using the available tweets and their corresponding human or machine annotator labels to explore differences between the two subsets of tweets. We conducted an exploratory data analysis and trained a BERT-based classifier to identify sampling biases between the two subsets. The exploratory data analysis found no significant differences between the samples and our best classifier achieved a precision of 0.52 and a recall of 0.51, indicating near-random performance. This confirms the lack of sampling biases between the two sets of tweets and is thus a valid dataset for a task designed to assess the authorship of annotations by humans versus machines.

1 Introduction

Tweeting has become a significant way for people to connect and communicate with each other individually and as a community. In 2021, 23% of adults in the United States reported to using Twitter (Dinesh and Odabaş, 2023). This type of social media is used for a variety of reasons, including news, entertainment, communication between family and friends, information on brands, and for developing a professional network. For example, COVID-19 spurred Twitter users to share their personal experiences, emotions, and beliefs during a turbulent and confusing time of a global epidemic (Cuomo et al., 2021).

Given the prevalence of its use, tweet content can provide information for understanding public sentiment and identifying specific areas of concern. However, sifting through the volumes of tweets can be an arduous task. Machine learning provides an automated approach for analyzing tweets. Leveraging its tremendous processing speed, a machine learning model, such as BERT and GPT-3, can be trained to identify specific patterns and annotate features of interest quickly compared to manual annotation by humans (Ding et al., 2023).

The #SMM4H 2024 Shared Task 7 builds over the dataset from Task 3 of the 2023 Social Media Mining for Health. Task 3 involved a dataset of 10,150 tweets describing COVID-related symptoms (Klein et al., 2023). The dataset was annotated by human experts who were medical doctors and also native speakers of Latin-American Spanish. The 2023 challenge involved training a machine learning model that could identify and extract symptoms in the tweets that were either personally described or mentioned by a third party. The leading model achieved an F1 score of 0.94 for identifying the character offsets of COVID-19 symptoms.

The dataset provided for the #SMM4H 2024 Shared Task 7 utilized some of the tweets from #SMM4H 2023 that were annotated by either human or machine (Xu et al., 2024). The intended aim was to identify whether the tweet was annotated by a human or a machine. Due to the lack of additional tweet annotation data for this classification task, we first conducted an exploratory data analysis to investigate if the two subset of data – those annotated by humans versus machines – were fundamentally different. Then, we trained a BERT-based classifier to explore differences in content between the two subsets of tweets. Our primary research goal was to confirm the absence of sampling biases that could affect the training of a classifier designed to distinguish between human and machine annotations.

Proceedings of The 9th Social Media Mining for Health Applications (#SMM4H) Workshop and Shared Tasks, pages 149–152
August 15, 2024

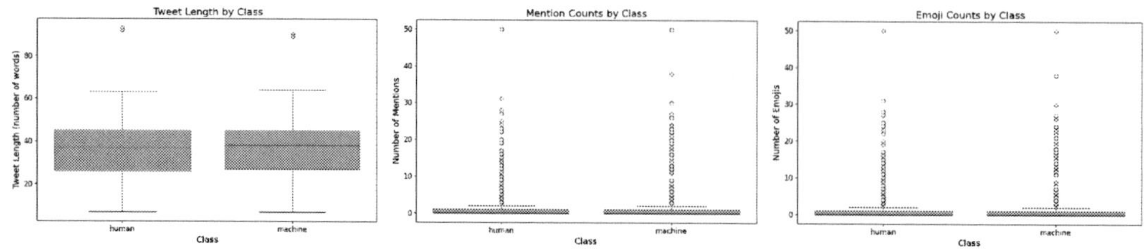

(a) (b) (c)

Figure 1: Comparative analysis of tweet characteristics by class and (a) tweet length, (b) mentions, and (c) emojis.

2 Exploratory Data Analysis

The Task 7 dataset included only the tweet text and an associated label indicating the author of the annotation, without any additional text annotations. While our initial understanding of the task was that the model should distinguish between annotations generated by humans versus machines, the absence of any generated annotations confirmed the overall purpose of the task. We framed our work to leverage the available data and examine the differences between tweets annotated by humans and those annotated by machines. We trained a binary tweet classifier to verify that there are no significant differences between the tweets in the two annotation groups. This aims to ensure that the tweets that were sampled for human annotation vs. machine annotation did not suffer from selection bias, and such bias did not affect the classifier trained to predict annotation authorship, as intended in the original Task 7.

We first approached the challenge by inspecting the provided dataset. There was a total of 4,603 tweets of which 2,288 were labelled "human" and 2,315 labeled "machine." The counts were sufficiently balanced. Then, we performed an exploratory data analysis to dive deeper into the two subset features.

As seen in Fig. 1, a comparison of the tweet lengths, number of mentions, and emojis did not show any significant differences between those labeled as humans or machines. This initial comparison suggests that both groups of tweets have similar textual features.

Word cloud analysis on the human and machine-annotated tweets were also similar, indicating that the overall words and their frequencies were similar between both groups (Fig. 2). Finally, analyzing the bi-grams, which include pairs of consecutive words, also didn't show any significant differences

Figure 2: Comparative analysis of tweet word cloud by class

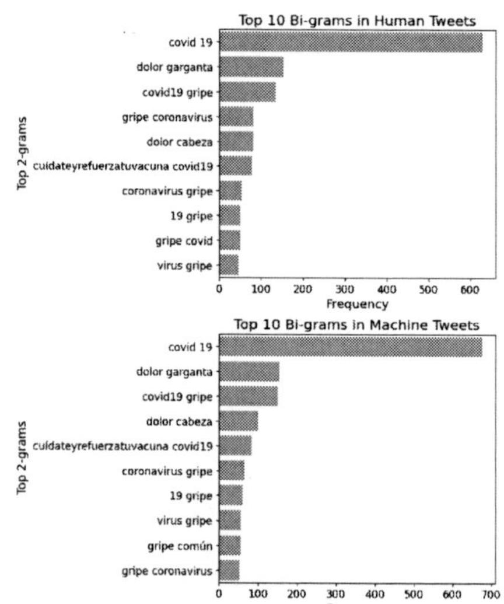

Figure 3: Comparative analysis of tweet bi-grams by class

between the two groups (Fig. 3). This further supports the hypothesis that the language content of the tweets was similar regardless of whether humans or machines annotated them.

3 BERT Model Selection and Fine-tuning

Data Pre-processing: After conducting the exploratory data analysis, we found that the number of emojis and mentions was similar between the two groups. Therefore, we removed emojis and mentions from the tweets to enable the model to focus on the core textual content. We retained the stop words to preserve the context of the tweets, which is beneficial for training a large language model.

Model training As the tweets were written in Spanish, we trained a large language model (LLM) using BETO, a BERT-based model for Spanish text (Cañete et al., 2020). This LLM was also referenced in the #SMM4H 2023 task overview (Klein et al., 2023). We fine-tuned BETO by adjusting its hyperparameters for optimal results.

For our initial submission, we found that the optimal parameters for the fine-tuned BETO model were as follows: training epochs=5, train batch size=8, evaluation batch size=32, learning rate=5e-5, and weight decay=0.01. In the following sections, this is referred to as Configuration 1.

After the evaluation period ended, we tried to further improve the model's performance and consistency by training it with various parameters and defining the seeds for the Transformers, PyTorch, and Numpy libraries. Additionally, to gain more robust metrics performance, we trained the model three times and submitted each of the runs to get their precision and recall scores. The best alternate configuration was as follows: training epochs=15, train batch size=16, evaluation batch size=32, warmup steps=100, learning rate=1e-6, and weight decay=0.01. This is referred to as Configuration 2.

4 Results

As shown in Table 1, Configuration 1 yielded a precision of 0.52 and a recall of 0.51 on the test set. Based on the scores scared by the organizers, this placed our submission at the top of the leaderboard.

In the post-evaluation phase, we found that Configuration 2 yielded better performance and consistency, with an average precision of 0.51 and a recall of 0.54, which was a slight increase compared to the initial submission. Tuning other hyperparameters did not yield any significant improvement.

Configuration	Precision	Recall	F1 score
Configuration 1	0.52	0.51	0.51
Configuration 2	0.51	0.54	0.52

Table 1: Model performance based on two parameter configurations

5 Discussion

Despite initial misunderstanding of the task, we realized that the actual aim of the shared task was to determine whether the dataset used in previously published work on generating machine-based annotations incurred any training bias. The dataset provided for training and validation did not contain any additional information aside from the tweet text and label. Particularly, there was no information about the specific COVID symptoms from the human or machine annotator that may have been informative in distinguishing between the human and machine annotations.

Based on our results, our models were unable to find any key distinguishing factors separating human-annotated and machine-annotated tweets. This strengthens the conclusions of the baseline system mentioned in the challenge, which achieved an 82% classification accuracy for a tweet annotation classifier. Our best models, formulated as tweet authorship classifiers, only achieved a random-chance performance, indicating that the tweets labeled by machines were indistinguishable from those labeled by humans. This aligns with our exploratory data analysis, which also revealed that the features of the two subsets of tweets are very similar.

6 Conclusion

Our best fine-tuned BETO model achieved an overall F1 score of 0.52, and was slightly better than our submitted run, which achieved an F1 score of 0.51. Our submitted run achieved the highest performance on #SMM4H 2024 Task 7 to distinguish between tweets that were annotated by humans vs. machines. Our findings from exploratory data analysis and training a classifier between the two subsets of tweets indicate no sampling biases and help confirm the applicability of this dataset in training a tweet annotation authorship classifier.

Limitations

Our study faced several limitations that should be noted. Firstly, the dataset was exclusively in Latin-American Spanish, which limited the author's understanding because none of them were native Spanish speakers. Additionally, we encountered the absence of detailed tweet annotations, particularly specific COVID-19 symptoms identified by human or machine annotators, further limiting our ability to distinguish between human and machine annotations effectively. These constraints may have influenced our model's performance and its alignment with the intended research objective of the #SMM4H shared task. Future research should incorporate detailed tweet annotations to enhance task clarity and model effectiveness.

Ethics Statement

In the Task 7 dataset of the #SMM4H 2024, which includes tweets referencing COVID-19 symptoms, the authors recognize these tweets as private data that is publicly accessible under the ongoing consent provided by Twitter/X's User Agreement and Terms and Conditions. The tweets for this study were sourced through the #SMM4H 2024 shared task coordinators and were accessed under the guidance of an academic advisor strictly for research purposes. The data was exclusively utilized for participation in #SMM4H 2024 Task 7 and for no other uses.

References

José Cañete, Gabriel Chaperon, Rodrigo Fuentes, Jou-Hui Ho, Hojin Kang, and Jorge Pérez. 2020. Spanish pre-trained bert model and evaluation data. In *Practical ML for Developing Countries Workshop @ ICLR 2020*, pages 1–9.

Raphael E. Cuomo, Vidya Purushothaman, Jiawei Li, Mingxiang Cai, and Tim K. Mackey. 2021. A longitudinal and geospatial analysis of COVID-19 tweets during the early outbreak period in the United States. *BMC Public Health*, 21(1):793. DOI:10.1186/s12889-021-10827-4.

Shradha Dinesh and Meltem Odabaş. 2023. 8 facts about Americans and Twitter as it rebrands to X. *Pew Research Center*.

Bosheng Ding, Chengwei Qin, Linlin Liu, Yew Ken Chia, Boyang Li, Shafiq Joty, and Lidong Bing. 2023. Is GPT-3 a good data annotator? In *Proceedings of the 61st Annual Meeting of the Association for Computational Linguistics (Volume 1: Long Papers)*, pages 11173–11195, Toronto, Canada. Association for Computational Linguistics.

Ari Z. Klein, Juan M. Banda, Yuting Guo, Ana Lucia Schmidt, Dongfang Xu, Jesus Ivan Flores Amaro, Raul Rodriguez-Esteban, Abeed Sarker, and Graciela Gonzalez-Hernandez. 2023. Overview of the 8th social media mining for health applications (#SMM4H) shared tasks at the AMIA 2023 annual symposium. *medRxiv*. Preprint. PMID: 37986776; PMCID: PMC10659479.

Dongfang Xu, Guillermo Lopez Garcia, Lisa Raithel, Rolland Roller, Philippe Thomas, Eiji Aramaki, Shuntaro Yada, Pierre Zweigenbaum, Sai Tharuni Samineni, Karen O'Connor, Yao Ge, Sudeshna Das, Abeed Sarker, Ari Klein, Lucia Schmidt, Vishakha Sharma, Raul Rodriguez-Esteban, Juan Banda, Ivan Flores Amaro, Davy Weissenbacher, and Graciela Gonzalez-Hernandez. 2024. Overview of the 9th social media mining for health applications (#SMM4H) shared tasks at ACL 2024. In *Proceedings of The 9th Social Media Mining for Health Research and Applications Workshop and Shared Tasks*, Bangkok, Thailand. Association for Computational Linguistics.

TLab at #SMM4H 2024: Retrieval-Augmented Generation for ADE Extraction and Normalization

Jacob Berkowitz[1], Apoorva Srinivasan[1],
Jose Miguel Acitores Cortina[1], Nicholas P Tatonetti[1]
[1]Department of Computational Biomedicine – Cedars-Sinai
jacob.berkowitz2@cshs.org, nicholas.tatonetti@cshs.org

Abstract

SMM4H 2024 Task 1 is focused on the identification and standardization of Adverse Drug Events (ADEs) in tweets. We introduce a novel Retrieval-Augmented Generation (RAG) method, leveraging the capabilities of Llama 3, GPT-4, and the SFR-embedding-mistral model, along with few-shot prompting techniques, to map colloquial tweet language to MedDRA Preferred Terms (PTs) without relying on extensive training datasets. Our method achieved competitive performance, with an F1 score of 0.359 in the normalization task and 0.392 in the named entity recognition (NER) task. Notably, our model demonstrated robustness in identifying previously unseen MedDRA PTs (F1=0.363) greatly surpassing the median task score of 0.141 for such terms.

1 Introduction

Social media is a potential wealth of information for contemporary public health monitoring, offering real-time insights into the effects of medications as reported by patients (Aichner et al., 2021). X (formerly Twitter), with its continuous flow of user-generated content, is a rich but challenging source for identifying Adverse Drug Events (ADEs), largely due to the informal and diverse language present in tweets (Klein et al., 2023).

The task of extracting standardized ADEs from such unstructured text is complex, as conventional methods typically depend on large training datasets. These not only require heavy computational resources, but also may fail to adapt to the nature of social media language (Yang et al., 2024). Here, we present a Retrieval-Augmented Generation (RAG) methodology to incorporate only relevant examples, side-stepping the need for large-scale data annotation and training processes (Gao et al., 2023).

2 Methods

2.1 ADE Tweet Classification

In the task of classifying tweets for the presence of ADEs, we used Google's BERT-large-uncased model (Devlin et al., 2019). This model is a widely recognized transformer-based neural network pretrained on a large corpus of uncased English text. Our choice of BERT-large-uncased was motivated by its proven capability in text classification tasks and its relative efficiency, making it suitable for processing the high volume of data typically found on social media platforms (Huang et al., 2022; Sakhovskiy et al., 2021).

To adapt BERT to our specific classification task, we fine-tuned the model on the provided training dataset, which included annotated tweets with a binary label showing the presence of an ADE. We fine-tuned the model over 3 epochs, a number selected to balance training time and computational resources. Future work could involve a more systematic exploration of this parameter to potentially enhance model performance.

2.2 Named Entity Recognition (NER)

NER is necessary for identifying and extracting specific text spans from the tweets that have been classified as containing at least one ADE. Our approach involves retrieving similar examples from a vector database to enhance the model's understanding and accuracy. We used the SFR-embedding-mistral model to create embeddings for the tweets in our training dataset. As of the Task 1 submission date, SFR-embedding-mistral leads the Massive Text Embedding Benchmark (MTEB) Leaderboard (Muennighoff et al., 2022). These embeddings capture the semantic similarities between tweets and are stored in a vector database. This storage allows for efficient retrieval of relevant examples.

When processing a new tweet, the same SFR-embedding-mistral model projects the tweet into

Proceedings of The 9th Social Media Mining for Health Applications (#SMM4H) Workshop and Shared Tasks, pages 153–157
August 15, 2024 ©2024 Association for Computational Linguistics

the embedding space. The resulting embedding is then compared against the stored embeddings in the tweet vector database using cosine similarity, defined as

$$\text{cosine similarity} = \frac{\sum_{i=1}^{n} t_{ji} * r_{ki}}{\sqrt{\sum_{i=1}^{n} t_{ji}^2}\sqrt{\sum_{i=1}^{n} r_{ki}^2}}$$

for a tweet t_j and reference tweet r_k, with an embedding dimension of n. The top 10 most similar tweets to t_j are retrieved based on these similarity scores. Known as few-shot prompting (Logan-IV et al., 2021), these similar examples provide contextual references and demonstrate the correct response format.

We pass this added information to Meta's Llama 3 7b, and ask it to extract the ADEs from t_j. The prompts fed to the model are shown in Table 1. We selected Llama 3 through consideration of the cost barrier when using GPT-4 with large quantities of tokens.

2.3 Text Normalization

Text normalization is the process of transforming text into a standardized and structured form (Aliero et al., 2023). Here, we map the extracted ADE terms to standardized MedDRA PTs. We used a RAG framework to compare the identified terms to MedDRA PTs and select the most appropriate standardized term.

To normalize an extracted ADE term, we generated its embedding using the SFR-embedding-mistral model and calculated cosine similarity scores (1) between this embedding and the stored embeddings of MedDRA PTs. We selected the top 10 most semantically similar MedDRA PTs based on these similarity scores. These similar terms provide a set of potential matches for the extracted ADE term.

With the retrieved MedDRA PTs, we used GPT-4(version=gpt-4-0125-preview, temperature=0, max new tokens=256), to process the full tweet along with the potential matches. Here, since we are retrieving individual terms rather than phrases, the cost barrier is less drastic than in the NER step. The prompts for this task are shown in Table 2. This strategy allows GPT-4 to consider both the context of the tweet and the extracted term when selecting the most appropriate MedDRA PT. By leveraging the semantic context provided by the full tweet, GPT-4 can map the colloquial language of the tweet to the standardized medical terminology of MedDRA.

3 Results

We evaluated our approach using standard metrics of precision (P), recall (R), and F1 score across three different tasks: named entity recognition (NER), normalization (Norm), and normalization on unseen data (Norm-Unseen). In Table 3, we compare our results with the mean and median scores of the task to demonstrate the effectiveness of our approach. Additionally, we include our results on the provided validation (Val) dataset.

3.1 Performance on Named Entity Recognition (NER) Task

For the NER task, our model was responsible for identifying and extracting specific ADE terms from tweets. The F1-NER score for our model was 0.392, compared to the task mean of 0.327 and median of 0.376. Our precision in the NER task was 0.437 while the recall was 0.355.

3.2 Performance on Normalization Task

In the normalization task, our approach mapped colloquial tweet language to standardized MedDRA terminology. Our method achieved an F1-Norm score of 0.359, which compares to the task mean of 0.283 and median of 0.293. The precision and recall scores of our approach were 0.400 and 0.326, respectively.

3.3 Performance on Normalization Task for Unseen Data

The ability to generalize to unseen data is important for the practical application of any model. In evaluating our approach on MedDRA PTs not seen in the training data, we achieved an F1-Norm-Unseen score of 0.363, compared to the task mean of 0.209 and median of 0.209. This demonstrates our model's robustness and ability to effectively handle the challenge of normalizing terms that were not present in the training data. The precision and recall for unseen data were 0.360 and 0.365, respectively.

4 Discussion

The results of our study highlight the effectiveness of our RAG approach in the context of the SMM4H 2024 Task 1. Our approach, which integrates the capabilities of Llama 3, GPT-4, and the SFR-embedding-mistral model, has demonstrated a particularly strong performance in handling unseen MedDRA Preferred Terms, an important aspect of real-world applicability given the evolving nature of language used in social media. This capability can enhance the timeliness and reliability of ADE detection from social media sources, contributing to faster and more informed public health responses.

For future use, there are certain areas of our pipeline that would benefit from experimental validation. For example, we arbitrarily selected the number 10 for both the ADE extraction and text normalization to provide our models with an adequate amount of information without overloading them. In addition to refining our pipeline, future work may aim to explore additional few-shot learning techniques and retrieval methods to enhance the model's ability to adapt to the continuously evolving language on social media. If willing to trade cost for a potential performance boost, it would be worthwhile to assess GPT-4 for the NER task. Or, with a large quantity of training data, the system may perform better with a fine-tuned local LLM (such as Llama 3) for both the NER and normalization steps.

References

T. Aichner, M. Grünfelder, O. Maurer, and D. Jegeni. 2021. Twenty-five years of social media: A review of social media applications and definitions from 1994 to 2019. *Cyberpsychology, Behavior, and Social Networking*, 24(4):215–222.

A. Ahmad Aliero, B. Sulaimon Adebayo, H. Olanrewaju Aliyu, A. Gogo Tafida, B. Umar Kangiwa, and N. Muhammad Dankolo. 2023. Systematic review on text normalization techniques and its approach to non-standard words.

J. Devlin, M.-W. Chang, K. Lee, K. T. Google, and A. I. Language. 2019. Bert: Pre-training of deep bidirectional transformers for language understanding.

Y. Gao et al. 2023. Retrieval-augmented generation for large language models: A survey.

C. Huang et al. 2022. A generic pre-trained bert-based framework for social media health text classification.

A. Z. Klein et al. 2023. Overview of the 8th social media mining for health applications (#smm4h) shared tasks at the amia 2023 annual symposium.

Robert L. Logan-IV, Ivana Balazevic, Eric Wallace, Fabio Petroni, Sameer Singh, and Sebastian Riedel. 2021. Cutting down on prompts and parameters: Simple few-shot learning with language models. *CoRR*, abs/2106.13353.

N. Muennighoff, N. Tazi, L. Magne, and N. Reimers. 2022. Mteb: Massive text embedding benchmark.

A. Sakhovskiy, Z. Miftahutdinov, and E. Tutubalina. 2021. Kfu nlp team at smm4h 2021 tasks: Cross-lingual and cross-modal bert-based models for adverse drug effects.

H. Yang, Y. Zhang, J. Xu, H. Lu, P. A. Heng, and W. Lam. 2024. Unveiling the generalization power of fine-tuned large language models.

A Appendices

A.1 LLM Prompts

A.1.1 Example of Adverse Drug Event Extraction

Role	Content
System	You are a helpful assistant whose job is to, given the user tweet, determine the adverse drug events mentioned. Return a JSON with a key 'uniqueTermCount', as well as a 'terms' key for each adverse drug event including the term name and its minimal text span.
User	@USER_ that sucks. Humira made me so sick. Remicade helped a but, but then seizures put that out. Now waiting, just prednisone now.
Assistant	{"uniqueTermCount": 2, "Terms": [{"Name": "Malaise", "textSpan": "sick"}, {"Name": "Seizure", "textSpan": "seizures"}, {"Name": "Seizure", "textSpan": "seizures"}]}
User	@USER________ @USER_______ Me too, Remicade for a year, Humira, and had an allergic reaction to Cimzia, almost killed me in November.
Assistant	{"uniqueTermCount": 2, "Terms": [{"Name": "Hypersensitivity", "textSpan": "allergic reaction"}, {"Name": "Apparent death", "textSpan": "almost killed me"}]}
User	@USER_ to it, and now I'm on Humira with just the bloat to battle. Things sometimes get super shitty in every sense. But stay strong xx
Assistant	{"uniqueTermCount": 1, "Terms": [{"Name": "Abdominal distension", "textSpan": "bloat"}]}
User	@USER_______ #humira helped my #crohns but then #respiratory issues ensued & now i can't take anything & have major lung issues
Assistant	{"uniqueTermCount": 2, "Terms": [{"Name": "Respiratory disorder", "textSpan": "Respiratory issues"}, {"Name": "Lung disorder", "textSpan": "lung issues"}]}
User	@USER_ what problems did you have with humira, if you don't mind me asking. I tried infliximab it destroyed my liver.
Assistant	{"uniqueTermCount": 1, "Terms": [{"Name": "Liver injury", "textSpan": "destroyed my liver"}]}
User	@USER_______ @USER_ i feel your pain :(just got 1st #fistula urgh ! humira worked for me for just 3months then got painful reactions
Assistant	{"uniqueTermCount": 2, "Terms": [{"Name": "Fistula", "textSpan": "fistula"}, {"Name": "Pain", "textSpan": "painful reactions"}]}
User	@USER______ i found the humira to fix all my crohn's issues, but cause other issues. i went off it due to issues w nerves/muscle spasms

Table 1: Example input for Llama 3 in our pipeline, with the last user message being the tweet to evaluate.

A.1.2 Example of Text Normalization

Role	Content
System	Given a tweet, a highlighted span, the medical terminology, and a reference list of MedDRA PTs, identify the most relevant MedDRA PT for the medical terminology. Return only the integer ID of the MedDRA PT.
User	Tweet: @USER________ i found the humira to fix all my crohn's issues, but cause other issues. i went off it due to issues w nerves/muscle spasms, Span: issues w nerves, Term: Neuralgia, Reference List: ['Neuralgia: 10029223', 'Facial neuralgia: 10061594', 'Morton's neuralgia: 10052288', 'Trigeminal neuralgia: 10044652', 'Trigeminal neuritis: 10074054', 'Post-traumatic neuralgia: 10076781', 'Trigeminal nerve disorder: 10060890', 'Occipital neuralgia: 10068106', 'Glossopharyngeal neuralgia: 10018391', 'Trigeminal nerve injection: 10044651']

Table 2: Example input for GPT-4 in our pipeline, with the retrieved list of MedDRA PT passed as the reference list.

A.2 Performance Metrics

Metric	Named Entity Recognition				Text Normalization			
	Val Results	**Task Mean**	**Task Median**	**Team Approach**	**Val Results**	**Task Mean**	**Task Median**	**Team Approach**
F1	0.620	0.327	0.376	0.392	0.623	0.283	0.293	0.359
P	0.631	0.356	0.437	0.437	0.634	0.292	0.339	0.400
R	0.609	0.340	0.374	0.355	0.612	0.334	0.326	0.326
	Unseen Text Normalization							
	Val Results	**Task Mean**	**Task Median**	**Team Approach**				
F1	0.200	0.209	0.141	0.363				
P	0.133	0.205	0.144	0.360				
R	0.400	0.287	0.365	0.365				

Table 3: Performance Metrics for Named Entity Recognition, Text Normalization, and Text Normalization of Unseen Terms

BIT@UA at #SMM4H 2024 Tasks 1 and 5: finding adverse drug events and children's medical disorders in English tweets

Luís Carlos Afonso 0009-0005-6728-3089, **João Rafael Almeida** 0000-0003-0729-2264, **Rui Antunes** 0000-0003-3533-8872, and **José Luís Oliveira** 0000-0002-6672-6176

IEETA/DETI, LASI, University of Aveiro, Aveiro, Portugal

Abstract

In this paper, we present our proposed systems, for Tasks 1 and 5 of the #SMM4H-2024 shared task (Social Media Mining for Health), responsible for identifying health-related aspects in English social media text. Task 1 consisted of identifying text spans mentioning adverse drug events and linking them to unique identifiers from the medical terminology MedDRA, whereas in Task 5 the aim was to distinguish tweets that report a user having a child with a medical disorder from tweets that merely mention a disorder.

For Task 1, our system, composed of a pre-trained RoBERTa model and a random forest classifier, achieved 0.397 and 0.295 entity recognition and normalization F1-scores respectively. In Task 5, we obtained a 0.840 F1-score using a pre-trained BERT model.

1 Introduction

Social media text, such as tweets from Twitter, holds a vast amount of textual information and can be a solid source for clinical findings (Dreisbach et al., 2019). Several research initiatives have been pursued to promote the development of data mining solutions from social media to foster healthier lives (Weissenbacher et al., 2018, 2019). Text from social media has been used by biomedical NLP (Natural Language Processing) researchers for different purposes including sentiment analysis (Yang et al., 2016), disease normalization (Tutubalina et al., 2018), suicide attempt prediction (Coppersmith et al., 2016), and classification of depression users (Trifan et al., 2020).

The 9th Social Media Mining for Health Research and Applications (#SMM4H-2024) Workshop continues this research endeavor promoting seven different tasks (Xu et al., 2024). In this work, we describe our participation in the #SMM4H 2024 shared task where we present our developed systems for Tasks 1 and 5, both consisting in mining English tweets from Twitter.

In Sections 2 and 3 we present the datasets in use and the methodology followed, and discuss the results obtained during the official challenge for Tasks 1 and 5 respectively. Finally, we draw some conclusions in Section 4 and present future lines of research.

2 Task 1: adverse drug events

In Task 1, participants had to develop systems to automatically identify mentions of Adverse Drug Events (ADEs) and link them to unique identifiers from the standard terminology MedDRA (Fescharek et al., 2014). Past research work also explored the identification of ADEs from medical case reports and social media (Gurulingappa et al., 2012; Liu and Chen, 2015).

2.1 Dataset

The dataset is composed of a few files annotated with entity mentions, representing adverse drug events, linked with unique MedDRA identifiers. All of the ADE annotations have an associated text span (character start and end offsets) as well as the unique MedDRA identifier. A unique tweet identifier is also used to specify a tweet within the dataset.

The organizers split the dataset into three subsets—training, development, and testing. During the challenge, participants had access to the training and development subsets, containing tweets annotated with ADE mentions, to develop their systems. Table 1 presents dataset statistics.

The training subset contains 18 185 tweets of which only 1 239 tweets contain ADE annotations, and 16 946 tweets do not contain any entity mention. It is annotated with a total of 1 711 entities meaning that some of the tweets have more than one ADE mention. The development subset fol-

Proceedings of The 9th Social Media Mining for Health Applications (#SMM4H) Workshop and Shared Tasks, pages 158–162
August 15, 2024 ©2024 Association for Computational Linguistics

Table 1: Task 1 dataset statistics.

	Training	Development	Testing*
# Tweets	18 185	965	11 799
with entities	1239	65	—
# Entities	1711	87	—

* Participants had no access to the gold standard entities in the testing subset during the challenge.

lows a similar distribution containing a total of 965 tweets and 87 entity annotations where only 65 tweets have at least one ADE annotation. For the challenge official evaluation, participants had to submit their predictions on the blind testing subset, composed of 11 799 tweets.

2.2 Method

In this subsection we detail our approach for detecting adverse drug events in tweets. Our strategy was based on a two-phase workflow where we first identify the spans of ADE mentions and then link them to MedDRA identifiers.

2.2.1 Named entity recognition

We treated the problem of detecting ADE mentions in tweets as a sequence labeling task. For simplicity, each token is attributed an I (Inside) or O (Outside) tag to specify if a token belongs to an ADE mention or not. Contrarily to the standard IOB (Inside, Outside, Beginning) tagging scheme, we employed this IO simpler scheme since the mentions were scarce and no adjacent entities were found in the training and development subsets.

Then, we experimented different machine learning classifiers for performing this token-level classification and obtained the best preliminary results using a RoBERTa model[1] that was (i) pre-trained in PubMed abstracts, (ii) trained on datasets annotated with mentions of diseases (Liu et al., 2019), and (iii) then fine-tuned, by us, using the training data for detecting ADEs.

We also experimented applying a filtering stage as a binary document classification task, before the NER module, to remove documents that did not contain ADEs but this did not prove beneficial and was therefore discarded.

2.2.2 Entity normalization

The final step in our pipeline involved entity normalization, responsible for linking the detected ADEs to unique MedDRA identifiers. This task was tackled as a classification problem where each detected entity needed to be assigned the correct MedDRA identifier.

Prior to classification, we applied standard preprocessing techniques to the textual data such as lowercasing, stop words removal, stemming, and lemmatization. We used a combination of features for the classification task, including token counts and TF-IDF (term frequency–inverse document frequency) features calculated using the the scikit-learn library (Pedregosa et al., 2011).

After experimenting with various machine learning models, also from scikit-learn, our final system employed a random forest classifier with 10 estimators. This choice was based on its superior performance on the development subset compared to other tested classifiers, and offered a good balance between computational efficiency and classification accuracy for this particular task. The classifier was trained on the provided training data, learning to map textual representations of ADEs to their corresponding MedDRA identifiers.

2.3 Results and discussion

During the implementation phase, we obtained an F1-score of 0.453 for entity normalization on the development subset. From Table 2 we observe that our normalization result, on the testing subset, deteriorated considerably (0.453 vs 0.295). We suspect that one of the reasons for this performance drop was because our model was only able to assign identifiers that were seen during the training phase.

We noticed that one of the main challenges in mapping detected entities to MedDRA identifiers was dealing with the variability in how ADEs are expressed in social media text and informal writing. This includes handling synonyms, abbreviations, and misspelled terms that may all refer to the same underlying medical concept.

Table 2: Task 1 official results on the testing subset. F1-score metric is employed. Norm.: entity normalization.

	NER	Norm
Our submission	0.397	0.295
Mean*	0.327	0.283
Median*	0.376	0.293
Baseline* (Magge et al., 2021)	0.481	0.439

* The organizers shared the results of a baseline model, and the mean and median of all submissions by the participating teams.

[1] https://huggingface.co/raynardj/ner-disease-ncbi-bionlp-bc5cdr-pubmed

3 Task 5: children's medical disorders

In Task 5, participants were asked to develop a system to identify tweets that mention a user having a child with a medical disorder, from tweets solely referring a disorder. This was considered a binary text classification task.

3.1 Dataset

The dataset is composed of a few files with tweets classified with a gold standard label that is either 0 or 1. A label of 1 represents a *true case* for this task, meaning that the linked tweet mentions a user having a child with a medical disorder, and a label of 0 represents the opposite scenario (*negative case*). Each tweet also has associated an unique identifier that represents the tweet uniquely within the dataset.

The organizers split the dataset in three subsets, publicly distributing the gold standard labels for the training and development subsets, while keeping the labels for the testing subset unknown for the participants.

Some statistics can be seen in Table 3 about each split of the original dataset where 'Positive' and 'Negative' refer to tweets with a label of 1 and 0 respectively. Participants had to submit their predictions for the unlabeled testing subset containing 10 000 tweets.

3.2 Method

Here we detail our approach for detecting tweets posted by users that report having a child with a disorder. Our strategy was based on the assumption that the task can be tackled as a simple text classification problem.

We experimented with different traditional models from the scikit-learn library (Pedregosa et al., 2011)—naive Bayes, SVM with a linear kernel, Logistic Regression—and XGBoost (Chen and Guestrin, 2016). For text representation we tried two different well-known approaches, available in scikit-learn, for converting a collection of text documents:

Table 3: Task 5 dataset statistics.

	Training	Development	Testing*
# Tweets	7398	389	10 000
Positive	2280	135	—
Negative	5118	254	—

* Participants had no access to the gold standard labels in the testing subset during the challenge.

1. Count vectorizer—to obtain a matrix of token counts; and
2. TF-IDF vectorizer—to obtain a matrix of term frequency–inverse document frequency features.

In a later stage we employed a pre-trained BERT variant[2] to inspect how a more complex model would compare (Devlin et al., 2019).

3.3 Results and discussion

The results for all the aforementioned models, varying the vectorizer (text representation features) for the traditional classifiers and the number of epochs for the BERT model, are presented in Table 4 and Table 5 respectively. As one can observe, the BERT model achieved the best results after being fine-tuned for at least 4 epochs after which the performance changes were not significant.

For the official submission, with the final predictions on the blind testing subset, we employed the BERT model fine-tuned for 8 epochs since it achieved the highest preliminary result on the development subset (0.8968 F1-score).

Table 6 presents the official challenge results on the blind testing subset where we observe that

[2] https://huggingface.co/google-bert/bert-base-uncased

Table 4: Task 5 results with traditional classifiers by applying 5-fold cross-validation on the training subset.

Classifier	Vectorizer	F1-score	Accuracy
Naive Bayes	Count	0.6833	0.7693
	TF-IDF	0.4145	0.6934
SVM	Count	0.7117	0.7814
	TF-IDF	0.7057	0.7830
Logistic Regression	Count	0.7255	0.7776
	TF-IDF	0.7085	0.7813
XGBoost	Count	0.7431	0.7922
	TF-IDF	0.7358	0.7862

Table 5: Task 5 results with a BERT model, fine-tuned for different numbers of epochs, by applying 5-fold cross-validation on the training subset.

Epochs	F1-score	Accuracy
2	0.8216	0.8592
4	0.8499	0.8778
8	0.8428	0.8628
16	0.8467	0.8708
32	0.8520	0.8755

Table 6: Task 5 official results on the testing subset.

	F1-score
Our submission	0.840
Mean*	0.822
Median*	0.901

* The organizers shared the mean and median results of all submissions by the participating teams.

despite our system achieved an F1-score 1.8 percentage points above the mean result it lags behind the median result, showing that there is significant room for improvement.

4 Conclusions

In this work, we presented machine learning models to detect ADEs (Task 1) and users reporting having children with medical disorders in English tweets (Task 5). We obtained more competitive results in Task 1 being slightly above the median. In Task 5, our classification F1-score was 6.1 percentage points below the median result demonstrating that our approach still holds great potential for improvement.

In Task 1, the most relevant aspect to enhance would be for our system to be able to link ADEs to identifiers that were not seen during the training phase. Such system should be able to consult the full MedDRA terminology and normalize any ADE mention to the respective identifier. We also hypothesize that the adoption of the BIO tagging scheme for NER (Task 1) could be beneficial and a more careful hyperparameter optimization through grid search could improve results on both tasks.

From our experiments, we conclude that BERT-based models achieved the best performance in entity recognition and document classification proving to be on par with the state-of-the-art.

5 Funding

This work has received support from the EU/EFPIA Innovative Medicines Initiative 2 Joint Undertaking under grant agreement No 806968. Rui Antunes is funded under the project UIDB/00127/2020[3].

[3]https://doi.org/10.54499/UIDB/00127/2020

References

Tianqi Chen and Carlos Guestrin. 2016. XGBoost: a scalable tree boosting system. In *22nd ACM SIGKDD International Conference on Knowledge Discovery and Data Mining*, pages 785–794, San Francisco, California, USA. ACM.

Glen Coppersmith, Kim Ngo, Ryan Leary, and Anthony Wood. 2016. Exploratory analysis of social media prior to a suicide attempt. In *Third Workshop on Computational Linguistics and Clinical Psychology*, pages 106–117, San Diego, California, USA. Association for Computational Linguistics.

Jacob Devlin, Ming-Wei Chang, Kenton Lee, and Kristina Toutanova. 2019. BERT: pre-training of deep bidirectional transformers for language understanding. In *2019 Conference of the North American Chapter of the Association for Computational Linguistics: Human Language Technologies, Volume 1 (Long and Short Papers)*, pages 4171–4186, Minneapolis, Minnesota, USA. Association for Computational Linguistics.

Caitlin Dreisbach, Theresa A. Koleck, Philip E. Bourne, and Suzanne Bakken. 2019. A systematic review of natural language processing and text mining of symptoms from electronic patient-authored text data. *International Journal of Medical Informatics*, 125:37–46.

Reinhard Fescharek, Jürgen Kübler, Ulrich Elsasser, Monika Frank, and Petra Güthlein. 2014. Medical dictionary for regulatory activities (MedDRA). *International Journal of Pharmaceutical Medicine*, 18(5):259–269.

Harsha Gurulingappa, Abdul Mateen Rajput, Angus Roberts, Juliane Fluck, Martin Hofmann-Apitius, and Luca Toldo. 2012. Development of a benchmark corpus to support the automatic extraction of drug-related adverse effects from medical case reports. *Journal of Biomedical Informatics*, 45(5):885–892.

Xiao Liu and Hsinchun Chen. 2015. A research framework for pharmacovigilance in health social media: Identification and evaluation of patient adverse drug event reports. *Journal of Biomedical Informatics*, 58:268–279.

Yinhan Liu, Myle Ott, Naman Goyal, Jingfei Du, Mandar Joshi, Danqi Chen, Omer Levy, Mike Lewis, Luke Zettlemoyer, and Veselin Stoyanov. 2019. RoBERTa: A robustly optimized BERT pretraining approach. *arXiv:1907.11692*.

Arjun Magge, Elena Tutubalina, Zulfat Miftahutdinov, Ilseyar Alimova, Anne Dirkson, Suzan Verberne, Davy Weissenbacher, and Graciela Gonzalez-Hernandez. 2021. DeepADEMiner: a deep learning pharmacovigilance pipeline for extraction and normalization of adverse drug event mentions on Twitter. *Journal of the American Medical Informatics Association*, 28(10):2184–2192.

F. Pedregosa, G. Varoquaux, A. Gramfort, V. Michel, B. Thirion, O. Grisel, M. Blondel, P. Prettenhofer, R. Weiss, V. Dubourg, J. Vanderplas, A. Passos, D. Cournapeau, M. Brucher, M. Perrot, and E. Duchesnay. 2011. Scikit-learn: machine learning in Python. *Journal of Machine Learning Research*, 12:2825–2830.

Alina Trifan, Rui Antunes, Sérgio Matos, and Jose Luís Oliveira. 2020. Understanding depression from psycholinguistic patterns in social media texts. In *42nd European Conference on Information Retrieval*, pages 402–409, Online. Springer Nature.

Elena Tutubalina, Zulfat Miftahutdinov, Sergey Nikolenko, and Valentin Malykh. 2018. Medical concept normalization in social media posts with recurrent neural networks. *Journal of Biomedical Informatics*, 84:93–102.

Davy Weissenbacher, Abeed Sarker, Arjun Magge, Ashlynn Daughton, Karen O'Connor, Michael J. Paul, and Graciela Gonzalez-Hernandez. 2019. Overview of the Fourth Social Media Mining for Health (SMM4H) Shared Tasks at ACL 2019. In *Fourth Social Media Mining for Health Applications (#SMM4H) Workshop & Shared Task*, pages 21–30, Florence, Italy. Association for Computational Linguistics.

Davy Weissenbacher, Abeed Sarker, Michael J. Paul, and Graciela Gonzalez-Hernandez. 2018. Overview of the Third Social Media Mining for Health (SMM4H) Shared Tasks at EMNLP 2018. In *2018 EMNLP Workshop SMM4H: The 3rd Social Media Mining for Health Applications Workshop & Shared Task*, pages 13–16, Brussels, Belgium. Association for Computational Linguistics.

Dongfang Xu, Guillermo Lopez Garcia, Lisa Raithel, Roland Roller, Philippe Thomas, Eiji Aramaki, Shuntaro Yada, Pierre Zweigenbaum, Sai Tharuni Samineni, Karen O'Connor, Yao Ge, Sudeshna Das, Abeed Sarker, Ari Klein, Lucia Schmidt, Vishakha Sharma, Raul Rodriguez-Esteban, Juan Banda, Ivan Flores Amaro, Davy Weissenbacher, and Graciela Gonzalez-Hernandez. 2024. Overview of the 9th Social Media Mining for Health (#SMM4H) Research and Applications Workshop and Shared Tasks at ACL 2024. In *Proceedings of The 9th Social Media Mining for Health Research and Applications Workshop and Shared Tasks*, Bangkok, Thailand. Association for Computational Linguistics.

Fu-Chen Yang, Anthony J.T. Lee, and Sz-Chen Kuo. 2016. Mining health social media with sentiment analysis. *Journal of Medical Systems*, 40(11):236.

FORCE: A Benchmark Dataset for Foodborne Disease Outbreak and Recall Event Extraction from News

Sudeshna Jana
TCS Research
sudeshna.jana@tcs.com

Manjira Sinha
TCS Research
sinha.manjira@tcs.com

Tirthankar Dasgupta
TCS Research
dasgupta.tirthankar@tcs.com

Abstract

The escalating prevalence of food safety incidents within the food supply chain necessitates immediate action to protect consumers. These incidents encompass a spectrum of issues, including food product contamination and deliberate food and feed adulteration for economic gain leading to outbreaks and recalls. Understanding the origins and pathways of contamination is imperative for prevention and mitigation. In this paper, we introduce FORCE (**F**oodborne disease **O**utbreak and **ReC**all **E**vent extraction from openweb). Our proposed model leverages a multi-tasking sequence labeling architecture in conjunction with transformer-based document embeddings. We have compiled a substantial annotated corpus comprising relevant articles published between 2011 and 2023 to train and evaluate the model. The dataset will be publicly released with the paper. The event detection model demonstrates fair accuracy in identifying food-related incidents and outbreaks associated with organizations, as assessed through cross-validation techniques.

1 Introduction

The escalating number of food safety concerns remains a source of significant apprehension(Amico et al., 2018; Kase et al., 2017; Boatemaa et al., 2019; Nerín et al., 2016; Kase et al., 2017). Globally, foodborne diseases continue to plague populations and stand as leading contributors to both illness and mortality(Bouzembrak and Marvin, 2016; Potter et al., 2012; Pádua et al., 2019; Lüth et al., 2019). Recent estimates have identified norovirus and Campylobacter as the most common reason behind foodborne illnesses, while fatalities have been associated with non-typhoidal Salmonella enterica, Salmonella Typhi, Taenia solium, hepatitis A virus, and aflatoxin(Djekic et al., 2017; Kleter et al., 2009; Bouzembrak and Marvin, 2019).

One of the repercussions of food safety issues is the necessity for food recalls, which pose substantial economic threats to both businesses and nations alike(Deng et al., 2016). This underscores the imperative of uncovering the root causes behind these incidents and the factors contributing to contamination (Zhou et al., 2020). Cross-contamination in food and beverages is a multifaceted issue that can transpire at various stages of the food processing chain, including external raw food contamination, transportation, cleaning processes, heating, food packaging, and even during food storage. Contamination events resulting in outbreaks can manifest at any point before, during, or after food processing (Scallan and Mahon, 2012; Gupta et al., 2004).

Consequently, the pivotal task of identifying the origins of contamination or the triggers for recalls is of paramount importance (Hall et al., 2013). It is essential to gain insights into potential sources and pathways of contamination leading to foodborne outbreaks and product recalls, and to devise effective measures for prevention(Tao et al., 2020; Zhou et al., 2021; Jin et al., 2020; Marvin et al., 2017). It is worth noting that there is a dearth of substantial work in the development of computational and/or analytical models addressing these concerns(Allard et al., 2016; Moumni Abdou et al., 2019). This scarcity of research can largely be attributed to the limited availability of data concerning the contributory factors associated with food safety incidents. As such, there is a pressing need to develop an automated tool that can mine reported food safety incidents and recalls to bridge this knowledge gap.

Applications of language technologies and data science for food-borne risk assessment are gaining ground(Harris et al., 2017; Altenburger and Ho, 2019; Maharana et al., 2019; Deng et al., 2021). data-intensive systems play important roles in tracking food-borne illness cases and agents(Pujahari and Khan, 2022; Oldroyd et al., 2021; Nychas et al., 2021; Gupta et al., 2004; Scallan and Mahon, 2012; Wang et al., 2021). Examples at the US federal

Proceedings of The 9th Social Media Mining for Health Applications (#SMM4H) Workshop and Shared Tasks, pages 163–169
August 15, 2024 ©2024 Association for Computational Linguistics

level include PulseNet (Swaminathan et al., 2001), the National Antimicrobial Resistance Monitoring System (NARMS) (Gupta et al., 2004), FoodNet (Scallan and Mahon, 2012), and the National Outbreak Reporting System (Hall et al., 2013). Implementation of whole-genome sequencing (WGS) in surveillance and outbreak investigation has fueled an explosion of publicly available foodborne pathogen genomes in new systems such as GenomeTrakr (Allard et al., 2016), EnteroBase (Zhou et al., 2020), and the National Center for Biotechnology Information's Pathogen Detection. These works are primarily focused on a) identifying the sentiment polarity of the documents, and b) identifying the occurrences of a set of predefined types of entities and events. However, none of the above works perform deep linguistic analysis of the textual contents to identify food-related incidents based only on the linguistic structure of the text. The model presented in this paper is distinct from the earlier approaches since it is capable of detecting novel and heretofore unknown incidents from reports based only on semantic and linguistic analysis of content.

Keeping in mind the above-mentioned requirements of end users, in this work we propose an event detection model that can identify food safety-related insights, recalls, and outbreaks mentioned in regulatory reports and social media platforms. The proposed model uses a multi-tasking sequence labeling architecture that works with transformer-based document embeddings. We have created a large annotated corpus containing relevant articles published by multiple regulatory agencies over twelve years (2011 - 2023) for training and evaluating the model. The dataset will be publicly released with this paper. The model has been thereafter applied to recent publications. Aggregate analysis of these extracted insights reveals interesting trends.

2 Dataset Creation

The dataset comprises regulatory news articles from two sources namely, a) A corpus of 6000 regulatory articles comprising around 121080 sentences, under Outbreak(O) and Recall(R) categories published between 2011 and 2023 by Food Safety News (FSN)[1] and b) A collection of around 2200 news articles from United States Food and Drug Administration (FDA) recall and outbreak announcements[2]. All together there are 8100 articles.

All the news articles were manually annotated to mark the various target entities. The annotation was done by multiple annotators following a rigorous procedure to ensure acceptable inter-annotator agreement.

The entities and events to be identified by the annotators are as follows:

Target Organization (TO): Of the many organization names that may appear in a document and are detected by the NER earlier, the task during annotation is to identify and tag the organization whose product has been recalled.

Product Name (P): The name of the product that has been recalled or caused the outbreak.

Infection Name (I): The name of the bacterial infection mentioned in the report that causes the outbreak/recall.

Safety Incident (SI) - Annotators have to tag phrases or sequences of words that collectively are indicative of the food safety incident.

Cause of Incident (CI) - Annotators have to tag phrases or sequences of words that indicate the primary cause of the food safety incident.

Number of People Affected (N): Annotators have to tag phrases or sequences of words that collectively are indicative of a number of people affected due to the outbreak.

To help the annotators, each document is first processed using the Stanford NER (Manning et al., 2014) to obtain the organization names, locations, and currency values as named entities. This helps in quick localization of the first four elements, if present in the document. The annotators are domain experts who are knowledgeable about the domain.

16 annotators took part in the annotation, with each expert annotating 700 documents using the Stanford simple manual annotation tool[3]. This included 200 documents, which were sent to all the annotators to compute the inter-annotator agreement later. The average length of a document is around 23 sentences. The experts read each document and performed the following tasks,

Task-1: - Label sentences of each document as **Food Recall** - if the document reported events of a product recall or **Disease Outbreak** - if the document mentions events that report an foodborne disease outbreak or **NEUTRAL**- in case none of the above factors hold.

Task-2: - This task had two components: (a)

[1]https://www.foodsafetynews.com/

[2]https://www.fda.gov/safety

[3]https://nlp.stanford.edu/software/

From among the named entities, the target organization, product name, infection, and locations were marked, if any, and (b) Mark phrases in the text that indicate food safety incidents, and its cause. At the end of the annotation, each word in the document is assigned a label TO, P, I, SI, CI, N, or None. Table 3 in A.2 illustrates the annotation with some example News texts. For the sake of understanding, we have shown labels of only the phrases that belong to any one of the following classes {TO, P, I, SI, CI, L, or N}.

Using the annotations obtained for 200 common documents, we measured the inter-annotator agreement using the Fleiss Kappa (Fleiss et al., 1981) measure (κ). This is computed as: $\kappa = \frac{\bar{P}-\bar{P}_e}{1-\bar{P}_e}$ The factor $1 - \bar{P}_e$ gives the degree of agreement that is attainable above chance, and $\bar{P} - \bar{P}_e$ gives the degree of agreement achieved above chance. It was observed that the inter-annotator score for Task-1 was 0.83, which is appreciably high. For Task 2, it was found to be 0.71. The scores are computed using word-label matches assigned by different annotators. The very high scores indicate that all experts were marking fairly uniformly and therefore, the expert annotated dataset is reliable to be used for training incident detection systems. Out of the 165,080 sentences from 82000 documents, around 27500 sentences were found to contain words belonging to at least one type mentioned in the incident knowledge schema. Altogether, we obtained 13000 safety incidents, 12100 causes, 2223 Target Organizations, 3300 locations, 4695 product names, and 4101 infection names. The entire annotated data will be publicly released with this paper.

Model No.	Model Name	Sequence Classification		
		P	R	F1
I.	Single task CNN-BiLSTM	0.63	0.59	0.62
II.	Single task PreTrained BERT	0.75	0.72	0.73
III.	Single-task-BERT-CNN-BiLSTM	0.72	0.75	0.73
IV.	LLAMA-2	0.75	0.79	0.78
V.	Mistral7B	0.73	0.79	0.77
VI.	Multi-task BERT-CNN-BiLSTM	**0.83**	**0.89**	**0.86**

Table 1: Results reporting accuracy of classifying food safety events as *Food Recall* or *Disease Outbreak*.

3 A Multi-tasking Neural Model for Food Safety Knowledge Extraction

The proposed model works on each sentence at a time to detect elements of interest that are defined in the incident knowledge schema. Multi-task

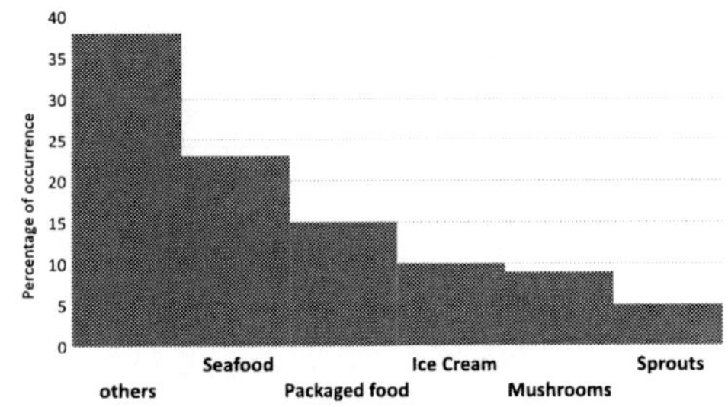

Figure 1: Distribution of occurrences of recalled products.

learning utilizes the correlation between related tasks to improve classification by learning tasks in parallel. In the present work, the two related tasks are *task-1:* classifying a sentence into either *Recall* or *Outbreak* classes as discussed earlier and *task-2:* labeling appropriate phrases in the text as per the incident knowledge schema.

A cascaded CNN-BiLSTM layer for the combined tasks of sentence classification and sequence label prediction, using the fine-tuned BERT for creating the sequence embeddings.

To obtain the multi-tasking model for dual tasks of sequence classification and sequence labeling, the $BERT-CNN-BiLSTM$ layers have been trained with two separate loss functions L_1 and L_2. Where, $L_1(\theta) = -\sum_{t=1}^{M}\sum_{k=1}^{K}\bar{y}_t^k log(y_t)$ and $L_2(\theta) = -\sum_{t=1}^{N}\sum_{j=1}^{J}\bar{q}_t^{i,j} log(q_t^i)$.

Here, q_t is the vector representation of the predicted output of the model for the input word w_t^i. K and J are the number of class labels for each task. The model is fine-tuned end-to-end by minimizing the cross-entropy loss.

We define the joint loss function using a linear combination of the loss functions of the two tasks:

$$L_{joint}(\theta) = \lambda * L_1(\theta) + (1-\lambda) * I_{[y_{sentence}==1]} * L_2(\theta) \quad (1)$$

Where λ controls the contribution of losses of the individual tasks in the overall joint loss. $I_{[y_{sentence}==1]}$ is an indicator function that activates the loss only when the corresponding sentence classification label is 1 since we do not want to back-propagate sequence labeling loss when the corresponding sequence classification label is 0.

4 Evaluation and Results

The performance of the proposed model has been compared with a number of baseline models used for single-objective document classification and sequence labeling tasks as well as large language

Model	Sequence labeling task																	
No.	TO			P			I			SI			CI			N		
	P	R	F	P	R	F	P	R	F	P	R	F	P	R	F	P	R	F
I.	0.76	0.78	0.77	0.71	0.67	0.69	0.77	0.78	0.77	0.72	0.77	0.74	0.77	0.8	0.78	0.72	0.78	0.72
II.	0.77	0.78	0.78	0.69	0.74	0.71	0.79	0.77	0.78	0.78	0.75	0.76	0.74	0.75	0.76	0.78	0.80	0.79
III.	0.80	0.81	0.80	0.71	0.75	0.73	0.76	0.86	0.80	0.78	0.79	0.78	0.75	0.76	0.75	0.78	0.77	0.78
IV.	0.82	0.87	0.84	0.75	0.79	0.77	0.82	0.86	0.82	0.78	0.79	0.78	0.78	0.72	0.75	0.84	0.89	0.86
V.	0.82	0.89	**0.85**	0.80	0.82	**0.81**	0.82	0.87	0.83	0.76	0.79	0.78	0.78	0.79	0.79	0.92	0.90	**0.91**
VI.	0.81	0.89	0.84	0.79	0.83	0.80	0.82	0.87	**0.84**	0.82	0.90	**0.84**	0.85	0.91	**0.88**	0.86	0.88	0.88

Table 2: Results reporting the performance of the food safety incident and entity extraction task.

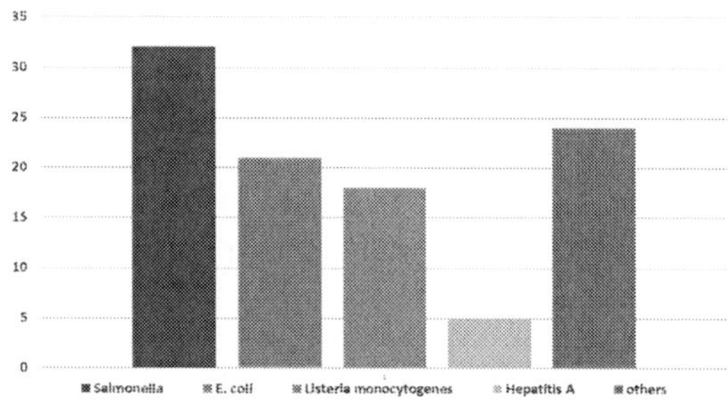

Figure 2: Distribution of germs found in the recalled products.

models like LLAMA-2 7B and fined-tuned Mistral 7B. Table 1 presents the precision, recall, and F1 scores of classifying food safety events as *food recall* or *disease outbreak*. We have obtained the highest F1 score of 0.86 with a high precision of 0.89.

Table 2 presents the accuracy of subsequent labeling of word sequences within a sentence by their respective categories - *TO, P, I, SI, CI, or N*, as described earlier. For both the cases, the performance of the proposed multi-objective architecture has been compared with several baseline state-of-the-art models designed with single objective functions. It was observed that for most of the classes like, S, SI and CI, the Multi-task BERT-CNN-BiLSTM model significantly outperforms the baseline models. On the other hand, mistral-7B model trained over the given dataset performs better recognizing the TO, P and N classes.

The primary reason for the poor performance of LLAMA-2 as well as mistral-7B can be attributed due to two reasons: a) lack of environmental domain knowledge due to which critical domain concepts like, *Salmonella*, *Listeria monocytogenes* and *E.Coli* gets ignored. b) Unable to identify phrase boundaries. We observe that despite in most of the cases LLAMA-2 correctly identified the safety incident and cause phrases, but the span of the phrases are either too long or too short. as a results of which outputs of the model get penalized. Similar observations were made for mistral 7B, however, since the mistral model is fine-tuned over the current dataset, problems related to domain concept mismatch were relatively less. However, the output word span detection for incident and causes still remains a challenge.

Apart from the classification and extraction of food Safety incidents, it is equally important to perform some basic analytics on the dataset. Figure 1 shows the distribution of the top five causes of *food recall* across different locations in the United States. In general, we have observed that infections such as *Salmonella, Listeria monocytogenes* and *E.Coli* are the biggest causes of food recall in the United States. Figure 2 depicts the top food products that contain the aforementioned germs.

5 Conclusion

In this paper we present resource creation and extraction of critical information related to food safety and food-borne infection from regulatory reports and social media platforms. The proposed model, founded on a multi-tasking sequence labeling architecture integrated with transformer-based document embeddings, demonstrates its effectiveness in this task. To develop and evaluate our model, we meticulously curated an annotated corpus comprising pertinent articles. Our initial analysis demonstrate the proposed multi-task model surpasses the performance of almost all the baseline models including LLMs such as LLAMA-27B and finetuned mistral-7B.

References

Marc W Allard, Errol Strain, David Melka, Kelly Bunning, Steven M Musser, Eric W Brown, and Ruth Timme. 2016. Practical value of food pathogen traceability through building a whole-genome sequencing network and database. *Journal of clinical microbiology*, 54(8):1975–1983.

Kristen M Altenburger and Daniel E Ho. 2019. Is yelp actually cleaning up the restaurant industry? a re-analysis on the relative usefulness of consumer reviews. In *The World Wide Web Conference*, pages 2543–2550.

Priscilla D' Amico, Daniele Nucera, Lisa Guardone, Martino Mariotti, Roberta Nuvoloni, and Andrea Armani. 2018. Seafood products notifications in the eu rapid alert system for food and feed (rasff) database: Data analysis during the period 2011–2015. *Food Control*, 93:241–250.

Sandra Boatemaa, McKenna Barney, Scott Drimie, Julia Harper, Lise Korsten, and Laura Pereira. 2019. Awakening from the listeriosis crisis: Food safety challenges, practices and governance in the food retail sector in south africa. *Food Control*, 104:333–342.

Yamine Bouzembrak and Hans JP Marvin. 2016. Prediction of food fraud type using data from rapid alert system for food and feed (rasff) and bayesian network modelling. *Food Control*, 61:180–187.

Yamine Bouzembrak and Hans JP Marvin. 2019. Impact of drivers of change, including climatic factors, on the occurrence of chemical food safety hazards in fruits and vegetables: A bayesian network approach. *Food control*, 97:67–76.

Xiangyu Deng, Shuhao Cao, and Abigail L Horn. 2021. Emerging applications of machine learning in food safety. *Annual Review of Food Science and Technology*, 12:513–538.

Xiangyu Deng, Henk C den Bakker, and Rene S Hendriksen. 2016. Genomic epidemiology: whole-genome-sequencing–powered surveillance and outbreak investigation of foodborne bacterial pathogens. *Annual review of food science and technology*, 7:353–374.

Ilija Djekic, Danijela Jankovic, and Andreja Rajkovic. 2017. Analysis of foreign bodies present in european food using data from rapid alert system for food and feed (rasff). *Food control*, 79:143–149.

Joseph L Fleiss, Bruce Levin, Myunghee Cho Paik, et al. 1981. The measurement of interrater agreement. *Statistical methods for rates and proportions*, 2(212-236):22–23.

Amita Gupta, Jennifer M Nelson, Timothy J Barrett, Robert V Tauxe, Shannon P Rossiter, Cindy R Friedman, Kevin W Joyce, Kirk E Smith, Timothy F Jones, Marguerite A Hawkins, et al. 2004. Antimicrobial resistance among campylobacter strains, united states, 1997–2001. *Emerging infectious diseases*, 10(6):1102.

Aron J Hall, Mary E Wikswo, Karunya Manikonda, Virginia A Roberts, Jonathan S Yoder, and L Hannah Gould. 2013. Acute gastroenteritis surveillance through the national outbreak reporting system, united states. *Emerging infectious diseases*, 19(8):1305.

Jenine K Harris, Jared B Hawkins, Leila Nguyen, Elaine O Nsoesie, Gaurav Tuli, Raed Mansour, and John S Brownstein. 2017. Research brief report: Using twitter to identify and respond to food poisoning: The food safety stl project. *Journal of Public Health Management and Practice*, 23(6):577.

Cangyu Jin, Yamine Bouzembrak, Jiehong Zhou, Qiao Liang, Leonieke M Van Den Bulk, Anand Gavai, Ningjing Liu, Lukas J Van Den Heuvel, Wouter Hoenderdaal, and Hans JP Marvin. 2020. Big data in food safety-a review. *Current Opinion in Food Science*, 36:24–32.

Julie Ann Kase, Guodong Zhang, and Yi Chen. 2017. Recent foodborne outbreaks in the united states linked to atypical vehicles—lessons learned. *Current Opinion in Food Science*, 18:56–63.

Diederik P Kingma and Jimmy Ba. 2014. Adam: A method for stochastic optimization. *arXiv preprint arXiv:1412.6980*.

GA Kleter, ALDO Prandini, LAURA Filippi, and HJP Marvin. 2009. Identification of potentially emerging food safety issues by analysis of reports published by the european community's rapid alert system for food and feed (rasff) during a four-year period. *Food and chemical toxicology*, 47(5):932–950.

Stefanie Lüth, Idesbald Boone, Sylvia Kleta, and Sascha Al Dahouk. 2019. Analysis of rasff notifications on food products contaminated with listeria monocytogenes reveals options for improvement in the rapid alert system for food and feed. *Food Control*, 96:479–487.

Adyasha Maharana, Kunlin Cai, Joseph Hellerstein, Yulin Hswen, Michael Munsell, Valentina Staneva, Miki Verma, Cynthia Vint, Derry Wijaya, and Elaine O Nsoesie. 2019. Detecting reports of unsafe foods in consumer product reviews. *JAMIA open*, 2(3):330–338.

Christopher D. Manning, Bauer Surdeanu, Mihai, Finkel John, Bethard Jenny, Steven J., and David. McClosky. 2014. The stanford corenlp natural language processing toolkit. In *52nd Annual Meeting of the Association for Computational Linguistics: System Demonstrations*, pages 55–60.

Hans JP Marvin, Esmée M Janssen, Yamine Bouzembrak, Peter JM Hendriksen, and Martijn Staats. 2017.

Big data in food safety: An overview. *Critical reviews in food science and nutrition*, 57(11):2286–2295.

Houda Moumni Abdou, Ilham Dahbi, Mohammed Akrim, Fatima Zahra Meski, Yousef Khader, Mohammed Lakranbi, Hind Ezzine, and Asmae Khattabi. 2019. Outbreak investigation of a multipathogen foodborne disease in a training institute in rabat, morocco: case-control study. *JMIR public health and surveillance*, 5(3):e14227.

Cristina Nerín, Margarita Aznar, and Daniel Carrizo. 2016. Food contamination during food process. *Trends in food science & technology*, 48:63–68.

George-John Nychas, Emma Sims, Panagiotis Tsakanikas, and Fady Mohareb. 2021. Data science in the food industry. *Annual Review of Biomedical Data Science*, 4:341–367.

Rachel A Oldroyd, Michelle A Morris, and Mark Birkin. 2021. Predicting food safety compliance for informed food outlet inspections: a machine learning approach. *International Journal of Environmental Research and Public Health*, 18(23):12635.

Inês Pádua, André Moreira, Pedro Moreira, Filipa Melo de Vasconcelos, and Renata Barros. 2019. Impact of the regulation (eu) 1169/2011: Allergen-related recalls in the rapid alert system for food and feed (rasff) portal. *Food control*, 98:389–398.

Antony Potter, Jason Murray, Benn Lawson, and Stephanie Graham. 2012. Trends in product recalls within the agri-food industry: Empirical evidence from the usa, uk and the republic of ireland. *Trends in food science & technology*, 28(2):77–86.

Rakesh Mohan Pujahari and Rijwan Khan. 2022. Applications of machine learning in food safety. In *Artificial Intelligence Applications in Agriculture and Food Quality Improvement*, pages 216–240. IGI Global.

Elaine Scallan and Barbara E Mahon. 2012. Foodborne diseases active surveillance network (foodnet) in 2012: a foundation for food safety in the united states. *Clinical Infectious Diseases*, 54(suppl_5):S381–S384.

Bala Swaminathan, Timothy J Barrett, Susan B Hunter, Robert V Tauxe, and CDC PulseNet Task Force. 2001. Pulsenet: the molecular subtyping network for foodborne bacterial disease surveillance, united states. *Emerging infectious diseases*, 7(3):382.

Dandan Tao, Pengkun Yang, and Hao Feng. 2020. Utilization of text mining as a big data analysis tool for food science and nutrition. *Comprehensive reviews in food science and food safety*, 19(2):875–894.

Hanxue Wang, Wenjuan Cui, Yunchang Guo, Yi Du, Yuanchun Zhou, et al. 2021. Machine learning prediction of foodborne disease pathogens: Algorithm development and validation study. *JMIR medical informatics*, 9(1):e24924.

Qinqin Zhou, Hao Zhang, and Suya Wang. 2021. Artificial intelligence, big data, and blockchain in food safety. *International journal of food engineering*, 18(1):1–14.

Zhemin Zhou, Nabil-Fareed Alikhan, Khaled Mohamed, Yulei Fan, Mark Achtman, Derek Brown, Marie Chattaway, Tim Dallman, Richard Delahay, Christian Kornschober, et al. 2020. The enterobase user's guide, with case studies on salmonella transmissions, yersinia pestis phylogeny, and escherichia core genomic diversity. *Genome research*, 30(1):138–152.

A Appendix

A.1 Fine-tuning the BERT language model

The basic $BERT_{base}$ model was first fine-tuned with a portion of the food safety corpus using over-sampling, to create FoodSafety-BERT, referred to as FS-BERT hereafter. A labeled document is broken into multiple smaller chunks, such that each chunk can be fed as a unit to $BERT_{base}$ to create its corresponding vector. Each chunk is associated with a label that is the same as its parent document. A classification task is now defined with these chunks during which the basic BERT model is fine-tuned while training. This model is designed as a fully connected layer over the BERT base model, with softmax as the activation function. Training was done with learning rate set to 2×10^{-5} using the Adam optimizer (Kingma and Ba, 2014). The model is fine-tuned for a few epochs (3-4) only to avoid over-fitting. The chunk representations are saved from the CLS token embeddings created during the process. The fine-tuned BERT model, FS-BERT, is subsequently used for document and incident recognition tasks.

A.2 Example Annotation Sentences

[ORGName]/**TO** of Poughkeepsie, NY, is *[recalling its 2-lb., 5-lb. and 15-lb. boxes]*/**SI** of "*[Abady Highest Quality Maintenance & Growth Formula for Cats]*/**P**" because they have the *[potential to be contaminated with Salmonella]*/**CI**.
[ORGName]/**TO** of Brampton, Ontario, *[recalled 36 pounds of fully cooked pork baby back ribs in* *recalled 36 pounds of fully cooked pork baby back ribs in]* /**SI** today because they were not presented for inspection at the U.S. border. The problem was discovered when U.S. Department of Agriculture Food Safety and Inspection Service import staff reviewed records and discovered that the independent third-party carrier *[did not present a product for USDA inspection at the U.S.-Canadian border]*/**CI**. According to a Public Health Alert released by FSIS, being recalled are 18-pound cases containing 1.5-pound packages of "*[Cobblestone Farms Fully Cooked Pork Baby Back Ribs in Honey Garlic Barbeque Sauce]*/**P**" bearing package code "Sell By 2015-AL-08" and case code "15201" bearing the Canadian mark of inspection with establishment number "624." The product was distributed to a retailer in *[New York]*/**L**. In its announcement, FSIS stated that it is working on solutions to prevent future failure-to-present episodes from occurring, including outreach to industry, foreign food-safety agencies, and importers.
A Listeria outbreak in the *[Midwest]*/**L** linked to *[one death]*/**N** and a miscarriage likely was caused by *[contamination during the cheese-making process]*/**CI**, according to a new report from the U.S. Centers for Disease Control and Prevention. The Minnesota Department of Agriculture tested samples of the cheese from two retail outlets, revealing the outbreak strain to be *[Listeria monocytogenes]*/**I**.
About *[96,000 pounds]*/**SI** of *[Oscar Mayer Classic Wieners]*/**P** *[were recalled]*/**SI** Sunday by *[ORGName]*/**TO** of Columbia, MO, because of a *[packaging error]*/**CI**.
[ORGName]/**TO** of Detroit, MI, is *[recalling approximately 1.8 million pounds]*/**SI** of *[ground beef products]*/**P** that may be *[contaminated with E. coli O157:H7]*/**CI**, the U.S. Department of Agriculture's Food Safety and Inspection Service (FSIS) announced Monday. At the time that the recall was issued, there were *[11 illnesses]*/**N** linked to the recalled product.

Table 3: Sample Food Safety News texts with the respective annotated entities and events. Note that all the target organization names were intentionally masked by the token [ORGName] to maintain anonymity.

Overview of #SMM4H 2024 – Task 2: Cross-Lingual Few-Shot Relation Extraction for Pharmacovigilance in French, German, and Japanese

Lisa Raithel[1,2,3], Philippe Thomas[3], Bhuvanesh Verma[3], Roland Roller[3], Hui-Syuan Yeh[4], Shuntaro Yada[5], Cyril Grouin[4], Shoko Wakamiya[5], Eiji Aramaki[5], Sebastian Möller[2,3], Pierre Zweigenbaum[4]

[1]BIFOLD – Berlin Institute for the Foundations of Learning and Data, Germany;
[2]Quality & Usability Lab, Technische Universität Berlin, Germany;
[3]German Research Center for Artificial Intelligence (DFKI), Berlin, Germany;
[4]Université Paris-Saclay, CNRS, LISN, Orsay, France;
[5]Nara Institute of Science and Technology, Nara, Japan;

Abstract

This paper provides an overview of Task 2 from the Social Media Mining for Health 2024 shared task (#SMM4H 2024), which focused on Named Entity Recognition (NER, Subtask 2a) and the joint task of NER and Relation Extraction (RE, Subtask 2b) for detecting adverse drug reactions (ADRs) in German, Japanese, and French texts written by patients. Participants were challenged with a few-shot learning scenario, necessitating models that can effectively generalize from limited annotated examples. Despite the diverse strategies employed by the participants, the overall performance across submissions from three teams highlighted significant challenges. The results underscored the complexity of extracting entities and relations in multi-lingual contexts, especially from user-generated content's noisy and informal nature.

1 Introduction

An adverse drug reaction (ADR) is defined as a "harmful or unpleasant reaction, resulting from an intervention related to the use of a medicinal product" (Edwards and Aronson, 2000). ADRs pose a significant challenge in pharmacovigilance. No medication is devoid of side effects, and despite clinical trials for each drug, the trial populations often fail to represent the entirety of real-world patients in terms of age, gender, health status, or ethnicity (Hazell and Shakir, 2006). Moreover, post-release surveillance efforts may miss patients experiencing issues with the medication (Hazell and Shakir, 2006), emphasizing the need for continuous monitoring of medication usage and effects.

Natural language processing can support this process by extracting potentially *novel* ADRs from text sources. Clinical texts and scientific literature are valuable resources containing information about ADRs. Still, they are either difficult to access for researchers outside a hospital or are published multiple weeks or even months after the occurrence/detection of an adverse effect. Social media, such as X (formerly known as Twitter) or patient forums, in which patients share and discuss their sorrows, concerns, and potential ADRs, instead became an alternative and up-to-date text source (Leaman et al., 2010; Segura-Bedmar et al., 2014; Zolnoori et al., 2019). Not all information from social media is necessarily reliable from a medical point of view. Still, it directly reflects the patient's perspective and at a much faster speed than well-curated scientific text.

To produce robust enough systems to process the vast amount of online data automatically, various datasets have been introduced in this context (Karimi et al., 2015; Klein et al., 2020; Tutubalina et al., 2021; Sboev et al., 2022, inter alia), shared tasks have been conducted (Magge et al., 2021a; Weissenbacher et al., 2022; Klein et al., 2024), and models have been published (Magge et al., 2021b; Portelli et al., 2022). However, like other text processing domains, most datasets exist only for English. To raise interest in this critical topic – particularly for non-English languages, which are not well-represented in this domain (Névéol et al., 2018) – we provide a Shared Task at #SMM4H 2024 (Xu et al., 2024): *Cross-Lingual Few-Shot Relation Extraction for Pharmacovigilance in French, German, and Japanese*. This paper describes the results and findings of this task. It targets joint cross-lingual named entity recognition and relation extraction in a multi-lingual setting. The data consists of French, German, and Japanese texts written by patients on social media such as X and patient fora, and is a subset of the KEEPHA corpus (Raithel et al., 2024).

Proceedings of The 9th Social Media Mining for Health Applications (#SMM4H) Workshop and Shared Tasks, pages 170–182
August 15, 2024 ©2024 Association for Computational Linguistics

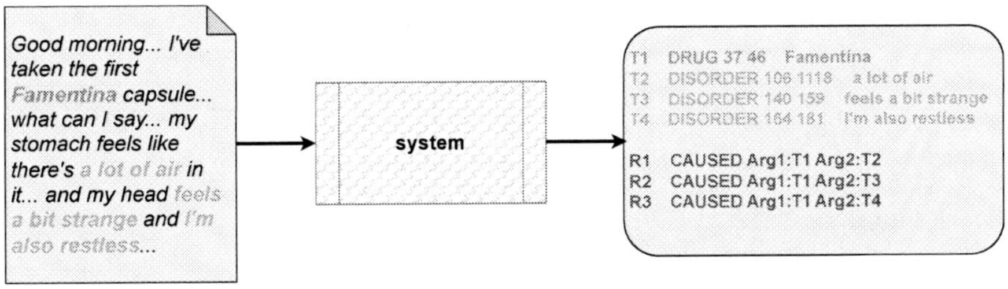

Figure 1: Visualization of input (a document) and expected output for both tasks. The output is a text file with predictions in brat format and shows an identifier (e.g., T1), a label (e.g., DRUG), the offsets of the entity, and the actual string (e.g., *Famenita*) in the case of NER. For RE, the annotations/predictions are extended by relations (identified with, e.g., R1), the relation type (e.g., CAUSED), and the head (Arg1) and tail (Arg2) arguments, referring to entity identifiers.

2 Shared Task

In this section, we present the task details and schedule, the data used for the challenge, our baseline system, and the participants' approaches.

2.1 Task

#SMM4H 2024 Task 2 targets extracting drug and disorder/body function mentions (Subtask 2a) and relations between those entities (Subtask 2b). The task is set up in a cross-lingual few-shot scenario: Training data consists mainly of Japanese and German data plus four French documents (see Table 1). The submitted systems are evaluated on Japanese, German, and French data.

Figure 1 visualizes the general process: Given a text document, a system should first predict entities and, subsequently, relations between these. The output format of the predictions is expected to be in `brat` format. The participants were asked to submit multi-lingual systems (FR + DE + JA) for one or all of the following tasks:

a) Named Entity Recognition (NER): Recognize mentions that belong to the classes `DRUG`, `DISORDER`, or `FUNCTION`.

b) Joint Named Entity and Relation Extraction (joint NER+RE): Recognize the entities mentioned above and the relations `TREATMENT_FOR` or `CAUSED` between them without relying on gold entities.

The participants could also submit predictions for only one or two languages.

2.2 Data

The data used for this challenge are a subset (in terms of fewer entities and relations) of the KEEPHA dataset (Raithel et al., 2024). It originates from different (non-parallel) social media sources (online patient fora and X) and is available in German, French, and Japanese. The choice of languages is due to the native languages spoken in the countries in which the labs involved in the data creation are located.[1] To diversify the data, the authors tried to use as many sources as possible, to get different populations of patients and also different types of text, e.g., short texts from X versus longer messages in fora. The German (training and test) data is from an online patient forum, whereas the Japanese documents are from X (training) and a patient forum (test). The French data, finally, is a translation of German documents from the same patient forum as the German data. The translation was necessary because there was no French patient forum (or other resource) that permitted access to the postings of its users. The translated French documents do not overlap with the German originals.

All data were annotated based on the same annotation guidelines[2], with a focus on the detection and extraction of adverse drug reactions, modeled by associating medication mentions (`DRUG`) with disorders (medical signs and symptoms, `DISORDER`) and body function mention (`FUNCTION`) using cause-consequence relations (`CAUSED`) to represent side effects elicited by medication intake, and treatment relations (`TREATMENT_FOR`) to represent medication used to treat medical symptoms. Figure 2 shows an example annotation. The tool used for annotation was `brat` (Stenetorp et al., 2012) (see an

[1]TU Berlin & DFKI in Berlin, Germany; LISN & Université Paris-Saclay in Orsay, France; NAIST in Nara, Japan, and RIKEN in Tokyo, Japan.

[2]`https://shorturl.at/BBHOS`

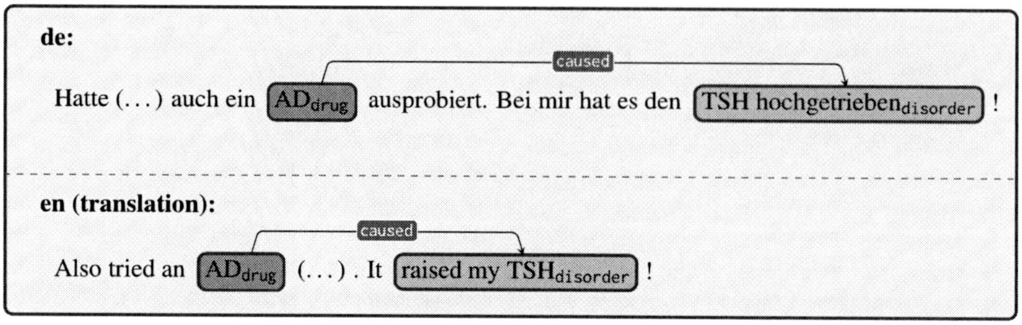

Figure 2: An annotated example. Top: original German text (shortened), bottom: English translation with projected annotations.

example in Figure 1, output of the system). The relation distribution is imbalanced. The number of `treatment_for` relations is much lower than that of `caused` relations, adding to the task's difficulty.

corpus	lang	src	#doc	#ent	#rel
	de	patient forum	70	1,207	476
train	ja	X	392	2,416	619
	fr	patient forum	4	69	32
	de	patient forum	23	424	141
dev	ja	patient forum	168	930	266
	fr	—	0	0	0

Table 1: Number of documents (#doc) with the number of entities (#ent), and relations (#rel) of each type for each language (lang) and source (src) of the training and development data.

2.3 Schedule

Table 2 shows the schedule of #SMM4H 2024 Task 2. The task was announced via several mailing lists (e.g., corpora-list, ML-news) and social media (e.g., X, LinkedIn) in two calls. In the beginning, 12 teams from diverse countries (Switzerland, India, France, China, USA, and Georgia) registered for Task 2.

Training data available	January 10, 2024
CodaLab available	January 17, 2024
Practice predictions due	April 3, 2024
Test data available	April 17, 2024
Evaluation end	April 24, 2024

Table 2: The schedule of Task 2 at #SMM4H 2024.

The CodaLab environment for the task submission was published in mid-January[3], shortly after the training and development data was released. We further provided the opportunity to test the prediction format until the beginning of April, which was, however, not used by many participants. The evaluation period lasted six days in total. Ultimately, only three teams submitted predictions to CodaLab during the evaluation phase.

2.4 Baseline

To provide a meaningful comparison, we developed two baseline systems for the shared task, one for NER and one for joint NER + RE.

2.4.1 NER

The baseline for NER is set up using the PyTorch-IE framework (Binder et al., 2024)[4], which allows to prototype and test information extraction pipelines quickly. For NER, we employed a simple token classification model that encapsulates a (pre-trained) Transformer model from the HuggingFace library (Wolf et al., 2019). We first fine-tuned an NER model for German and Japanese data separately with different hyper-parameters and performed inference on the test data using the corresponding best-performing models. For German, we utilized a German version of BERT (Devlin et al., 2019)[5] as the pre-trained model, while for Japanese, we used multi-lingual XLM-RoBERTa (Conneau et al., 2019)[6]. For French, we used the best-performing Japanese XLM-RoBERTa model. Utilizing pooled output embeddings from the pretrained models and a classification head, we generate token-level predictions and convert the results back into `brat` format. We then combine the pre-

[3]`https://codalab.lisn.upsaclay.fr/competitions/17204`

[4]`https://github.com/ArneBinder/pytorch-ie`

[5]`dbmdz/bert-base-german-uncased`

[6]`FacebookAI/xlm-roberta-base`

dictions of all three languages to obtain the overall results.

2.4.2 Joint NER + RE

For joint NER+RE, we combined the NER system from above with a few-shot experiment using an LLM-based approach with Llama-3-8B-UltraMedical[7] (Zhang et al., 2024). We utilized the entities predicted during the NER task as input for the prompt given to the LLM. The specific details of the final prompt used in the experiment are provided in Appendix A.3. Specifically, we include definitions of the relations to be determined. We constructed three prompts, following the format outlined in Appendix A.3. We exclusively used German examples for the first prompt to predict relations within German data. Similarly, Japanese examples were used for the second prompt to predict relations within Japanese data. Finally, the third prompt was created using one German and Japanese example, aiming to predict relations across the entire dataset collectively. This last prompt was used for the French data.

2.5 Submitted Systems

The submitted systems are summarized in Table 3. For the leaderboard on CodaLab and the following tables, we selected the best three runs with more than three distinct submissions.

Yseop (Gupta, 2024) focused on NER for French and Japanese and on RE for Japanese only. For French NER, the authors utilized a combination of advanced language models, including the instruction LLM Mistral-7B (Jiang et al., 2023) and DrBERT-CASM2 (Labrak et al., 2023). For Japanese NER, they employed a multifaceted approach involving rule-based methods (medkit[8]), a Japanese-Multilingual Dictionary (JMdict[9]), and a Japanese medical language model based on RoBERTa (Liu et al., 2019), which was pre-trained on Japanese case reports and fine-tuned for NER using MedTxt-CR (Yada et al., 2022).[10] For the Japanese Relation Extraction, they re-used the RoBERTa model and fine-tuned it with the provided training data. Team Yseop submitted three runs for NER and joint NER + RE.

Team HBUT (Ke et al., 2024) concentrated solely on the named entity recognition task for all three languages. The methodology employed by the team focused on the use of LLMs. They explored three distinct prompting strategies to identify the most effective approach for NER. The team did not explore fine-tuning transformer architectures. The authors initially evaluated two different LLMs and selected GLM-3-Turbo (Zeng et al., 2023) as their preferred model. For the use with LLMs, the task had to be transferred into a generation task (instead of token classification), for which the authors designed specific prompts to get structured output. These outputs were then post-processed to result in `brat` format. Team HBUT submitted two runs to Subtask 2a and did not work on Relation Extraction.

Predictions of a third system were submitted to CodaLab. However, the predictions did not comply with the `brat` format and could not be evaluated.

2.6 Evaluation

The participants' submissions are ranked by non-weighted macro F_1 score ($\mathbf{F}_1$), precision (**P**), and recall (**R**) for both tasks. The evaluation script is a slightly modified version of 'brateval'[11] and can be found online[12]. The modifications were necessary to comply with the required output format for the evaluation platform CodaLab.

For Subtask 2a, we use an exact match of entities to calculate the previously mentioned scores. In the evaluation script, this corresponds to the parameters "-span-match exact".

For Subtask 2b, joint entity and relation extraction, note that both entity boundaries and types and relation types and arguments must match precisely. In the evaluation script, this corresponds to the parameters "-type-match exact -span-match exact". We also provide scores for relaxed (lenient) entity evaluation (for both subtasks) and results per language.

3 Results

In the following, we briefly describe the results for the two subtasks.

3.1 Subtask 2a – Named Entity Recognition

Table 4 presents the overall results for NER across languages using a strict match of entities. The re-

[7] https://huggingface.co/TsinghuaC3I/Llama-3-8B-UltraMedical

[8] https://medkit.readthedocs.io/en/stable/index.html

[9] https://www.edrdg.org/jmdict/j_jmdict.html

[10] https://huggingface.co/daisaku-s/medtxt_ner_roberta

[11] https://github.com/READ-BioMed/brateval

[12] https://github.com/Erechtheus/brateval

Team	Task	Language	P	R	F_1	System Summary
Yseop	NER	fr, ja	58.31	42.14	48.92	fr: Mistral-7B + DrBERT-CASM2; ja: rule-based system + dictionary + RoBERTa
HBUT	NER	de, fr, ja	60.52	26.54	36.90	GLM-3-Turbo + post-processing
Yseop	RE	ja	02.24	01.63	01.89	ja: RoBERTa fine-tuned

Table 3: Summary of the submitted systems with the scores on CodaLab (exact macro F_1, precision, and recall).

Team	Run	Overall			DISORDER			DRUG			FUNCTION		
		P	R	F_1	P	R	F_1	P	R	F_1	P	R	F_1
baseline	run 1†	47.55	58.83	**52.60**	47.36	58.86	**52.48**	55.08	67.43	60.63	29.49	36.70	**32.70**
Yseop	run 1	23.02	11.10	14.98	49.23	09.09	15.34	15.77	19.09	17.27	00.00	00.00	00.00
	run 2	22.36	09.39	13.22	**57.66**	08.17	14.31	14.01	15.25	14.61	00.00	00.00	00.00
	run 3†	58.31	42.14	48.92	56.65	42.86	48.80	71.03	50.10	58.76	**30.13**	18.35	22.81
HBUT	run 1	55.04	24.21	33.63	47.69	23.60	31.57	67.95	34.75	45.98	00.00	00.00	00.00
	run 2†	**60.52**	26.54	36.90	54.89	27.89	36.98	**71.24**	34.44	46.43	00.00	00.00	00.00
∪	—	42.00	71.10	52.81	41.10	71.66	52.24	50.78	81.12	62.46	25.16	42.82	31.69

Table 4: Named Entity Recognition (NER) results on the test set for all participants using exact match evaluation across languages. † denotes the system with the best overall performance for each team. The overall score is the unweighted micro average across the three languages. ∪ shows the results when combining all predictions of the best (†) systems.

sults indicate the difficulty of our task. The best score on overall NER is an F_1 score of 52.60. While the detection of DRUG appears to be slightly more attainable (F_1 up to 60), the detection of FUNCTION achieves overall the lowest scores. Since team Yseop targeted only Japanese and French, and team HBUT did not find any FUNCTION entities, it is no surprise that the baseline system achieves the best results. Interestingly, the joint NER+RE approach yielded better results for the baseline system than the NER approach described in Section 2.4.1. Therefore, we only show the results of one baseline system for both subtasks.

Note that all systems achieve similar results concerning overall precision but that the submitted systems only produce very low recall, i.e., they fail to catch many of the gold entities. In contrast, precision and recall of the baseline system seem to be relatively balanced, with recall being slightly higher than precision.

A language-specific overview of the NER results is provided in Table 5. The approach of Yseop achieves the best overall results in the few-shot scenario for French. In contrast, HBUT achieves the best performance considering the detection of DRUG mentions in French documents. For Japanese, team Yseop performs similarly as the baseline but substantially drops in performance for FUNCTION.

Finally, Table 6 presents the relaxed scores, i.e., resulting scores for entities that do not exactly match but have some overlap with the gold data. The results highlight (similarly as in Table 4) that while the baseline was optimized for recall, the systems of the team Yseop and the team HBUT both achieve good scores in terms of precision.

3.2 Subtask 2b – Joint Entity and Relation Extraction

Table 7 presents the joint named entity and relation extraction task results, highlighting the task's difficulty. To extract relations correctly, the entities need to be detected accurately in the first place. Here, we differ between exact (where entities are detected correctly according to exact boundaries) and relaxed (where entity boundaries have to overlap at least partially), which allows a more flexible mapping of the corresponding entities. The performance of the NER system directly and strongly

Lang	Team	Overall			DISORDER			DRUG			FUNCTION		
		P	R	F_1	P	R	F_1	P	R	F_1	P	R	F_1
de	baseline	41.26	**55.56**	**47.35**	42.59	**52.09**	**46.86**	51.28	**64.52**	**57.14**	**20.00**	**46.15**	**27.91**
	Yseop	00.00	00.00	00.00	00.00	00.00	00.00	00.00	00.00	00.00	00.00	00.00	00.00
	HBUT	**62.28**	27.51	38.17	**54.63**	27.44	36.53	**76.27**	36.29	49.18	00.00	00.00	00.00
fr	baseline	33.05	**46.43**	38.61	29.97	**36.63**	32.97	45.76	**71.15**	55.70	**08.16**	**16.33**	**10.88**
	Yseop	**60.68**	31.33	**41.32**	49.54	26.26	34.32	**76.52**	48.35	**59.26**	00.00	00.00	00.00
	HBUT	60.57	31.33	41.30	**55.90**	32.25	**40.90**	68.81	38.19	49.12	00.00	00.00	00.00
ja	baseline	**61.57**	**67.79**	**64.53**	**59.76**	**75.38**	**66.67**	67.90	**65.34**	**66.60**	**57.14**	**43.51**	**49.41**
	Yseop	57.52	59.03	58.27	58.98	64.05	61.41	68.22	64.50	66.31	30.13	28.87	29.49
	HBUT	60.00	23.15	33.41	54.12	25.05	34.25	**72.20**	31.09	43.47	00.00	00.00	00.00

Table 5: Language-specific named entity recognition results on the test set for all participants (best system) using exact match evaluation. The overall score is the non-weighted micro average for each language.

Team	Run	Overall			DISORDER			DRUG			FUNCTION		
		P	R	F_1	P	R	F_1	P	R	F_1	P	R	F_1
baseline	run 1	64.61	**79.94**	**71.46**	68.18	**84.74**	**75.57**	69.66	**85.27**	**76.68**	35.26	**43.88**	**39.10**
Yseop	run 1	46.38	22.36	30.17	73.07	13.49	22.77	38.99	47.20	42.70	00.00	00.00	00.00
	run 2	48.65	20.42	28.77	**86.29**	12.23	21.42	39.75	43.26	41.43	00.00	00.00	00.00
	run 3	75.19	54.34	63.08	75.60	57.20	65.13	85.15	60.06	70.44	**43.23**	26.33	32.73
HBUT	run 1	71.74	31.55	43.83	66.86	33.09	44.27	80.32	41.08	54.36	00.00	00.00	00.00
	run 2	**79.85**	35.02	48.68	75.14	38.17	50.63	**88.84**	42.95	57.90	00.00	00.00	00.00
∪	—	51.16	86.60	64.32	51.69	90.11	65.69	58.44	93.36	71.88	31.09	52.93	39.17

Table 6: Named Entity Recognition (NER) results on the test set for all participants using relaxed match evaluation across languages. For each team, the system with the best overall performance is highlighted with †. The overall score is the unweighted micro average across the three languages. ∪ shows the results when combining all predictions of the best (†) systems.

influences the overall results. Therefore, as the previous results on NER were low, it is unsurprising that participants and baselines result in an F_1 score below 10. The fact that team Yseop did not target all languages also influenced the results.

Language-specific results are shown in Table 8. Team Yseop predicted only relations for the Japanese dataset.

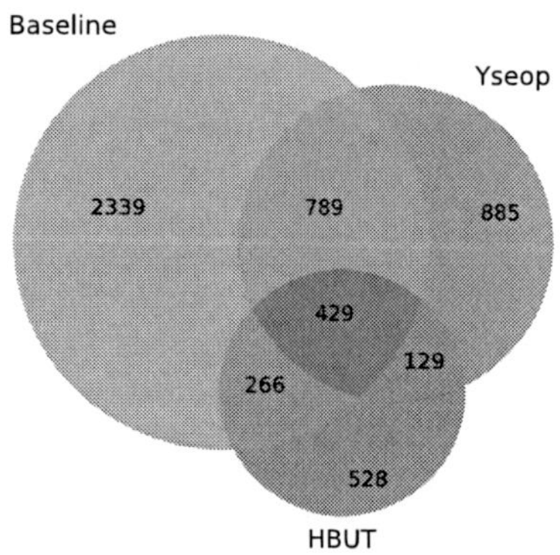

Figure 3: Exact overlap of entity predictions (Subtask 2a) for the best performing submission (†) for each team.

4 Analysis

We provide a brief analysis of the achieved results and the challenges the participants and their systems faced. This section is meant for further versions of the shared task and showcases common pitfalls.

4.1 Common Mistakes and Challenges

Several teams had difficulty providing predictions in the required `brat` format. At first glance, the `brat` format seems quite simple since the predictions are written in a text file, one extracted entity or relation per line, separated by either whitespace or tabular space, as shown in Figure 1. However, it seems that LLMs cannot consistently produce the correct `brat` format. Therefore, LLMs' output must be validated and/or pre-processed to ensure the correct format.

We also noticed that many offsets in the prediction files did not correspond to the actual string (i.e., the extracted entity) and that some spans started with -1, which resulted in invalid entity spans. Finally, we found several relations in the prediction files associated with non-existing entity mentions and, therefore, were ignored during automatic evaluation.

Overall, it seemed that not only was the development of the actual system the challenge in this task, but also, the post-processing of the systems' outputs provided some difficulty, especially when an LLM returned it.

4.2 Overlapping Entities

Figure 3 presents a Venn diagram of detected entities and their overlap between the best-performing submissions for each team. While the baseline system tends to have a high recall, the participants seem to have targeted a high precision. Therefore, it is unsurprising that the baseline detected the most significant number of entities. However, it is interesting to see that each team detected a large number of entities that the others did not detect.

We tested this by building the union of the predictions of the best system for each team. As shown in Table 4 and Table 6, the recall increases substantially, demonstrating that a large proportion of the three entities was indeed found by at least one system.

4.2.1 `FUNCTION` Mentions

`FUNCTION` entities were one of the more difficult mentions to detect. For instance, team HBUT did not find any mention correctly, and the baseline only reached an (exact) F_1 score of 32.7. This might be due to these mentions having a more difficult underlying concept: `FUNCTION` can be simply nouns or verbs (*"... I can sleep too"*), but they can also encompass more complicated phrases (*"I still had a relatively regular cycle."*). Also, the distinction between a `FUNCTION` and an actual `DISORDER` (which might be a negated body function) is often ambiguous. Detecting `FUNCTION` mentions worked much better for the Japanese data than for German and French. This could be because the boundaries of body functions might be easier to detect in the Japanese script than in the Latin script.

5 Discussion & Conclusion

In Task 2 of #SMM4H 2024, the participants had to tackle a difficult task. Starting with a small, multilingual, and layperson dataset and only a few examples for French plus no English data support, their systems had to distinguish three medical entities and two different relations, which are determined by temporal order and medical knowledge: The order of `DRUG` mentions, and `DISORDER`/`FUNCTION` mentions decides if the relation between the expressions is a "cause" or a "treatment" relation.

Team	Run	Exact			Relaxed		
		P	R	F_1	P	R	F_1
baseline	run 1	**04.25**	**06.81**	**05.23**	**07.82**	**12.53**	**09.63**
Yseop	run 3	02.24	01.63	01.89	03.17	02.32	02.68

Table 7: Relation extraction results on the test set for all participants using exact and relaxed match evaluation. Relaxed evaluation allows two entities to match if their boundaries overlap. The overall score is the non-weighted micro average across the three languages.

Lang	Team	Overall		
		P	R	F_1
de	baseline	**08.33**	**10.87**	**09.43**
	Yseop	00.00	00.00	00.00
fr	baseline	**03.88**	**06.38**	**04.83**
	Yseop	00.00	00.00	00.00
ja	baseline	**03.07**	**05.24**	**03.87**
	Yseop	02.26	04.49	03.01

Table 8: Language-specific relation extraction results on the test set for all participants (best system) using exact match evaluation. The overall score is the non-weighted macro average for each language.

Based on our baseline development using an LLM and the participants' submissions, a lot of post-processing seems necessary to be applied to the LLM output. It cannot be taken as is since the desired output format might not be consistently returned. We also noticed that it matters in which language the prompts are given to an LLM and that sometimes, for example, with an English prompt, the returned entities are correct but in English and, therefore, do not match the gold entities. Additionally, even if the LLM approach worked better than a transformer-based approach, the results were still unsatisfactory, especially for relation extraction.

Of course, combining different languages in different scripts and colloquial texts on which the models were not trained is somewhat tricky. However, given the success of LLMs in other domains or other genres of text, e.g., scientific documents, we were surprised that none of the teams beat the baseline. We, therefore, think that there is still a lot of work to be done in the medical domain concerning patient-generated texts, especially for non-English speaking patients, and that even LLMs seem to be only a part of a potential solution.

It is worth looking into the details of the single systems' benefits in future work. Despite the low number of participating teams, there were still many different approaches to Task 2 (rule-, transformer-, and LLM-based). Inspecting the detected entities and relation of each system might yield further insights and lead to a more successful system.

Acknowledgements

We thank the anonymous reviewers for their constructive feedback on this paper. Our work was supported by the Cross-ministerial Strategic Innovation Promotion Program (SIP) on "Integrated Health Care System" Grant Number JPJ012425, by JPMJCR20G9, ANR-20-IADJ-0005-01, and DFG-442445488 under the trilateral ANR-DFG-JST AI Research project KEEPHA, and by the German Federal Ministry of Education and Research under the grant BIFOLD24B.

References

Arne Binder, Leonhard Hennig, and Christoph Alt. 2024. Pytorch-ie: Fast and reproducible prototyping for information extraction. *Preprint*, arXiv:2406.00007.

Alexis Conneau, Kartikay Khandelwal, Naman Goyal, Vishrav Chaudhary, Guillaume Wenzek, Francisco Guzmán, Edouard Grave, Myle Ott, Luke Zettlemoyer, and Veselin Stoyanov. 2019. Unsupervised cross-lingual representation learning at scale. *CoRR*, abs/1911.02116.

Jacob Devlin, Ming-Wei Chang, Kenton Lee, and Kristina Toutanova. 2019. BERT: Pre-training of Deep Bidirectional Transformers for Language Understanding. In *Proceedings of the 2019 Conference of the North American Chapter of the Association for Computational Linguistics: Human Language Technologies*, Minneapolis, Minnesota, USA. Association for Computational Linguistics.

I. Ralph Edwards and Jeffrey K. Aronson. 2000. Adverse drug reactions: Definitions, diagnosis, and management. *The Lancet*, 356(9237):1255–1259.

Anubhav Gupta. 2024. "a team at #smm4h 2024: Pharmacovigilance shared task in english, french and japanese". In *Proceedings of The 9th Social Media Mining for Health Research and Applications Workshop and Shared Tasks*, Bangkok, Thailand. Association for Computational Linguistics.

Lorna Hazell and Saad A. W. Shakir. 2006. Under-reporting of adverse drug reactions : A systematic review. *Drug Safety*, 29(5):385–396.

Albert Q Jiang, Alexandre Sablayrolles, Arthur Mensch, Chris Bamford, Devendra Singh Chaplot, Diego de las Casas, Florian Bressand, Gianna Lengyel, Guillaume Lample, Lucile Saulnier, et al. 2023. Mistral 7b. *arXiv preprint arXiv:2310.06825*.

Sarvnaz Karimi, Alejandro Metke-Jimenez, Madonna Kemp, and Chen Wang. 2015. Cadec: A corpus of adverse drug event annotations. *Journal of Biomedical Informatics*, 55:73–81.

Yuanzhi Ke, Zhangju Yin, Xinyun Wu, and Caiquan Xiong. 2024. "hbut at #smm4h 2024 task2: Cross-lingual few-shot medical entity extraction using a large language model". In *Proceedings of The 9th Social Media Mining for Health Research and Applications Workshop and Shared Tasks*, Bangkok, Thailand. Association for Computational Linguistics.

Diederik P. Kingma and Jimmy Ba. 2014. Adam: A Method for Stochastic Optimization. *Internation Conference on Learning Representations*.

Ari Klein, Ilseyar Alimova, Ivan Flores, Arjun Magge, Zulfat Miftahutdinov, Anne-Lyse Minard, Karen O'Connor, Abeed Sarker, Elena Tutubalina, Davy Weissenbacher, and Graciela Gonzalez-Hernandez. 2020. Overview of the Fifth Social Media Mining for Health Applications (#SMM4H) Shared Tasks at COLING 2020. In *Fifth Social Media Mining for Health Applications(#SMM4H) Shared Tasks at COLING 2020*, page 10.

Ari Z Klein, Juan M Banda, Yuting Guo, Ana Lucia Schmidt, Dongfang Xu, Ivan Flores Amaro, Raul Rodriguez-Esteban, Abeed Sarker, and Graciela Gonzalez-Hernandez. 2024. Overview of the 8th Social Media Mining for Health Applications (#SMM4H) shared tasks at the AMIA 2023 Annual Symposium. *Journal of the American Medical Informatics Association*, 31(4):991–996.

Yanis Labrak, Adrien Bazoge, Richard Dufour, Mickaël Rouvier, Emmanuel Morin, Béatrice Daille, and Pierre-Antoine Gourraud. 2023. Drbert: A robust pre-trained model in french for biomedical and clinical domains. In *Proceedings of the 61st Annual Meeting of the Association for Computational Linguistics (Volume 1: Long Papers)*, pages 16207–16221.

Robert Leaman, Laura Wojtulewicz, Ryan Sullivan, Annie Skariah, Jian Yang, and Graciela Gonzalez. 2010. Towards Internet-Age Pharmacovigilance: Extracting Adverse Drug Reactions from User Posts in Health-Related Social Networks. In *Proceedings of the 2010 Workshop on Biomedical Natural Language Processing*, pages 117–125, Uppsala, Sweden. Association for Computational Linguistics.

Yinhan Liu, Myle Ott, Naman Goyal, Jingfei Du, Mandar Joshi, Danqi Chen, Omer Levy, Mike Lewis, Luke Zettlemoyer, and Veselin Stoyanov. 2019. RoBERTa: A robustly optimized BERT pretraining approach. *arXiv:1907.11692 [cs]*.

Arjun Magge, Ari Klein, Antonio Miranda-Escalada, Mohammed Ali Al-garadi, Ilseyar Alimova, Zulfat Miftahutdinov, Eulalia Farre-Maduell, Salvador Lima Lopez, Ivan Flores, Karen O'Connor, Davy Weissenbacher, Elena Tutubalina, Abeed Sarker, Juan M Banda, Martin Krallinger, and Graciela Gonzalez-Hernandez, editors. 2021a. *Proceedings of the Sixth Social Media Mining for Health (#SMM4H) Workshop and Shared Task*. Association for Computational Linguistics, Mexico City, Mexico.

Arjun Magge, Elena Tutubalina, Zulfat Miftahutdinov, Ilseyar Alimova, Anne Dirkson, Suzan Verberne, Davy Weissenbacher, and Graciela Gonzalez-Hernandez. 2021b. DeepADEMiner: A deep learning pharmacovigilance pipeline for extraction and normalization of adverse drug event mentions on Twitter. *Journal of the American Medical Informatics Association*, 28(10):2184–2192.

Aurélie Névéol, Hercules Dalianis, Sumithra Velupillai, Guergana Savova, and Pierre Zweigenbaum. 2018. Clinical Natural Language Processing in languages other than English: Opportunities and challenges. *Journal of Biomedical Semantics*, 9(1):12.

Beatrice Portelli, Simone Scaboro, Emmanuele Chersoni, Enrico Santus, and Giuseppe Serra. 2022. AILAB-Udine@SMM4H'22: Limits of Transformers and BERT Ensembles. In *Proceedings of The Seventh Workshop on Social Media Mining for Health Applications, Workshop & Shared Task*, pages 130–134, Gyeongju, Republic of Korea. Association for Computational Linguistics.

Lisa Raithel, Hui-Syuan Yeh, Shuntaro Yada, Cyril Grouin, Thomas Lavergne, Aurélie Névéol, Patrick Paroubek, Philippe Thomas, Tomohiro Nishiyama, Sebastian Möller, Eiji Aramaki, Yuji Matsumoto, Roland Roller, and Pierre Zweigenbaum. 2024. A Dataset for Pharmacovigilance in German, French, and Japanese: Annotating Adverse Drug Reactions across Languages. In *Proceedings of the 2024 Joint International Conference on Computational Linguistics, Language Resources and Evaluation (LREC-COLING 2024)*, Turin, Italy.

Alexander Sboev, Sanna Sboeva, Ivan Moloshnikov, Artem Gryaznov, Roman Rybka, Alexander Naumov, Anton Selivanov, Gleb Rylkov, and Vyacheslav Ilyin. 2022. Analysis of the Full-Size Russian Corpus of Internet Drug Reviews with Complex NER Labeling Using Deep Learning Neural Networks and Language Models. *Applied Sciences*, 12(1):491.

Isabel Segura-Bedmar, Ricardo Revert, and Paloma Martínez. 2014. Detecting drugs and adverse events from Spanish social media streams. In *Proceedings of the 5th International Workshop on Health Text Mining and Information Analysis (Louhi)*, pages 106–115, Gothenburg, Sweden. Association for Computational Linguistics.

Pontus Stenetorp, Sampo Pyysalo, Goran Topić, Tomoko Ohta, Sophia Ananiadou, and Jun'ichi Tsujii. 2012. Brat: A Web-based Tool for NLP-Assisted Text Annotation. In *Proceedings of the Demonstrations at the 13th Conference of the European Chapter of the Association for Computational Linguistics*, pages 102–107, Avignon, France. Association for Computational Linguistics.

Elena Tutubalina, Ilseyar Alimova, Zulfat Miftahutdinov, Andrey Sakhovskiy, Valentin Malykh, and Sergey Nikolenko. 2021. The Russian Drug Reaction Corpus and neural models for drug reactions and effectiveness detection in user reviews. *Bioinformatics (Oxford, England)*, 37(2):243–249.

Davy Weissenbacher, Juan Banda, Vera Davydova, Darryl Estrada Zavala, Luis Gasco Sánchez, Yao Ge, Yuting Guo, Ari Klein, Martin Krallinger, Mathias Leddin, Arjun Magge, Raul Rodriguez-Esteban, Abeed Sarker, Lucia Schmidt, Elena Tutubalina, and Graciela Gonzalez-Hernandez. 2022. Overview of the Seventh Social Media Mining for Health Applications (#SMM4H) Shared Tasks at COLING 2022. In *Proceedings of The Seventh Workshop on Social Media Mining for Health Applications, Workshop & Shared Task*, pages 221–241, Gyeongju, Republic of Korea. Association for Computational Linguistics.

Thomas Wolf, Lysandre Debut, Victor Sanh, Julien Chaumond, Clement Delangue, Anthony Moi, Pierric Cistac, Tim Rault, Rémi Louf, Morgan Funtowicz, and Jamie Brew. 2019. Huggingface's transformers: State-of-the-art natural language processing. *CoRR*, abs/1910.03771.

Dongfang Xu, Guillermo Lopez Garcia, Lisa Raithel, Roland Roller, Philippe Thomas, Eiji Aramaki, Shuntaro Yada, Pierre Zweigenbaum, Sai Tharuni Samineni, Karen O'Connor, Yao Ge, Sudeshna Das, Abeed Sarker, Ari Klein, Lucia Schmidt, Vishakha Sharma, Raul Rodriguez-Esteban, Juan Banda, Ivan Flores Amaro, Davy Weissenbacher, and Graciela Gonzalez-Hernandez. 2024. Overview of the 9th social media mining for health (#SMM4H) research and applications workshop and shared tasks at ACL 2024. In *Proceedings of The 9th Social Media Mining for Health Research and Applications Workshop and Shared Tasks*, Bangkok, Thailand. Association for Computational Linguistics.

Shuntaro Yada, Shoko Wakamiya, Yuta Nakamura, and Eiji Aramaki. 2022. Real-MedNLP: Overview of REAL document-based MEDical Natural Language Processing Task. In *Proceedings of the 16th NTCIR Conference on Evaluation of Information Access Technologies*, Tokyo, Japan.

Aohan Zeng, Xiao Liu, Zhengxiao Du, Zihan Wang, Hanyu Lai, Ming Ding, Zhuoyi Yang, Yifan Xu, Wendi Zheng, Xiao Xia, Weng Lam Tam, Zixuan Ma, Yufei Xue, Jidong Zhai, Wenguang Chen, Peng Zhang, Yuxiao Dong, and Jie Tang. 2023. GLM-130B: An Open Bilingual Pre-trained Model. *Preprint*, arXiv:2210.02414.

Kaiyan Zhang, Ning Ding, Biqing Qi, Sihang Zeng, Haoxin Li, Xuekai Zhu, Zhang-Ren Chen, and Bowen Zhou. 2024. Ultramedical: Building specialized generalists in biomedicine. `https://github.com/TsinghuaC3I/UltraMedical`.

Maryam Zolnoori, Kin Wah Fung, Timothy B. Patrick, Paul Fontelo, Hadi Kharrazi, Anthony Faiola, Yi Shuan Shirley Wu, Christina E. Eldredge, Jake Luo, Mike Conway, Jiaxi Zhu, Soo Kyung Park, Kelly Xu, Hamideh Moayyed, and Somaieh Goudarzvand. 2019. A systematic approach for developing a corpus of patient reported adverse drug events: A case study for SSRI and SNRI medications. *Journal of Biomedical Informatics*, 90:103091.

A Baseline Details

A.1 Discontinuous Entities

The dataset contains fragmented spans where a single entity consists of two disjoint spans. To handle these fragmented spans, we split them into two separate spans and establish a temporary relation between the new spans. If a relationship involves a fragmented span, we create new relationships accordingly. For instance, if a `CAUSED` relation existed between a fragmented span (span 1) and a continuous span (span 2), we create new relations after splitting span 1 into span 11 and span 12: span 11 `CAUSED` span 2 and span 12 `CAUSED` span 2. This implies that we train models with simple entities containing only a single span. During prediction, entities linked by the temporary relation are combined to form a fragmented entity with two spans, but only if both entities have the same predicted label; otherwise, the temporary relations are ignored. If any of these entities had a relation with another entity, that relation is maintained after conversion to the fragmented entity.

A.2 Experimental Details

A.2.1 NER

For the NER task on German data, we utilized the dbmdz/bert-base-german-uncased pretrained model. For Japanese and French, we employed the FacebookAI/xlm-roberta-base model. We implemented a BIO encoding scheme for token labeling, resulting in a total of 7 classes. During training, we used a cross-entropy loss function and the Adam optimizer (Kingma and Ba, 2014) with a learning rate of 1e-05. To handle lengthy texts, we split them based on two parameters: max_length (set to 512) and stride (set to 64). Each split contains up to 512 tokens, with an overlap of 64 tokens between consecutive splits.

A.2.2 Joint NER+RE

For joint NER-RE inference, combinations of predicted entities are used for relation classification. However, this approach may introduce entity pairs that are widely separated in the text. To address this, we employed a max_window parameter (set to 512), which specifies the maximum allowed inner distance between entity pairs. Additionally, we included reversed gold relations by swapping the head and tail entities.

A.3 Prompt Format Descriptions

The NER prompt begins by defining the entities identified in the text, including `DRUG`, `DISORDER`, and `FUNCTION`. The prompt then provides examples to illustrate the task. Each example presents an input text followed by a list of identified entities, with explanations for their classification. Each identified entity is presented with its associated information, including the entity text, a True/False label indicating the accuracy of the entity classification, and an explanation justifying the classification. The explanation specifies the entity type (`DRUG`, `DISORDER`, or `FUNCTION`) and provides context for why the entity fits into that category.

The prompt for the joint NER + RE task specifies that we should classify relations between provided entities as `CAUSED` or `TREATMENT_FOR`. We then define the two types of relations. Following these definitions, the prompt includes multiple examples to illustrate the process. Each example is structured first to present the input text where the relations must be identified. Then, the identified entities from the text are listed and categorized into `DRUG`, `DISORDER`, or `FUNCTION`. Finally, the output section details the relationships identified between the entities. Each relationship is specified with the first entity, the second entity, a True/False label indicating the presence of the relationship, and an explanation justifying the prediction and identifying the relation type (`CAUSED` or `TREATMENT_FOR`).

A.4 Prompt used for NER task

```
Defn: The following are the definitions of the entities
DRUG: any mention of a medication name ("iburpofen"), brand ("Vick"), or agent
("Sorbitan"), including dietary supplements ("magnesium"), even when abbreviated ("AD"
for "anti depressive")
DISORDER: any disease, sign, or symptom related to the patient's health, including
mental issues. Sometimes a disorder may be expressed as a parameter in combination
with a value: high LDL
FUNCTION: all body functions and processes. Body functions are often represented in
biomarkers eg: HDL, WBC. It also includes mental functions
Difference between DISORDER and FUNCTION: We annotate adverse biological processes as
disorder and neutral/positive processes as function

<1>
Example 1: Ich nehme seit zwei Wochen Ibuprofen gegen meine Kopfschmerzen.

Entities:
1. Ibuprofen | True | as it refers to a specific medication (DRUG)
2. Kopfschmerzen | True | as it refers to a type of pain (DISORDER)
</1>

<2>
Example 2: Seitdem ich Magnesium nehme, fühle ich mich weniger müde.

Entities:
1. Magnesium | True | as it refers to a dietary supplement (DRUG)
2. müde | True | as it refers to fatigue, a symptom (DISORDER)
</2>

<3>
Example 3:
Text: {text}

Entities:
```

A.5 Prompt used for RE task

```
# Task: Use ONLY the provided entities to classify relations : [CAUSED, TREATMENT_FOR]

## Definition:
TREATMENT_FOR: This relation connects a DRUG and the targeted DISORDER, describing the
medication that was used to treat the patient's symptoms.
CAUSED: We only annotate a caused relation when the entities DRUG, DISORDER, or
FUNCTION are concerned. Explicit formulation of a <cause>-<consequence> relation,
Lexical semantics of nouns or verbs, e.g., <cause> provokes <consequence>

### Examples
<1>
Example 1:
Input: Ich nehme seit zwei Wochen Ibuprofen gegen meine Kopfschmerzen.
Entities:
DRUG - [Ibuprofen]
DISORDER - [Kopfschmerzen]
Output:
Relations:
Ibuprofen | Kopfschmerzen | True | Ibuprofen is used as a treatment for headaches
(TREATMENT_FOR)
</1>

<2>
Example 2:
Input: Seitdem ich Magnesium nehme, fühle ich mich weniger müde.
Entities:
DRUG - [Magnesium]
DISORDER - [müde]
Output:
Relations:
Magnesium | müde | True | Magnesium is used as a treatment for fatigue (TREATMENT_FOR)
</2>

<3>
Example 3:
Input: {text}
Entities:
{entities}
Output:
Relations:
```

Overview of the 9th Social Media Mining for Health Applications (#SMM4H) Shared Tasks at ACL 2024 – Large Language Models and Generalizability for Social Media NLP

Dongfang Xu[1], Guillermo Lopez-Garcia[1], Lisa Raithel[2, 3, 4], Roland Roller[2], Philippe Thomas[2], Eiji Aramaki[5], Shoko Wakamiya[5], Shuntaro Yada[5], Pierre Zweigenbaum[6], Karen O'Connor[7], Sophia Hernandez[8], Sai Tharuni Samineni[1], Yao Ge[9], Swati Rajwal[9], Sudeshna Das[9], Abeed Sarker[9], Ari Klein[7], Ana Lucia Schmidt[10], Vishakha Sharma[11], Raul Rodriguez-Esteban[10], Juan M. Banda[12], Ivan Flores Amaro[1], Davy Weissenbacher[1], Graciela Gonzalez-Hernandez[1]

[1]Cedars-Sinai Medical Center, Los Angeles, CA, USA
[2]German Research Center for Artificial Intelligence (DFKI), Berlin, Germany
[3] Quality & Usability Lab, Technische Universität Berlin, Berlin, Germany
[4] BIFOLD – Berlin Institute for the Foundations of Learning and Data, Berlin, Germany
[5]Nara Institute of Science and Technology, Nara, Japan
[6]Université Paris-Saclay, CNRS, LISN, Orsay, France
[7]University of Pennsylvania, Philadelphia, PA, USA
[8]University of Pittsburgh, Pittsburgh, PA, USA
[9]Emory University, Atlanta, GA, USA
[10]Roche Innovation Center, Basel, Switzerland
[11]Roche Diagnostics, Santa Clara, CA, USA
[12]Stanford Health Care, Newark, CA, USA
Correspondence: dongfang.xu@cshs.org

Abstract

For the past nine years, the Social Media Mining for Health Applications (#SMM4H) shared tasks have promoted community-driven development and evaluation of advanced natural language processing systems to detect, extract, and normalize health-related information in publicly available user-generated content. This year, #SMM4H included seven shared tasks in English, Japanese, German, French, and Spanish from Twitter, Reddit, and health forums. A total of 84 teams from 22 countries registered for #SMM4H, and 45 teams participated in at least one task. This represents a growth of 180% and 160% in registration and participation, respectively, compared to the last iteration. This paper provides an overview of the tasks and participating systems. The data sets remain available upon request, and new systems can be evaluated through the post-evaluation phase on CodaLab.

1 Introduction

The number of social media (SM) users continues to grow worldwide: 86% of US adults and 79% in Europe use SM (Center; Elliott and Sverdlov, 2012). Advances in automated data processing, machine learning and natural language processing (NLP) allow us to incorporate this massive real-time data source from around the world for biomedical and public health applications, providing researchers a venue to address the many methodological challenges unique to this media. The Social Media Mining for Health Applications (#SMM4H) Workshop, in its 9th annual iteration, brings together researchers interested in developing and sharing NLP methods that enable the systematic use of SM data for health research. The tasks of this year use data from various platforms (X, Reddit, and patient forums such as Lifeline[1] or YJQA[2]) and languages (English, Spanish, French, German, and Japanese), with a special focus on Large Language Models (LLMs) for Social Media NLP. Seven tasks organized by experienced research teams from around the world were selected for 2024. We prioritized tasks evaluating the generalizability of the proposed approaches by explicitly creating test sets with out-of-distribution data such as unseen concepts, multi-lingual and multi-source texts. These tasks are: extraction and normalization of adverse drug events in English tweets (Task 1), cross-lingual few-shot relation extraction for pharmacovigilance in French, German,

[1]https://fragen.lifeline.de/forum/
[2]https://chiebukuro.yahoo.co.jp/

Proceedings of The 9th Social Media Mining for Health Applications (#SMM4H) Workshop and Shared Tasks, pages 183–195
August 15, 2024 ©2024 Association for Computational Linguistics

and Japanese (Task 2), multi-class classification of effects of outdoor spaces on social anxiety symptoms in Reddit (Task 3), extraction of the clinical and social impacts of non-medical substance use from Reddit (Task 4), binary classification of English tweets reporting children's medical disorders (Task 5), self-reported exact age classification with cross-platform evaluation in English (Task 6), and identification of whether an LLM or a human domain expert annotated data in the context of health-related applications (Task 7).

Teams could register for a single task or multiple tasks. Teams were provided with gold-standard annotated training and validation sets to develop their systems and, subsequently, an unlabeled test set for the final evaluation. After receiving the test set, all teams were given 5 days to submit the predictions of their systems to CodaLab —a platform that facilitates data science competitions— for automatic evaluation, promoting a systematic performance comparison. Among the 84 teams that registered, 45 teams submitted at least one set of predictions: 11 teams for Task 1, 3 teams for Task 2, 15 teams for Task 3, 3 teams for Task 4, 20 teams for Task 5, 7 teams for Task 6, and 3 teams for Task 7. Teams that submitted predictions were invited to submit a short manuscript describing their system, and 38 of the 45 teams did. Each of these 38 system descriptions was peer-reviewed by at least 2 reviewers. In this article, we present the annotated corpora, the technical summaries of all the participating systems, and the performance results, providing insights into state-of-the-art methods for mining social media data for health informatics.

2 Tasks

2.1 Task 1: Extraction and normalization of adverse drug events in English tweets

Adverse drug events (ADEs) are harmful and undesired reactions attributed to the intake of a drug or medication. Active post-market surveillance is essential, given clinical trials may not detect all potential ADEs, particularly for vulnerable populations. Social media can complement traditional reporting systems, such as the FDA's Adverse Event Reporting System (FAERS) for pharmacovigilance (Leaman et al., 2010; Tricco et al., 2018). Task 1 involved automatically extracting ADE text spans in tweets and normalizing them to their standard preferred term IDs (ptIDs) in the Medical Dictionary for Regulatory Activities (MedDRA).

Dataset The dataset for Task 1 contains a total of 18,185 tweets with 1,650 adverse drug events (ADEs) labeled in the training set, 965 tweets with 85 ADEs in the development set, and 11,799 tweets with 1,232 ADEs in the test set. The training, development, and test splits include 1,239, 65, and 915 tweets reporting at least one ADE. Notably, 5.8% of the ADEs in the development set are unseen (i.e., they do not appear in the training set), and 22.0% of the ADEs in the test set are unseen (i.e., they do not appear in either the training or development sets). This was done explicitly to test the systems' generalizability (understood as their capacity to detect unseen mentions).

Evaluation We used three different evaluation metrics: the ADE normalization score for all ptIDs, the ADE normalization score for unseen ptIDs, and ADE extraction scores. The first two metrics are the same as those used in the shared task in SMM4H-2023 (Klein et al., 2024b), while the third metric was used in past ADE extraction tasks in SMM4H (Magge et al., 2021a; Weissenbacher et al., 2022a). We use precision, recall, and F_1 scores for all three evaluation metrics, where a true positive prediction means that for each tweet, the predicted annotation (either ADE ptID or ADE text span) matches the gold standard annotation. The CodaLab site for this task is `https://codalab.lisn.upsaclay.fr/competitions/18363`.

2.2 Task 2: Cross-Lingual few-shot relation extraction for pharmacovigilance in French, German, and Japanese

Task 2, like Task 1, focuses on information extraction for pharmacovigilance, but it evaluates a multilingual corpus of texts gathered from diverse sources, including patient forums, social media, and clinical reports in German, French, and Japanese. This task has two subtasks: (2a) named entity recognition (NER) for identifying mentions of drugs, disorders, and body functions, and (2b) joint NER and relation extraction (RE), evaluating both the extraction of entities and their relationships.

Dataset The training data consists of texts collected from both Twitter and a Q&A forum related to healthcare issues. It includes 392 documents in the training and 168 documents in the development set in Japanese, as well as texts in German collected from a health forum (70 documents in the training set and 23 documents in the development set). In addition, 4 French documents were added to the training set by automatically translating German documents collected from the same patient forum as German data and manually reviewed by a native French speaker. The test data comprised 118 Japanese documents, 25 German documents, and 96 French documents. Note that the translated French data did not overlap with the German data. All data were taken from the fine-grained KEEPHA corpus (Raithel et al., 2024b) and filtered for the aforementioned entity and relation types. All data were annotated with the same annotation guidelines, focusing on detecting and extracting adverse drug reactions, modeled by associating medication mentions with disorder (medical signs and symptoms) and body function mentions. The relation distribution is imbalanced (i.e., the number of "treatment_for" relations is much lower than that of "caused" relations), adding to the task's difficulty. The format of the annotations, and therefore the format of the desired predictions, is `brat` (Stenetorp et al., 2012).

Evaluation Participating systems were evaluated on CodaLab using macro precision, recall and F_1 score for both Subtask 2a and 2b across all languages in an exact match setup (only exact matching of entities are considered correct). For further analysis, single submissions were evaluated by language and with relaxed entity scores. The CodaLab site for this task is `https://codalab.lisn.upsaclay.fr/competitions/17204`.

2.3 Task 3: Multi-class classification of effects of outdoor spaces on social anxiety symptoms in Reddit

Social anxiety disorder (SAD), which is anxiety that occurs or is triggered by any situation involving other people where the individual may be judged or scrutinized, may affect up to 12% of the population at some point in their lives (Kessler et al., 2005). The onset of SAD occurs early in life, beginning by age 11 in 50% and by age 20 in 80% of patients (Stein and Stein, 2008). Individuals with SAD report experiencing symptoms for a decade before seeking treatment. However, some turn to social media platforms like Reddit to discuss their symptoms, share experiences, and seek advice for alleviating their condition. While outdoor activities in *green* -like gardens or parks- or *blue* -like rivers or lakes- outdoor spaces have been shown to benefit those with other anxiety disorders, research on their impact on SAD remains limited. To assess the perceived effects of outdoor environments on SAD, social media posts that reference these settings can be used to capture the patients' perspectives and sentiments towards them.

For this task, we challenged participants to develop a classifier to categorize posts that mention one or more words related to outdoor spaces into one of four categories: 1) positive effect, 2) neutral or no effect, 3) negative effect, or 4) unrelated, where the word mentioned is not referencing an actual outdoor space.

Dataset The data for this task was collected from the subreddit r/socialanxiety and includes only users between 12 and 25 years old (Schmidt et al., 2023). The posts from the collection were filtered to include only posts that contained at least one term from a list of about 80 keywords related to green spaces, blue spaces, and activities that take place in these spaces (e.g., running, baseball). Two annotators annotated these posts, categorizing each as nature-related or unrelated to nature. The nature-related posts were then categorized into one of the three effect categories following detailed annotation guidelines. For the subset of posts that were double annotated (n = 650), the inter-annotation agreement was k=0.796 for the initial binary annotation and k=0.72 for the multi-class annotation. The training, validation, and test sets contain 1800, 600, and 600 posts, respectively. The distribution of the classes is unbalanced, with 1,757 posts (58.6%) unrelated to nature, 298 posts (10%) reporting a positive effect attributed to the outdoor space, 214 reporting a negative effect (7%), and 731 (24.4%) labeled neutral. To prevent manual annotations during evaluation, an additional 600 decoy posts were included in the unlabeled test set provided to participants. The predictions for these decoy posts will not be evaluated.

Evaluation Participating systems were evaluated using the macro-averaged F_1 score for multi-class classification. The CodaLab site for this task is `https://codalab.lisn.upsaclay.fr/competitions/18305`.

2.4 Task 4: Extraction of the clinical and social impacts of nonmedical substance use from Reddit

Nonmedical opioid use, whether prescribed or illicit, has become a significant public health concern, leading to addiction, overdose, and associated health issues. Understanding the clinical and social impacts of nonmedical opioid use is essential for improving the treatment of opioid use disorder. It helps healthcare professionals develop more effective interventions and medications to address addiction. By studying these impacts, researchers can develop more effective prevention and education programs to reduce the occurrence of opioid abuse and its associated clinical and social consequences.

Dataset In this task, we focused on extracting two entity types from a social media dataset for analyzing clinical and social effects of substance use (Ge et al., 2024), which belonged to the category with the least number of samples: nonmedical use clinical impacts and nonmedical use social impacts. Instances in the "nonmedical use clinical impacts" category describe the clinical effects, consequences, or impacts of substance abuse or medication misuse on an individual's health, physical condition, or mental well-being. Instances in the category of "nonmedical use social impacts" describe the societal, interpersonal, or community-level effects, consequences, or impacts of substance abuse or medication misuse. These impacts may include social relationships, community dynamics, or broader social issues. The training, validation, and test sets contain 843, 259, and 278 posts, respectively. Around 27.8% of posts contain words or phrases marked as clinical or social impacts. Systems designed for this task should automatically distinguish between clinical impacts and social impacts in text data derived from Reddit, with specific spans.

Evaluation We used both token-level F_1vscore, entity-level strict F_1 score and entity-level relaxed F_1 score for evaluation. For entity-level relaxed F_1 score, we focused on the partial match between predictions and golden annotations and used the scripts from SemEval[3] to compute the score. The CodaLab site for this task is: `https://codalab.lisn.upsaclay.fr/competitions/16648`

2.5 Task 5: Binary classification of English tweets reporting children's medical disorders

Many children are diagnosed with disorders that can impact their daily lives and can last throughout their lifetime. For example, in the United States, 17% of children are diagnosed with a developmental disability (Zablotsky et al., 2019), and 8% of children are diagnosed with asthma (Zahran et al., 2018). Data sources for assessing the potential association of these outcomes with pregnancy exposures remain limited. Among users who reported their pregnancy on Twitter (Klein et al., 2023b), this binary classification task involved automatically distinguishing tweets that reported having a child with attention-deficit/hyperactivity disorder (ADHD), autism spectrum disorders (ASD), delayed speech, or asthma, from tweets that merely mentioned the disease. The technologies developed under this task could enable the potential use of Twitter not only for epidemiologic studies (Golder et al., 2019; Klein et al., 2022a,b, 2023a), but, more generally, to explore parents' experiences and directly target support interventions.

Dataset The training, validation, and test sets contained 7398 tweets, 389 tweets, and 1947 tweets, respectively: 3019 (31%) that reported having a child with a disorder and 6715 (69%) that did not. (Klein et al., 2024a). Inter-annotator agreement (Fleiss' kappa), based on 1000 tweets that were annotated by all three annotators, was 0.88.

Evaluation The evaluation metric for this task was the F_1 score for the class of tweets that reported having a child with a disorder. The CodaLab site for this task is: `https://codalab.lisn.upsaclay.fr/competitions/17310`.

2.6 Task 6: Self-reported exact age classification with cross-platform evaluation in English

Advancing the utility of social media data for research applications requires methods for automatically detecting demographic information, such as users' age, within social media study populations. Automatically identifying the exact self-reported age of social media users, rather than their age groups (the standard approach), enables large-scale use of social media data for applications that do not fit predefined age groupings of existing models. This can be particularly useful for linking specific age-related risk factors in observational studies. In this task, we focused on automatically extracting self-reported ages in posts of two social media platforms: Twitter (now X) and Reddit.

Dataset The training data consisted of 8800 labeled tweets (32% with a reported age and 68% without) (Weissenbacher et al., 2022b) and 100,000 unlabeled Reddit posts from the subreddit *r/AskDocs* that included 2-digit numbers. The validation data consisted of 2200 tweets (32% with a reported age, 68% without) and 2000 Reddit posts (53% with a reported age, 43% without) (Weissenbacher et al., 2022b). The testing data consisted

[3]`https://github.com/davidsbatista/NER-Evaluation`

of 2200 tweets (35% with a reported age, 65% without) and 6000 Reddit posts (60% with a reported age, 40% without). The Reddit posts were a combination of posts from two different sources, *r/socialanxiety* and *r/Dryeyes*, and the posts from the *r/socialanxiety* subreddit included only posts with reported ages on the 13 to 25 range. The inter-annotation agreement for the tweets yielded a Fleiss's kappa of 0.80, while the Reddit posts had a Cohen's Kappa inter-annotator agreement of 0.939 for the dry eye posts and 0.967 for the social anxiety posts.

Evaluation The evaluation metric was the F_1 score for the class of tweets/posts that contained the user's self-reported age. The CodaLab site for this task is: `https://codalab.lisn.upsaclay.fr/competitions/17452`.

2.7 Task 7: Identification of LLM or human domain-expert data annotations in the context of health-related applications.

The current widespread adoption of LLMs, like ChatGPT, for data annotation tasks has the NLP field at odds. While some researchers are embracing it due to their performance in many types of annotation tasks and certain domains, others are skeptical due to the potential underlying biases and the 'hallucinations' commonly reported with the models. It will become of paramount importance to be able to identify data annotated by LLMs and distinguish it from data annotated by humans. In this task, we provide one dataset of Tweets in Latin American Spanish containing COVID-19 symptoms with labels of annotation being made by human domain experts and by an LLM (GPT-4).

Dataset We augmented the domain-expert annotated dataset used in Task 3 of SMM4H 2023 with an equally sized dataset, consisting of non-overlapping tweets, annotated using GPT-4 with some prompt engineering. The total size of the dataset is 10,150, which was split into 4,603 tweets for training, 3,437 for validation, and 2,110 for testing.

Evaluation The evaluation metric for this task is the classification accuracy of our 'human' and 'machine' labels. The CodaLab site for this task is: `https://codalab.lisn.upsaclay.fr/competitions/17405`

3 Results

3.1 Task 1

Of the 27 teams registered for Task 1, 11 submitted system predictions to the CodaLab server, and 9 submitted system description papers. Table 1 shows the performance of the best submission from each team compared to the baseline system, DeepADEMiner (Magge et al., 2021b). DeepADEMiner uses a pipeline approach with BERT-based models for three sequential subtasks: 1) ADE classification, which is a binary classifier to identify whether a tweet contains ADEs; 2) ADE extraction, a sequence labeling classifier to detect ADE text spans; and 3) ADE normalization, a multi-class classifier to map the extracted ADEs to their corresponding ptIDs from MedDRA. Among all the participating teams, three teams (TLab, LHS712 and RIGA) followed the same three-step pipeline strategy, while the remaining teams only performed ADE extraction and normalization. Only two teams (SRCB and zongxiong) outperformed the baseline for ADE extraction (F_1-NER) and normalization on the overall ADEs (F_1-Norm). Additionally, three teams (SRCB, zongxiong, and TLab) outperformed the baseline for normalization on unseen ADEs (F_1-unseen).

For the ADE classification task, all three teams (Tlab, LHS712, and RIGA) fine-tuned BERT-style binary classifiers to detect the presence of ADEs in tweets. For the ADE extraction task, six teams (SRCB, Yseop, RIGA, BIT@UA, ADE Oracle, and PolyUCBS) used the BIO tagging schema and fine-tuned BERT-style language models for token classification. Meanwhile, three teams (Tlab, LHS712, and HBUT) applied prompt tuning and LLMs to directly generate ADE text spans. In the ADE normalization task, two teams (LHS712 and BIT@UA) fine-tuned a BERT-style model and a random forest multi-class classifier, respectively, to map extracted ADEs to MedDRA ptIDs. Five teams (Yseop, RIGA, HBUT, ADE Oracle, and PolyUCBS) used vector space models (VSMs) and semantic search to identify the most similar MedDRA term for each ADE. Additionally, two teams (SRCB and Tlab) employed a generate-and-rank framework, initially using VSMs to find candidate MedDRA terms and then a ranker to select the most similar term.

A total of 6 teams applyed LLMs for this task. SRCB, achieving the highest F_1-Norm of 0.536, applied LLMs for data augmentation. Specifically, the team prompted two different LLMs (GLM-4 and GPT-3.5) to rewrite ADE mentions and tweet contexts, generate synthetic tweets with diverse ADE expressions, rewrite tweets to avoid informal grammar, and generate explanations for MedDRA terms during the ranker step. They further trained an ensemble of multiple BERT-style LMs on the combination of original and augmented data for ADE extraction and normalization and showed that both two tasks benefited from the augmented data. The remaining 5 teams mainly leverage the in-context learning of LLMs for the extraction and normalization task. For instance, Tlab team used retrieved tweets as few-shot examples as input to Llama 3 to extract ADEs, and a retrieval augmented generation framework to take the candidate MedDRA terms and the original tweet as input to GPT4, and generate the best MedDRA candidate term. Similarly, the LHS712 team also used a one-shot example as input and experimented with 10 different prompts for ADE extraction, but they achieved worse performance. The RIGA team used GPT4 with prompt tuning to find potential ADE text spans, which, along with the original tweets, were then fed as input for classification and extraction.

In conclusion, Task 1 underscored the challenges existing pipeline systems face in ADE extraction and normalization tasks. While LLMs have been widely adopted for these tasks, using them directly as in-context few-shot learners yielded poorer performance than BERT-style models. However, leveraging LLMs for data augmentation can enhance the performance of BERT-style models.

3.2 Task 2

Out of the 13 teams registered for Task 2, only 3 teams submitted system predictions to the CodaLab server, and only 2 teams submitted system description papers. Furthermore, these two teams did not participate in all subtasks or cover all languages. The results for Task 2a (NER) and Task 2b (joint NER and RE) in detecting Adverse Drug Reactions (ADRs) across German, French, and Japanese are displayed in Table 2.

Our baseline system first employed the Pytorch-IE framework (Binder et al., 2024)to predict entities with BERT-style token classification models. For the German data, we fine-tuned a German version of BERT (Devlin et al., 2019)); for the Japanese data, we fine-tuned the multilingual XLM-RoBERTa (Conneau et al., 2019)); and for the French data, we reused the model fine-tuned on Japanese without additional fine-tuning. The predicted entities from these models were then used as input for prompt templates to generate the relationships between the detected entities. These prompts included examples from the training data, brief definitions of the desired entities and relationships, and a requirement for the model to explain its decisions. The model we used for prompting was the open Llama-3-8B-UltraMedical (Zhang et al., 2024).

Task 2a: Team Yseop participated in the NER task for Japanese and French, while team HBUT submitted predictions for all three languages.

For French NER, Team Yseop used a combination of advanced language models, including the large language models Mistral-7B (Jiang et al., 2023) and DrBERT-CASM2 (Labrak et al., 2023), along with the medkit framework. For Japanese NER, they employed a Japanese-Multilingual Dictionary (JMdict) and a Japanese medical language model based on RoBERTa (Liu et al., 2019), pretrained on Japanese case reports and fine-tuned for NER using MedTxt-CR (Yada et al., 2022). They achieved a macro F_1 score of 48.92 across the three languages (including German, for which they did not provide predictions).

Team HBUT focused solely on the NER task for all three languages. Their methodology also employed LLMs. They explored three distinct prompting strategies to identify the most effective approach for NER. After evaluating two different LLMs, they selected GLM-3-Turbo (Zeng et al., 2023) as their preferred model. To adapt the task for LLMs, they designed specific prompts to obtain structured output. These outputs were then post-processed into the desired `brat` format. Evaluating these predictions against the gold entities resulted in an F_1 score of 36.9 for Team HBUT.

Task 2b: For the Japanese Relation Extraction task, Team Yseop reused the Japanese XLM-RoBERTa model and fine-tuned it with the provided training data. Since Team Yseop was the only team to submit predictions for the RE task, they were automatically declared the winners of the challenge, achieving an F_1 score of 1.89.

Summary: Table 2 presents the best-performing system from each team. Notably, neither team surpassed the baseline in any of the tasks, but Team HBUT achieved higher precision in the NER task compared to the other team and the baseline. Ex-

Team	F_1-Norm	F_1-NER	F_1-Unseen	System Summary
SRCB	**0.536**	**0.521**	**0.494**	Ensemble of BERT-style models for extraction and normalization, LLM-based data augmentation
zongxiong	0.528	0.513	0.492	–
Baseline	0.439	0.481	0.323	BERT-style models in 3-step pipeline system
Yseop	0.400	0.472	0.295	BERT-style model for extraction, BERT-style VSM for normalization
TLab	0.359	0.392	0.363	BERT-style model for classification, Llama3 for extraction, GPT-4 for normalization
LHS712	0.354	0.338	0.259	BERT-style models for classification and normalization, GPT-4 for extraction, BioBERT for normalization
RIGA	0.318	0.403	0.212	GPT-4 for preprocessing, and BERT-style models for classification and extraction, OpenAI Embeddings as VSM for normalization
BIT@UA	0.295	0.397	0	RoBERTa-base for span extraction, random forest classifier for normalization
Proddis	0.221	0.001	0.098	–
HBUT	0.205	0.216	0.106	GLM for extraction, ensemble of BERT-style models for normalization
ADE Oracle	0.082	0.132	0.014	Spacy tool for extraction, BERT-style VSM for normalization
PolyuCBS	0.044	0.010	0	LLM for preprocessing ,BERT-style model for extraction, SapBERT for normalization

Table 1: System summaries and micro-averaged F_1-scores for **task 1**. F_1-*Norm* is the ADE normalization score for all ptIDs, F_1-*NER* is the ADE extraction score, and F_1-*Unseen* is the ADE normalization score for unseen ptIDs. '-' in the *System Summary* indicates no system description paper was submitted.

Team	Task	Languages	P	R	F_1	System Summary
Yseop	NER	fr, ja	58.31	42.14	<u>48.92</u>	fr: Mistral-7B + DrBERT-CASM2 + medkit; ja: dictionary + RoBERTa
HBUT	NER	de, fr, ja	**60.52**	26.54	36.90	GLM-3-Turbo prompting + post-processing
baseline	NER	de, fr, ja	47.55	**58.83**	**52.60**	BERT, XLM-RoBERTa
Yseop	RE	ja	02.24	01.63	<u>01.89</u>	ja: RoBERTa fine-tuned
baseline	RE	de, fr, ja	**04.25**	**06.81**	**05.23**	language-specific prompting with Llama-3-8B-UltraMedical

Table 2: Summary of the submitted systems for **Task 2**. Subtask 2a: Named Entity Recognition (NER). Subtask 2b: joint NER and Relation Extraction (RE). The scores show exact macro F_1 score ($\mathbf{F_1}$), precision (**P**), and recall (**R**) as reported in CodaLab. The underlined scores belong to the winning systems/teams of the challenge.

cluding the baseline, Team Yseop won both subtasks in Task 2. For a more detailed description of the task and its results, we refer the reader to Raithel et al. (2024a).

In conclusion, participants employed diverse methods such as dictionary-based approaches, transformers, and LLMs, yet the task remains highly challenging. This difficulty likely stems from the task's multi-language nature, the presence of noisy, user-generated texts, and the limited number of training instances. Challenges also included generating correct output formats, identifying valid entity spans, and establishing accurate relations (some predictions included relations to non-existent entities). Team Yseop's approach, which combined outputs from multiple models in an ensemble manner, appears promising. Understanding how Team HBUT achieved high precision for entities would be particularly insightful. Combining their approach with our baseline prompting strategy could potentially enhance overall performance. Despite the use of LLMs, none of the participating teams fully solved the task of joint multilingual named entity recognition and relation extraction. This underscores the need for further exploration and possible integration of different methodologies.

3.3 Task 3

Of 37 teams registered for Task 3, 15 submitted system predictions, and 12 submitted task description papers. Table 3 displays the macro-average F1 score, precision, and recall for the top-performing submission from each team. The top 5 teams achieved closely matched scores, with only a 0.05 difference between the first and fifth positions. Team CTYUN-AI attained the high-

Team	F_1	P	R	System Summary
CTYUN-AI	**0.692**	**0.704**	**0.686**	Qwen-72B-Chat, data augmentation
1024m	0.679	0.677	0.682	BART-Large (2-stage)
PCIC	0.655	0.687	0.636	XLNet-large, data augmentation (traditional + paraphrasing (T5-base))
Dilab	0.654	0.654	0.661	RoBERTa-Large with Fuzzy string matching
Dolomites	0.642	0.67	0.623	Mistral-7B, fine-tuning (QLoRA), Multi-Task Learning Data Augmentation (In-domain + Drills)
AAST-NLP	0.635	0.631	0.644	RoBERTa-Large
ThangDLU	0.627	0.62	0.644	BART-base
Golden_Duck	0.596	0.603	0.601	RoBERTa-base: Concatenation of Mean pooling, CLS, and Attention Head
IMS_medicALY	0.563	0.629	0.534	SocBERT
LAMA	0.545	0.633	0.536	Pipeline: MentalBERT (binary classification: related or unrelated) + RoBERTa (multiclass: pos, neg or neu)
gcortal	0.414	0.425	0.418	–
Transformers	0.413	0.431	0.52	RoBERTa-Large, under-sampling
interrupt-driven	0.358	0.365	0.411	Relevance-weighted sentiment analysis model, Passive-Aggressive Regressor
Omkar_Khade	0.233	0.683	0.287	–
TeamZSA_codalab	0.196	0.167	0.27	–

Table 3: System summaries and macro-averaged F_1-score (F_1), precision (P), and recall (R) for **Task 3**: multi-class classification of effects of outdoor spaces on social anxiety symptoms in Reddit.

Team Name	Relaxed F_1	Strict F_1	Token-level F_1	System Summary
UKYNLP	**0.462**	0.171	**0.531**	Span-based encoder-only model leveraging BERT and ALBERT, combined with a BiLSTM layer for classification of entity types.
Dolomites	0.448	**0.208**	0.496	Experimented with two MTL-DA techniques to fine-tune Mistral (7B) with QLoRA for low-resource settings
LHS712 NV	0.314	0.008	0.052	Fine-tuning pre-trained BERT

Table 4: Brief system approaches and evaluation metric results for **Task 4**: Extraction of the clinical and social impacts of nonmedical substance use from Reddit.

est macro-average F1 score (0.692) by employing the LLM Qwen-72b-Chat, which was pre-trained and fine-tuned for classification. They also implemented data augmentation to balance classes, involving the random shuffling of strings within the original post based on delimiters.

The team with the second-highest macro-average F1 score (0.679), 1024m, built its system around a BART-large model using a two-stage strategy: initially, posts are classified as either class 0 or not, followed by categorizing non-0 posts into one of the remaining three classes. The third-highest scoring team, PCIC, achieved a macro-average F1 score of 0.655 by employing XLNet-large. Their approach included using a combined loss function, implementing data augmentation to balance class distributions, and increasing the token window size to 256.

Only two teams, CTYUN-AI and Dolomites, achieved their highest performance using an LLM. Dolomites (F1 score: 0.642) utilized Mistral-7B and employed multi-task learning data augmentation (MTL-DA) techniques to fine-tune the LLM. These techniques included in-domain augmentation from similar resources such as Reddit, as well as drills that decomposed the task into smaller subtasks to generate replicated data. This approach outperformed GPT-4 with zero-shot or few-shot learning on the validation set. The remaining teams achieved their best results using transformer-based models such as BART, RoBERTa, and XLNet.

In conclusion, Task 3 underscored the effectiveness of transformer models in classification tasks, with only a limited number of teams effectively leveraging LLMs.

3.4 Task 4

Out of 23 teams registered for Task 4, only 3 teams ultimately submitted task description papers. Table 4 presents the performances of three different teams. UKYNLP achieved the highest scores, with a *Relaxed* F_1 score of 0.462 and a *Token-level* F_1 score of 0.531, demonstrating excellence in both broad recognition and fine-grained token-level accuracy. UKYNLP experimented with both BERT and ALBERT models combined with a BiLSTM

Team	F_1	P	R	System Summary
CTYUN-AI	**0.956**	**0.954**	0.959	Qwen-72B (fine-tuned), multi-task learning (Task 6)
LT4SG	0.938	0.930	0.946	BERTweet-Large ensemble
PolyuCBS	0.935	0.954	0.917	Llama-2-7B (fine-tuned), LoRA
UTRad-NLP	0.933	0.932	0.934	DeBERTa-V3-Large, GPT-4 data augmentation
Golden_Duck	0.928	0.919	0.937	RoBERTa-Large
Chaai	0.927	0.907	0.949	Twitter-RoBERTa, GPT-4 (zero-shot), under-sampling
Baseline	0.927	0.923	0.930	RoBERTa-Large
UKYNLP	0.924	0.924	0.924	DeBERTa-V3-Large
KUL	0.923	0.906	0.940	BERTweet, data augmentation, R-Drop
1024m	0.918	0.923	0.912	BART-Large
SMC	0.901	0.885	0.917	MentalBERT, PsychBERT, TwHIN-BERT, DistilBERT, data augmentation
Transformers	0.900	0.854	0.950	RoBERTa-Large, under-sampling
DILAB	0.898	0.883	0.914	Twitter-RoBERTa-Base-Sentiment ensemble
HALELab-NITK	0.868	0.858	0.879	RoBERTa-Base
Thang-DLU	0.841	0.844	0.839	T5-Small, GPT data augmentation
BIT@UA	0.840	0.829	0.851	BERT-Base
PhoenixTrio_918	0.823	0.721	0.959	RoBERTa-Large, 5-fold cross-validation
MUET	0.671	0.508	**0.988**	-
HULAT-UC3M	0.633	0.702	0.576	-
Be Better Health	0.522	0.630	0.445	-
Z-AGI Labs	0.357	0.321	0.401	-

Table 5: System summaries and F_1-score (F_1), precision (P), and recall (R) for **Task 5**: Binary classification of English tweets reporting children's medical disorders.

layer for NER. Their BERT model achieved the highest F_1 score of 0.462 on the test set, surpassing the ALBERT model with BiLSTM.

Dolomites utilized multi-task learning and data augmentation techniques, achieving moderate success overall with a strict F_1 score of 0.208. This suggests their approach is effective for precise entity matching. They employed a strategy of extracting smaller tasks (drills) from the target dataset and replicating the training set to include additional examples for these drills.

In contrast, team LHS712 NV showed the lowest performance across all metrics, indicating potential challenges in model implementation or task-specific tuning despite utilizing robust BERT-style models.

3.5 Task 5

Out of 48 teams registered for Task 5, 20 teams submitted system predictions to the Codalab server, and 16 teams ultimately submitted task description papers. Table 5 presents the F_1, precision, and recall scores for a RoBERTa-large baseline classifier (Klein et al., 2024a) and the best-performing system from each of the 20 teams for Task 5. CTYUN-AI achieved the highest F_1 score (0.956) and precision (0.954) using a Qwen-72B LLM and multi-task learning, improving upon the baseline (0.927) by approximately 0.03. Initially, they continued pre-training a Qwen-72B LLM using unlabeled Reddit posts from Task 6. Subsequently, they fine-tuned two additional Qwen-72B LLMs using labeled tweets and Reddit posts from Task 6, deploying them for binary classification of the unlabeled Reddit posts. They further refined this LLM by focusing on Reddit posts where the two classifiers agreed, combining them with labeled tweets and Task 6 Reddit posts for additional fine-tuning. Finally, they fine-tuned this LLM further using labeled tweets from Task 5. An ablation study is necessary to determine the precise impact of LLMs and multi-task learning on their performance.

Among the other top-performing teams that exceeded the baseline (0.927), PolyuCBS achieved an F_1 score of 0.935 using a Llama-2-7B LLM fine-tuned with LoRA, narrowly edged out by LT4SG's ensemble of BERTweet-Large models with an F_1 score of 0.938. Of these teams, only CTYUN-AI and PolyuCBS utilized LLMs in their approaches. Additionally, two other top teams used a GPT-4 LLM to augment training data for a DeBERTa-V3-Large classifier (UTRad-NLP) and validate predictions from a Twitter-RoBERTa classifier (Chaii). All teams that submitted system descriptions employed deep neural network architectures based on pre-trained transformer models.

Team	F1	P	R	System Summary
CTYUN-AI	**0.970**	**0.976**	0.963	fine-tuned Qwen-72B-Chat, data augmentation by random shuffling, ensemble labeling of unlabeled Reddit data and cross-task training
1024m	0.959	0.953	**0.965**	BART-Large
Dolomites	0.957	0.965	0.949	Mistral-7B, fine-tuning (QLoRA), Multi-Task Learning Data Augmentation (In-domain + Drills)
AAST-NLP	0.946	0.932	0.959	Ensemble of BERTweet, RoBERTa and Mistral-7b, rule-based data augmentation from unlabeled data
UTRad-NLP	0.936	0.947	0.926	DeBERTa-V3-Large, synthetic data augmentation with GPT-4
Baseline	0.900	0.902	0.897	RoBERTa-Large
SMM4H-TIET	0.900	0.916	0.884	BERTweet, back-translation augmentation of minority class, under-sampling of majority class
IITRoorkee	0.878	0.899	0.858	RoBERTa

Table 6: System summaries and F_1-scores (F_1), precision (P), and recall (R) for **Task 6**: Self-reported exact age classification with cross-platform evaluation in English.

Team Name	Accuracy	System Summary
Baseline	0.82	Fine-tuned COVID-Twitter-BERT w 500K silver-standard annotations
712forTask7	0.5166	Fine-tuned BETO
BrainStorm	0.5090	Fine-tuned two different BERT-style models, topical embeddings from BERTopic
Deloitte	0.5109	Zero-shot prompt tuning on GPT4

Table 7: System summaries and classification accuracy results for **Task 7**.

3.6 Task 6

Out of 27 teams registered for Task 6, 7 teams ultimately submitted task description papers. Table 6 displays the F_1 score, precision, and recall for the RoBERTa-Large baseline classifier (Klein et al., 2022c) and the top-performing system runs of these 7 teams. Among these teams, four utilized LLMs: CTYUN-AI applied Qwen-72B-Chat, Dolomites used Mistral-7B, and AAST-NLP integrated Mistral-7B into their ensemble models. Additionally, UTRad-NLP employed synthetic data generated with GPT-4 to augment their training dataset. The remaining teams focused on transformer models: team 1024m utilized BART-Large, team UTRad-NLP employed DeBERTa-V3-Large, team AAST-NLP used an ensemble of BERTweet, RoBERTa, and Mistral-7B, team SMM4H-TIET used BERTweet, and team IITRoorkee used RoBERTa.

The top-performing team, CTYUN-AI, achieved the highest F_1 score (0.970) and precision (0.976) by fine-tuning the Qwen-72B-Chat model. Their approach included data augmentation through random sentence shuffling, ensemble labeling of the unlabeled Reddit data provided, and cross-task training, significantly enhancing their model's performance. Only two teams employed different approaches to label the provided unlabeled Reddit posts: CTYUN-AI used an ensemble model, and AAST-NLP employed a rule-based approach. Notably, team 1024m achieved the highest recall, utilizing a BERT-style model.

Comparing the use of LLMs against traditional BERT-style models, LLMs generally demonstrated superior performance. For example, CTYUN-AI's use of Qwen-72B-Chat outperformed the baseline RoBERTa-Large model by 0.070 in terms of F_1 score. This trend was consistent across other teams' results, where LLMs, when fine-tuned and augmented with advanced techniques, consistently achieved higher precision and recall compared to traditional BERT-style models. However, BERT-style models also performed well, especially when used in ensembles or enhanced with data augmentation strategies.

3.7 Task 7

Out of 19 teams registered for Task 7, only 3 teams ultimately submitted system predictions on the test set and system description papers. Table 7 pdisplays the classification accuracy for the best system run from these 3 teams on the unseen test set. The baseline system utilized a fine-tuned COVID-Twitter-BERT (Müller et al., 2023) trained on the provided dataset and an additional 500K 'silver-standard' tweets sourced from Banda et al. (2021). These tweets were generated using weak supervision annotation for half and GPT-4 for the other half. We hypothesize that this extensive data augmentation played a significant role in the performance disparity between the participant scores and the baseline score.

Among the participants, two teams focused on traditional BERT-style models. One team employed BETO, a BERT-based model for Spanish text (Cañete et al., 2020) , while another team translated Spanish-language tweets into English and utilized BERTopic embeddings (Grootendorst, 2022). The third team, Deloitte, used GPT-4 to classify the provided tweets.

One team, 712forTask7, conducted an analysis of the training dataset, highlighting minimal differences among the tweets, which posed challenges for traditional approaches. The substantial data augmentation in the baseline system likely contributed significantly to its performance advantage. It would be intriguing to explore whether participant teams could achieve comparable performance using the same augmented dataset with their respective approaches.

4 Conclusion

This paper provides an overview of the SMM4H 2024 shared tasks. This year, seven tasks were proposed, reflecting the growing interest and participation in the SMM4H shared tasks. The top-performing teams predominantly used transformer-based models, including encoder-based LMs like DeBERTa-v3-large and decoder-based LLMs like GPT-4 or Qwen-72B-Chat. These teams frequently employed LLM-based data augmentation techniques to address issues such as data imbalance, unseen examples, and domain mismatch in social media data. Notably, out of 38 teams that submitted a system description paper, 11 participated in multiple tasks, often using the same systems fine-tuned for different tasks. This trend signifies an important effort within the community to develop high-performing classifiers and label sequencers that are both generalizable and reusable.

Acknowledgments

The work for SMM4H 2024 at Cedars-Sinai Medical Center was supported by the National Institutes of Health (NIH) National Library of Medicine (NLM) [grant number R01LM011176]. The content is solely the responsibility of the authors and does not necessarily represent the official views of the National Institutes of Health.

The work of the organizers of Task 2 was supported by the Cross-ministerial Strategic Innovation Promotion Program (SIP) on "Integrated Health Care System" Grant Number JPJ012425, by JPMJCR20G9, ANR-20-IADJ-0005-01, and DFG-442445488 under the trilateral ANR-DFG-JST AI Research project KEEPHA, and by the German Federal Ministry of Education and Research under the grant BIFOLD24B.

Creation of the dataset for task 4 was supported in part by the National Institute on Drug Abuse (NIDA) of the National Institutes of Health (NIH) under award number R01DA057599. The content is solely the responsibility of the authors and does not necessarily represent the official views of the National Institutes of Health.

The work of the organizer of Task 7 was supported by a Google Award for Inclusion Research (AIR).

The authors thank those who contributed to annotating the data, the program committee of the #SMM4H 2024 Workshop, and additional peer reviewers of the system description papers.

References

Juan M. Banda, Ramya Tekumalla, Guanyu Wang, Jingyuan Yu, Tuo Liu, Yuning Ding, Ekaterina Artemova, Elena Tutubalina, and Gerardo Chowell. 2021. A large-scale covid-19 twitter chatter dataset for open scientific research—an international collaboration. *Epidemiologia*, 2(3):315–324.

Arne Binder, Leonhard Hennig, and Christoph Alt. 2024. Pytorch-ie: Fast and reproducible prototyping for information extraction. *Preprint*, arXiv:2406.00007.

José Cañete, Gabriel Chaperon, Rodrigo Fuentes, Jou-Hui Ho, Hojin Kang, and Jorge Pérez. 2020. Spanish pre-trained bert model and evaluation data. In *PML4DC at ICLR 2020*.

Pew Research Center. SM Fact Sheet 2021. `https://www.pewresearch.org/internet/fact-sheet/social-media/`. Accessed 2nd August 2023.

Alexis Conneau, Kartikay Khandelwal, Naman Goyal, Vishrav Chaudhary, Guillaume Wenzek, Francisco Guzmán, Edouard Grave, Myle Ott, Luke Zettlemoyer, and Veselin Stoyanov. 2019. Unsupervised cross-lingual representation learning at scale. *CoRR*, abs/1911.02116.

Jacob Devlin, Ming-Wei Chang, Kenton Lee, and Kristina Toutanova. 2019. BERT: Pre-training of Deep Bidirectional Transformers for Language Understanding. In *Proceedings of the 2019 Conference of the North American Chapter of the Association for*

Computational Linguistics: Human Language Technologies, Minneapolis, Minnesota, USA. Association for Computational Linguistics.

N. Elliott and G. Sverdlov. 2012. Global SM adoption. the social marketing playbook: Master the next wave of social.

Yao Ge, Sudeshna Das, Karen O'Connor, Mohammed Ali Al-Garadi, Graciela Gonzalez-Hernandez, and Abeed Sarker. 2024. Reddit-impacts: A named entity recognition dataset for analyzing clinical and social effects of substance use derived from social media. *Preprint*, arXiv:2405.06145.

S. Golder, S. Chiuve, D. Weissenbacher, A. Klein, K. O'Connor, M. Bland, M. Malin, M. Bhattacharya, L. J. Scarazzini, and G. Gonzalez-Hernandez. 2019. Pharmacoepidemiologic evaluation of birth defects from health-related postings in social media during pregnancy. *Drug Saf*, 42(3):389–400.

Maarten Grootendorst. 2022. Bertopic: Neural topic modeling with a class-based tf-idf procedure. *arXiv preprint arXiv:2203.05794*.

Albert Q Jiang, Alexandre Sablayrolles, Arthur Mensch, Chris Bamford, Devendra Singh Chaplot, Diego de las Casas, Florian Bressand, Gianna Lengyel, Guillaume Lample, Lucile Saulnier, et al. 2023. Mistral 7b. *arXiv preprint arXiv:2310.06825*.

Ronald C. Kessler, Wai Tat Chiu, Olga Demler, and Ellen E. Walters. 2005. Prevalence, Severity, and Comorbidity of 12-Month DSM-IV Disorders in the National Comorbidity Survey Replication. *Archives of General Psychiatry*, 62(6):617–627.

A. Z. Klein, J. A. Gutiérrez Gómez, L. D. Levine, and G. Gonzalez-Hernandez. 2024a. Using Longitudinal Twitter Data for Digital Epidemiology of Childhood Health Outcomes: An Annotated Data Set and Deep Neural Network Classifiers. *J Med Internet Res*, 26:e50652.

A. Z. Klein, S. Kunatharaju, S. Golder, L. D. Levine, J. C. Figueiredo, and G. Gonzalez-Hernandez. 2023a. Association between COVID-19 during pregnancy and preterm birth by trimester of infection: a retrospective cohort study using longitudinal social media data. *medRxiv*.

A. Z. Klein, S. Kunatharaju, K. O'Connor, and G. Gonzalez-Hernandez. 2023b. Pregex: Rule-Based Detection and Extraction of Twitter Data in Pregnancy. *J Med Internet Res*, 25:e40569.

A. Z. Klein, K. O'Connor, and G. Gonzalez-Hernandez. 2022a. Toward using Twitter data to monitor COVID-19 vaccine safety in pregnancy: proof-of-concept study of cohort identification. *JMIR Form Res*, 6(1):e33792.

A. Z. Klein, K. O'Connor, L. D. Levine, and G. Gonzalez-Hernandez. 2022b. Using Twitter data for cohort studies of drug safety in pregnancy: proof-of-concept with B-blockers. *JMIR Form Res*, 6(6):e36771.

Ari Z Klein, Juan M Banda, Yuting Guo, Ana Lucia Schmidt, Dongfang Xu, Ivan Flores Amaro, Raul Rodriguez-Esteban, Abeed Sarker, and Graciela Gonzalez-Hernandez. 2024b. Overview of the 8th Social Media Mining for Health Applications (SMM4H) shared tasks at the AMIA 2023 Annual Symposium. *Journal of the American Medical Informatics Association*, 31(4):991–996.

Ari Z Klein, Arjun Magge, and Graciela Gonzalez-Hernandez. 2022c. Reportage: Automatically extracting the exact age of twitter users based on self-reports in tweets. *PloS one*, 17(1):e0262087.

Yanis Labrak, Adrien Bazoge, Richard Dufour, Mickaël Rouvier, Emmanuel Morin, Béatrice Daille, and Pierre-Antoine Gourraud. 2023. Drbert: A robust pre-trained model in french for biomedical and clinical domains. In *Proceedings of the 61st Annual Meeting of the Association for Computational Linguistics (Volume 1: Long Papers)*, pages 16207–16221.

Robert Leaman, Laura Wojtulewicz, Ryan Sullivan, Annie Skariah, Jian Yang, and Graciela Gonzalez. 2010. Towards Internet-age pharmacovigilance: Extracting adverse drug reactions from user posts in health-related social networks. In *Proceedings of the 2010 Workshop on Biomedical Natural Language Processing*, pages 117–125, Uppsala, Sweden. Association for Computational Linguistics.

Yinhan Liu, Myle Ott, Naman Goyal, Jingfei Du, Mandar Joshi, Danqi Chen, Omer Levy, Mike Lewis, Luke Zettlemoyer, and Veselin Stoyanov. 2019. RoBERTa: A robustly optimized BERT pretraining approach. *arXiv:1907.11692 [cs]*.

Arjun Magge, Ari Klein, Antonio Miranda-Escalada, Mohammed Ali Al-Garadi, Ilseyar Alimova, Zulfat Miftahutdinov, Eulalia Farre, Salvador Lima López, Ivan Flores, Karen O'Connor, Davy Weissenbacher, Elena Tutubalina, Abeed Sarker, Juan Banda, Martin Krallinger, and Graciela Gonzalez-Hernandez. 2021a. Overview of the sixth social media mining for health applications (#SMM4H) shared tasks at NAACL 2021. In *Proceedings of the Sixth Social Media Mining for Health (#SMM4H) Workshop and Shared Task*, pages 21–32, Mexico City, Mexico. Association for Computational Linguistics.

Arjun Magge, Elena Tutubalina, Zulfat Miftahutdinov, Ilseyar Alimova, Anne Dirkson, Suzan Verberne, Davy Weissenbacher, and Graciela Gonzalez-Hernandez. 2021b. DeepADEMiner: a deep learning pharmacovigilance pipeline for extraction and normalization of adverse drug event mentions on Twitter. *Journal of the American Medical Informatics Association*, 28(10):2184–2192.

Martin Müller, Marcel Salathé, and Per E. Kummervold. 2023. Covid-twitter-bert: A natural language processing model to analyse covid-19 content on twitter. *Frontiers in Artificial Intelligence*, 6.

Lisa Raithel, Philippe Thomas, Bhuvanesh Verma, Roland Roller, Hui-Syuan Yeh, Shuntaro Yada, Cyril Grouin, Shoko Wakamiya, Eiji Aramaki, Sebastian Möller, and Pierre Zweigenbaum. 2024a. Overview of #SMM4H 2024 – Task 2: Cross-lingual few-shot relation extraction for pharmacovigilance in french, german, and japanese. In *Proceedings of The Ninth Workshop on Social Media Mining for Health Applications, Workshop & Shared Task*, Bangkok, Thailand. Association for Computational Linguistics.

Lisa Raithel, Hui-Syuan Yeh, Shuntaro Yada, Cyril Grouin, Thomas Lavergne, Aurélie Névéol, Patrick Paroubek, Philippe Thomas, Tomohiro Nishiyama, Sebastian Möller, Eiji Aramaki, Yuji Matsumoto, Roland Roller, and Pierre Zweigenbaum. 2024b. A dataset for pharmacovigilance in German, French, and Japanese: Annotating adverse drug reactions across languages. In *Proceedings of the 2024 Joint International Conference on Computational Linguistics, Language Resources and Evaluation (LREC-COLING 2024)*, pages 395–414, Torino, Italia. ELRA and ICCL.

Ana Lucia Schmidt, Karen O'Connor, Graciela Gonzalez Hernandez, and Raul Rodriguez-Esteban. 2023. Studying social anxiety without triggering it: Establishing an age-controlled cohort of social media users for observational studies. *medRxiv*.

Murray B. Stein and Dan J. Stein. 2008. Social anxiety disorder. *Lancet*, 371(9618):1115–1125. London, England.

Pontus Stenetorp, Sampo Pyysalo, Goran Topić, Tomoko Ohta, Sophia Ananiadou, and Jun'ichi Tsujii. 2012. Brat: A Web-based Tool for NLP-Assisted Text Annotation. In *Proceedings of the Demonstrations at the 13th Conference of the European Chapter of the Association for Computational Linguistics*, pages 102–107, Avignon, France. Association for Computational Linguistics.

Andrea C Tricco, Wasifa Zarin, Erin Lillie, Serena Jeblee, Rachel Warren, Paul A Khan, Reid Robson, Ba' Pham, Graeme Hirst, and Sharon E Straus. 2018. Utility of social media and crowd-intelligence data for pharmacovigilance: a scoping review. *BMC medical informatics and decision making*, 18:1–14.

Davy Weissenbacher, Juan Banda, Vera Davydova, Darryl Estrada Zavala, Luis Gasco Sánchez, Yao Ge, Yuting Guo, Ari Klein, Martin Krallinger, Mathias Leddin, Arjun Magge, Raul Rodriguez-Esteban, Abeed Sarker, Lucia Schmidt, Elena Tutubalina, and Graciela Gonzalez-Hernandez. 2022a. Overview of the seventh social media mining for health applications (#SMM4H) shared tasks at COLING 2022. In *Proceedings of The Seventh Workshop on Social Media Mining for Health Applications, Workshop & Shared Task*, pages 221–241, Gyeongju, Republic of Korea. Association for Computational Linguistics.

Davy Weissenbacher, Juan Banda, Vera Davydova, Darryl Estrada Zavala, Luis Gasco Sánchez, Yao Ge, Yuting Guo, Ari Klein, Martin Krallinger, Mathias Leddin, et al. 2022b. Overview of the seventh social media mining for health applications (# smm4h) shared tasks at coling 2022. In *Proceedings of The Seventh Workshop on Social Media Mining for Health Applications, Workshop & Shared Task*, pages 221–241.

Shuntaro Yada, Shoko Wakamiya, Yuta Nakamura, and Eiji Aramaki. 2022. Real-MedNLP: Overview of REAL document-based MEDical Natural Language Processing Task. In *Proceedings of the 16th NTCIR Conference on Evaluation of Information Access Technologies*, Tokyo, Japan.

B. Zablotsky, L. I. Black, M. J. Maenner, L. A. Schieve, M. L. Danielson, R. H. Bitsko, S. J. Blumberg, M. D. Kogan, and C. A. Boyle. 2019. Prevalence and Trends of Developmental Disabilities among Children in the United States: 2009-2017. *Pediatrics*, 144(4).

H. S. Zahran, C. M. Bailey, S. A. Damon, P. L. Garbe, and P. N. Breysse. 2018. Vital Signs: Asthma in Children - United States, 2001-2016. *MMWR Morb Mortal Wkly Rep*, 67(5):149–155.

Aohan Zeng, Xiao Liu, Zhengxiao Du, Zihan Wang, Hanyu Lai, Ming Ding, Zhuoyi Yang, Yifan Xu, Wendi Zheng, Xiao Xia, Weng Lam Tam, Zixuan Ma, Yufei Xue, Jidong Zhai, Wenguang Chen, Peng Zhang, Yuxiao Dong, and Jie Tang. 2023. GLM-130B: An Open Bilingual Pre-trained Model. *Preprint*, arXiv:2210.02414.

Kaiyan Zhang, Ning Ding, Biqing Qi, Sihang Zeng, Haoxin Li, Xuekai Zhu, Zhang-Ren Chen, and Bowen Zhou. 2024. Ultramedical: Building specialized generalists in biomedicine. `https://github.com/TsinghuaC3I/UltraMedical`.

Author Index

Association for Computational Linguistics
209 N. Eighth Street
Stroudsburg, Pennsylvania 18360

ISBN 979-8-3313-0177-4

CURRAN ASSOCIATES INC.
proceedings.com